"Superb. . . . Barnet has added greatly to our understanding of the way human beings with a vision can change society for the better by pursuing their dreams."   —*St. Louis Post-Dispatch*

"Fascinating and deliciously detailed."   —*Dallas News*

"Wide-reaching and exciting. . . . Barnet is smart, engaging, and highly readable. With enthusiasm and eloquence . . . she tells the women's stories, while making her larger point that each was responsible for helping to change the world."   —*Lakeville Journal*

"With both resonant detail and purposeful distillation, Barnet tells [her subjects'] dramatic stories within the context of the counterculture of fifty years ago, charts the ongoing vitality and influence of their compassionate visions, and asks if we will yet accomplish what these four 'accidental revolutionaries' call on us to do to preserve the web of life."
                                                    —*Booklist* (starred review)

"What a perfect moment for this lucid, enlightening, wonderful book. These quite different and deeply independent women each worked alone. But now that Andrea Barnet has masterfully woven together their stories—four passionate, tenacious, groundbreaking outsiders shifting a paradigm and transforming the world during the 1960s and '70s—I'll always think of them as a team of superheroes."   —Kurt Andersen, author of *Fantasyland*

"Fascinatingly original—and enormously timely. Barnet unearths the fact that Jacobs, Carson, Goodall, and Waters improved the way we experience our global home and our daily lives with scintillating clarity."
                                        —Sheila Weller, author of *Girls Like Us*

"A fascinating cohort of four brilliant, impassioned women who singly and collectively transformed the way we view life on this planet."
                        —Linda Lear, author of *Rachel Carson: Witness for Nature*

"Each of Barnet's beautifully drawn portraits casts a light on the others, but what her shining prose illuminates most of all are the beliefs these women shared about the interconnectedness of the human community, animal species, natural world and built environment, convictions that have transformed the way we live now."
                        —Akiko Busch, author of *The Incidental Steward*

also by ANDREA BARNET

*All-Night Party: The Women of Bohemian
Greenwich Village and Harlem, 1913–1930*

Visionary WOMEN

how RACHEL CARSON, JANE JACOBS,
JANE GOODALL, *and* ALICE WATERS
CHANGED OUR WORLD

Andrea Barnet

An Imprint of HarperCollinsPublishers

for KIT *and* PHILIPPA

A hardcover edition of this book was published in 2018 by Ecco, an imprint of HarperCollins Publishers.

FIRST ECCO PAPERBACK EDITION PUBLISHED 2019.

Designed by Renata De Oliveira

Title page image credits: Rachel Carson: courtesy Library of Congress; Jane Jacobs: Fred McDarrah/Getty Images; Jane Goodall: © the Jane Goodall Institute/Judy Goodall; Alice Waters: courtesy of Alice Waters.

Library of Congress Cataloging-in-Publication Data has been applied for.

ISBN 978-0-06-231073-6

20 21 22 23 LSC 10 9 8 7 6 5 4 3 2

The grounds for hope are in the shadows, in the people who are inventing the world while no one looks, who themselves don't know yet whether they will have any effect.

—REBECCA SOLNIT

Like the standing wave in front of a rock in a fast-moving stream, a city is a pattern in time.

—JOHN HOLLAND

When we try to pick out anything by itself, we find it hitched to everything else in the universe.

—JOHN MUIR

The boy builds us a fire out of pinecones, puts on a kettle, and makes us tea. Then he produces a small piece of cheese and painstakingly cuts it into even smaller pieces, which he offers us gravely. . . .

He has given us everything he has, and he has done this with absolutely no expectation of anything in return. A small miracle of trust, and a lesson in hospitality that changed my life.

—ALICE WATERS

CONTENTS

....

PREFACE

Revolutions are sometimes sparked by unexpected characters, out-liers whose excellence and originality, driving energy, and perse-verance confound expectations. But these characters, it should be added, are rarely women, especially women who seemingly emerge from out of the blue, full-blown, defiantly themselves. This is the story of four remarkable women who changed the way we think about the world: four women linked not by friendship, or age, or even their fields, but by their monumental cultural impact, and the frequent, often surprising parallels in their thinking. All but one wrote iconic books that ignited social movements; all were "green" thinkers before the word had entered our vocabulary; all opposed the culture's blind obeisance to technology, seeing in its reckless quest to conquer and counterfeit nature an arrogant and deadly path forward. And all found their political voices in the 1960s, be-coming a kind of true north for the gathering counterculture, who heard their call to arms and drew upon their ideas and their activ-ism for their own.

They are Rachel Carson, who published *Silent Spring* in 1962 at age fifty-five, giving birth to the environmental movement; Jane Jacobs, in her forties in 1961 when she saved Greenwich Village from the wrecking ball and published *The Death and Life of Great*

American Cities, spawning a new consciousness about organized complexity and the life of cities; Jane Goodall, who in 1960 at age twenty-six discovered chimps using tools, altering mankind's understanding of the animal world overnight; and Alice Waters, who had her epiphanies about food in France as a student in 1965, and opened fresh, local-food-serving Chez Panisse in Berkeley, California, six years later, kicking off the sustainable food movement.

Like many of us, I suspect, I knew something of these women's thinking long before I had read their books. Growing up in the 1950s, I remember the fogging trucks that prowled the leafy streets of our suburban Massachusetts neighborhood, discharging deadly clouds of pesticides meant to kill the bugs that lived among us. I remember that several of our neighbors, "crackpots" my father called them, were building bomb shelters in their backyards, the idea being that they would somehow survive a nuclear attack in these homegrown bunkers. It was a notion, my father soberly explained, that was nonsense, without wading into the larger issue, which was that technology untethered from morality had a lethal, doomsday edge. At school there were nuclear drills: I recall lining up in our grade school corridors and then returning to our desks to practice duck-and-cover maneuvers, as if crouching under one's desk would be a sufficient response to an atomic event. My point is that even from a child's point of view, there was a certain peril, and a certain degree of willful denial that permeated the air of the 1950s. And as I would later discover in reading *Silent Spring*, it was this peril and purposeful avoidance of inconvenient truths that Carson was able to use to such powerful effect in her searing exposé of the identical, equally deadly dangers of chemical pesticides, exploding public complacency almost overnight.

I remember the awfulness of the food during the fifties too, the occasional TV dinner my mother served. My sister and I didn't like them. The food was gluey and tasteless we thought. Though we did secretly like the way the little foil trays separated each foodstuff into a neat compartment. McDonald's was beginning to appear in the landscape by then. I don't specifically remember stopping at one, but I do remember my mother marveling at the bargain-basement price for a burger—just nineteen cents, she said. And that the food was standardized, even to the point that the mustard and pickles had already been added. The kitchen was a mechanized assembly line; even the patties were made by a machine.

Later, long before I had read Jane Jacobs's *Death and Life*, I remember driving into Cambridge with my father, passing a bank of bleak public housing towers along the way, feeling their anomie from the car—there were no people anywhere—and then arriving in bustling Cambridge. I liked the narrow, winding streets immediately; the beautiful old-brick buildings there; the crowded one-of-a-kind bookstores and steamy coffee shops. I didn't know Jane Jacobs had written about the virtues of just such a human-scaled neighborhood; that she had used her own tatty block in Greenwich Village, where people stopped and talked, where even strangers looked out for one another, as a contrast to these sterile, government-engineered housing projects, which, amputated as they were from the body of the city, were socially dead, leached of the street life that connected a community, the incidental public spaces that served as social moorings.

Being a kid, of course, I couldn't see the lines of connection between all these things: engineered housing, wartime chemicals, industrialized agriculture, manufactured food. I didn't know that they were actually different faces of the same problem, part of the same cultural push to bend nature and natural systems to serve mankind's ends.

It wouldn't be until I was in college, in the 1970s, that I understood the extent to which much of what was passed off as food in the average supermarket had been manipulated to the point that it wasn't really food at all, but rather the spawn of some "food technologist" working in a gleaming chemistry lab somewhere. Or knew that the residues of industrial endeavors were turning up "in the shells of birds, in mother's milk, in the blood of children, in the body fat of all Americans." Or that DDT had nearly wiped out the bald eagle and the peregrine falcon, among other species, thinning their eggshells to the point that they could no longer properly hatch, just as Carson had warned. Or even that my own gut feelings about the life and cheer of a place like Cambridge, versus the sterility of the high-rise housing projects routinely pushed by urban planners, went beyond personal taste, that there were solid, sensible explanations behind these visceral responses.

By then I had joined my generation in its growing aversion to the technocratic direction of American life; its repudiation of America's war machine; its protest against the chemicalization of food and farming; its alarm over the plunder of nature, and the parallel plunder of America's cities; its quest for social justice and local solutions; its worry over the loss of place and community. I believed, inoculated with the arrogance of youth, but also as conventional history would have it, that these ideas had begun with us, with the counterculture.

Yet returning to that time, and to these four visionary women, I saw how little we did invent, that many of our most progressive ideals had begun not with us, but with them. Indeed, so much that I had attributed to my own time, and my own generation, had actually been shaped not so much by the 1960s but in response to the misguided values and priorities of the 1950s, priorities that each of these four women, in interestingly similar and adjacent ways, had spotlighted and pushed back against, unafraid to speak truth to power, to indict technologies that didn't consider "feedbacks" such as pollution, or the degraded quality of experience, but only the "narrow purposes" for which their innovations were designed.

There is a fifth woman, it perhaps should be added, who would seem to be a natural addition to this iconic group: Betty Friedan, who, in 1963, a year after the appearance of *Silent Spring*, published *The Feminine Mystique*, launching the modern women's movement. And in many respects Friedan does belong with this revolutionary band, but for one difference. While Carson, Jacobs, Goodall, and Waters were all trying to conserve endangered aspects of the culture, Friedan was trying to blow it apart wholecloth, seeing the system as it stood as pinching to the female psyche. While the other four were trying to preserve and restore systems they saw as threatened, Friedan was trying to dismantle a profoundly unjust one. Which is to say they were "green" thinkers, while Friedan was not.

And so the question: Why fold these four women's stories into a single narrative? There have been superb individual biographies written about each and certainly it will come as no surprise that each was remarkable in her own right. The reasons, I believe, are many and intriguing. Seeing these four lives in parallel reveals patterns that would otherwise be hidden: it brings into sharp focus the striking overlaps and consonances in these women's thinking, and their nearly simultaneous emergence as cultural touchstones in the early sixties, raising interesting questions about the inevitability of history versus the momentum of individual vision. It highlights the ways in which the threads of their respective and very different stories, when viewed together, reveal a second, more layered narrative about the seismic shifts of consciousness unfolding in what we now view as a watershed moment in the culture. It adds dimension and further angles to the complicated interplay between the private and the collective, the vision of the individual and the cultural tides of her time or, to put it another way—the extent to which these four pioneers were channeling the anxieties of their particular moment, versus the degree to which their towering accomplishments arose from something more elusive and unusual in their particular characters. None of these women knew each other, yet the philosophical roots of their thinking, even as it applied to their different fields or spheres of interest, were remarkably aligned. Some of this was individual prescience certainly, but some also had to do with their particular cultural time. History as traditionally told unfolds as a progression of events, one following the next in succession. But more revealing, I would contend, are the shifts in *consciousness* that drive those events. These four women, viewed together, catalyzed a radical shift in consciousness that rippled across not one line of thinking, but many, ultimately touching and transforming the entire culture. Reading their stories together only underscores just how monumental—and intertwined—their achievements were.

In closing I should add a brief word about this book's structure. Because this is a group biography and not meant to be an encyclopedic retelling of these women's lives, I have set their stories side by side with an eye to the genesis of their ideas, and the quirks of character and circumstance that shaped who they would become. For this reason, I have told their stories in some depth up to their breakthrough moments in the 1960s, and from there summarized the events of their later activism and work in a more cursory way. (The exception is Alice Waters, who, because younger, is taken further forward in time.) It is my hope that these bold, brave, and prophetic women will move and inspire the reader as powerfully as they have me.

introduction

THE AGE OF WRECKERS AND EXTERMINATORS

Bikini Atoll, Pacific Ocean, July 22, 1946

For weeks the forty-two thousand soldiers had been preparing. They had relocated the eleven local families who lived on Bikini Island, a lush tropical paradise with ice blue water, white sand beaches, and coconut trees. Bikini was one of a handful of pristine islands that circled the lagoon, part of the Pacific Ocean's Marshall Islands chain. The navy had decided it was the perfect target for testing atomic bombs. The 161 islanders, many of them simple fishermen, had been assured that the move would be temporary, which everyone believed.

Now the soldiers dressed in khakis stood on deck. They peered through their binoculars, their sights trained on the "ghost fleet" of ships they had towed into the harbor a few weeks earlier. They were nine miles offshore and the ships looked tiny, like toys. But the visibility was good today, even with the tinted goggles they had been issued to protect their eyes during the blast; the radio station they had set up to record their impressions was ready.

That morning their families back home had been assured by the vice admiral that the undersea bomb "will not start a chain reaction in the water, converting it all to gas . . . It will not blow out the bottom of the sea and let all the water run down the holes. It will not destroy gravity." The seamen were proud to be part of such an important operation. As the countdown began, they tensed in anticipation. And then they saw it: a blast of white light, followed immediately by a frothing, seething mushroom cloud of sea spray that shot into the sky. It was spectacular. The foaming, boiling column of water kept widening. It engulfed the tiny ships in the distance and, for a moment, blotted out the blue of the sky. Then quite suddenly the winds shifted. Instead of being out of range of the fallout, the troops found themselves directly downwind. Spray rained down on the deck and drenched the soldiers' clothes. They were hit with bits of coral and small stones and debris from the explosion. But no one was worried. The rain of rubble and lethal seawater was over quickly. And most soldiers had never even heard of the word "radioactive."

Within hours of the blast, the seamen were fast at work, steaming toward the island, eager to record the nuclear bomb's effects. Some of the ghost ships had sunk; others were blasted with holes, scorched tar black, their metal fittings melted into grotesque shapes. The lambs and pigs they had put onboard had oozing burns, lesions covering their bodies; many were dead and were already beginning to bloat.

For the rest of that summer it was wickedly hot. Often the mercury rose to 100 degrees Fahrenheit. The soldiers swam in the blue lagoon to cool off. They washed their clothes in it and used it to cook their food. Often when they worked, they shed all but their navy-issued khaki shorts.

Over the next eleven years, the American military conducted repeated atomic tests at Bikini, detonating hundreds more bombs, including one on March 1, 1954, that was "1000 times more powerful" than the bomb that was dropped on Hiroshima. On that day the winds had shifted again and the soldiers and islanders on neighboring atolls had been blanketed with fine white ash. The children had played in it as if it were snow. It was an innocence whose imminent loss would signal the beginning of a new consciousness. The potential for the extinction of all nature had arrived as blindly as that white windborne ash.

Maine, Summer 1962

IT WAS MIDNIGHT WHEN THE LONE, AUBURN-HAIRED WOMAN ARRIVED on the beach. Tall and stooped, just shy of fifty-five, Rachel Carson looked considerably older than her years. She swayed a moment as she sat, drank in the briny air. To feel the full wildness, she switched off her flashlight. Then, adjusting her eyes to the darkness, she turned her attention to the swell and roar of the sea. Tonight it was full of "diamonds and emeralds," flecks of phosphorescence that wave after wave hurled onto the sand. The individual sparks were huge. She could see them "glowing in the sand, or sometimes, caught in the in-and-out play of water," sluicing back and forth.

This is what Carson lived for: bearing witness to the natural world in all its mystery, attuning herself to the earth's rhythms and eternal cycles, feeling a part of the vast stream of time. It was why she'd spent the last four difficult years pushing so hard to complete *Silent Spring*. For all her travails, she had known from the moment she'd first read the field studies on the dangers of the synthetic pesticide DDT that she would feel "no future peace" until she shared with the world the gravity of what she saw. She had written the book because she wanted to change things, to alter the way people treated the natural world, to stop the mindless poisoning of it. Though Carson knew she had little time left to live, sitting on this beach tonight she had no regrets; she was filled with a sense that it had all been worth it: the years of isolation; the painstaking work; even her battle, now lost, against the cancer. The public's reception of the excerpts appearing all summer in *The New Yorker* had been immediate and enthusiastic, greater, even, than she had dared dream. Especially cheering had been E. B. White's kind note, commending her for—by now she had memorized the words—"the courage you showed in putting on the gloves and going in with this formidable opponent, and for your skill and thoroughness." *Silent Spring* would be "an *Uncle Tom's Cabin* of a book," he predicted, "the sort that will help turn the tide." Perhaps she could relax now. Finally, people were beginning to ask questions. They no longer "assumed that someone was looking after things," that the mass aerial spraying of DDT "must be all right, or it wouldn't be done." They were beginning to understand that once these pesticides entered the biosphere, they carried the same hazards as nuclear fallout, the same capacity to alter our genetic makeup in grave and irreversible ways; these chemicals not only killed bugs but also migrated up the food chain to poison birds and fish and eventually sicken humans.

Carson hadn't been surprised by the smear campaign the chemical industry was mounting that summer. She had anticipated their aggressive attacks on the book. But the defamation of her character—the charges that she was a Communist and a subversive, that her purpose in urging more care in the use of agricultural chemicals was to "jeopardize the nation's food supply"—this she had not expected. The most vulgar had come from the former secretary of agriculture, who had wondered aloud "why a spinster with no children was so concerned about genetics?"

But none of this mattered; she was winning in the court of popular opinion. Her meticulous care in presenting the science had paid off. It was 1962 and the world was changing. There was a new optimism in the air, a sense that things were opening up in response to the deep freeze of the Cold War. A fresh generation of young people had come of age. They were better educated than their elders, more idealistic and open, more willing to ask questions, to openly challenge the status quo. Their heroes were Pete Seeger and Joan Baez, Bob Dylan and a band of subversive dreamers who called themselves the Beats. Their anthems spoke of a different sort of America, one in which individualism and community might coexist, an America where social justice and personal fulfillment were not at odds. President John F. Kennedy was newly elected. He projected youth and daring, as did his young wife, Jackie, with her designer clothes and cosmopolitan chic, her fluency in French and dazzling sophistication. Kennedy had just invited Carson to a private gathering at the White House. He had referred to "Miss Carson's book" in a recent presidential press conference. In less than a month she was going to do an hour-long interview on television news, the new medium. The chemical companies were right to be worried. Their claim that now "no housewife would reach for a bug bomb without fear" was well founded. Women were already concerned about a host of contamination issues: "food additives, thalidomide, radioactive fallout." And now they had to worry about poisons in their vegetable gardens. The strict postwar division of the sexes, which had stranded a generation of women in the suburbs, their sole duty to be good mothers and consumers, was backfiring. The problems she was identifying were bigger and more irrefutable than even her critics understood. She had shined a spotlight on big business's carelessness toward the natural world, daring to make its indifference public. She had had the audacity to mount a critique of the "gospel of technological progress," forcing an open discussion of the notion that living things and their environment were intertwined.

Carson had never been the wild-eyed crusader her critics hoped to portray. Shy and self-effacing, considered in her speech, she exuded an inner stillness, a ladylike dignity that must have disarmed her foes. A loner at heart, almost pathologically private, devoted to her small, broken family and a handful of friends, attached to her cats and her beloved Maine cottage, she was happiest amidst the wild beauty of the wilderness, alert to the birdsong in the shadowed forest, the seagulls wheeling overhead, the swirling fog and mysterious tide pools along the shore.

She turned to the sea again. It was getting late; she had already stayed longer than she planned. Her eye caught a passing firefly now, his lamp blinking. "He was flying so low over the water that his light cast a long surface reflection, like a little headlight." She registered a pulse of joy. How ingenious nature was, how intricate the interrelationships between species, including humans. This too was what she wanted to share: the wonder of the natural world in all its variety and strangeness, the amazing interconnectedness of it all.

Her eyes swept the shoreline one last time, making a mental note of the rocks "crowned with foam," the long white crests running down the beach. She was leaving tomorrow. She hoped it wouldn't be the last time she could manage this walk to the beach. Flicking on her flashlight, she rose now with some effort and started toward the path, the funnel of her flashlight beam bobbing as she crept along. She took slow, halting steps, her breath labored. This too no longer mattered. She had finished the book. She had spoken out, and to her relief, the world seemed to be listening.

That same summer of 1962, several hundred miles south of where Carson sat, an impish, white-haired woman in a dark shift and a costume-jewelry necklace of oversize beads stood holding a placard on a street corner in Lower Manhattan. Tall and square-faced, with a Dutch boy haircut and thick, black-rimmed glasses perched atop an aquiline nose, she searched the crowd, smiled in recognition as she spotted her neighbor, the owner of the coffee bar down her block, which lately had become an ersatz community clubhouse, the place where she and others from Greenwich Village and Little Italy had been strategizing over martinis and cigarettes for weeks. Bighearted and affable, the neighborhood sage, tonight he was dressed as a skeleton and carried a placard shaped like a tombstone, on which was scrawled the words "Death of a Neighborhood." She shot him a quick, amused look, nodded in appreciation at his getup, then leaned in to confer for a moment with the congressman standing to her left, knitting her brow in concentration. She glanced at her watch and shook her head in agreement, sending her thatch of white hair flying. Then, moving with obvious deliberation, she threaded her way through the throng toward the podium, her quizzical face set despite the patter of applause that swelled through the crowd.

The object of this applause was Jane Jacobs, a magnetic forty-six-year-old writer and mother, who had recently become a celebrity and pariah to every urban planner in the land. A year before, in 1961, she had written an audacious little book called *The Death and Life of Great American Cities*, arguing that the men supposedly bettering America's cities were actually laying waste to them. The power and eloquence of the book had instantly hit a nerve, giving voice to what many had begun to sense—that the "poohbahs" who planned high-rise housing projects and made their decisions mostly behind closed doors were not always acting in the public's best interest. Almost overnight she had been hailed as an urban hero, a silver-tongued prophet of the people, her name synonymous with grassroots efforts to halt urban renewal projects that demolished vital existing neighborhoods. The *Village Voice*, the scrappy new alternative downtown paper, was calling her "the terror of every politico in town," gleefully claiming she had made more enemies than any American woman since Margaret Sanger. Diane Arbus had photographed her for *Esquire*. *Vogue* had paid its homage, dubbing her simply "Queen Jane."

On this particular steamy evening, Jacobs stood amidst an overflow crowd of residents from Greenwich Village, Little Italy, Manhattan's Lower East Side, and what would soon be dubbed Soho. The idea was to stage a mock, New Orleans–style funeral march down Broome Street into Little Italy, to protest the proposed Lower Manhattan Expressway, a monstrous, elevated superhighway that master builder Robert Moses was trying to string across Lower Manhattan, part of a vast "spaghetti dish"of expressways that would loop around and across the city. The hulking ten-lane highway would rip through low-slung blocks that still had a tatty, Paris-like feel as it made its way along Broome Street. It would mean bulldozing more than 400 buildings that housed 2,200 residents and 800 small businesses. It would wipe out the pastry shops and cozy restaurants of Little Italy, the lighting and restaurant supply shops on the Bowery, the shady park on Chrystie Street.

This was just the sort of top-down planning Jacobs deplored. It was arrogant and misguided, having nothing to do with what gave a city neighborhood its vitality and charm, its ability to adapt and remake itself as conditions changed. Cities were no different from oyster beds. Or "colonies of prairie dogs." They were living organisms; they thrived, just as in nature, on diversity and readaptation, not rigid order imposed on them from outside. This kind of progress killed cities. It slashed apart close-knit neighborhood communities, leaving desolate, gaping holes in the urban fabric. It was a monstrous mistake that went against everything Jacobs had observed about urban life from her own lively little block in Greenwich Village, whose intricate rhythms she had likened in her book to a kind of exquisite "sidewalk ballet," a dance that commenced every morning with the clatter of trash cans and the babble of children en route to school, expanding and reinventing itself throughout the day. What made her own neighborhood—and so many others like it—feel so vital was its short, bustling blocks and patchwork of old and new buildings, its crazy-quilt mix of commercial and residential uses: houses interspersed with stores and cafés, warehouses with restaurants and bodegas. At any hour of the day or night, there were people on the street: mothers pushing toddlers in strollers, longshoremen slipping into taverns at the end of their shifts; teenagers preening, checking their reflections in storefront windows, fathers strutting home after work; theatergoers scurrying off in evening clothes. There were always "eyes on the street," as she liked to describe it, which made everyone feel safe.

For decades, planners had ignored what occurred "tangibly and physically" on the street, which she found exasperating. They had sailed off on "metaphysical fancies" instead of asking people what kind of housing actually made them feel good, or why certain blocks felt inviting, while others breathed menace. This was the genius of her book. She had actually bothered to wade in and ask people, to walk the blocks of neighborhoods that worked—even those the experts deemed expendable—and then describe, in her trademark pungent prose, the intricate dance of particulars that made these districts thrive while others withered. This is what had made the so-called planning experts so mad—besides the fact that she wasn't college educated, let alone trained in urban planning: she had had the audacity to dismantle, point by point, all their airy abstract theories, drawing on a mix of intuition and her own firsthand observations to make her case.

But clearly that case had to be made again. This was why she and so many others were here tonight—Democrats and Republicans, shopkeepers and professionals, plumbers and artists, Catholics and Jews. To explain once more why this outrageous boondoggle of a road would rip the soul out of these vital neighborhoods. To put the city on notice that the residents here would not stand for it. That they would keep on fighting this ill-conceived plan until it was wiped off the map.

Jacobs was at the podium now. She clomped up the steps and paused, taking full measure of the crowd. This was good, she reflected. The turnout was large; people had gone all out. Many were outfitted with gas masks, to emphasize the soot and air pollution the highway would bring. The press photos would be theatrical, as she had hoped.

Leaning in to the microphone now, she graciously acknowledged the state senator and representative who had just preceded her. The crowd went silent, all eyes on the striking, white-haired woman in a sack dress and sandals who looked like a hausfrau, but seemed to command the respect of a queen. Then she began to speak.

"What kind of administration could even consider bulldozing the homes of more than two thousand families at a time like this?" she asked matter-of-factly. A camera flash popped and flared, briefly illuminating the platform where she stood, but she seemed not to notice. "With the amount of unemployment in the city, who would think of wiping out thousands of minority jobs?" she continued, widening her eyes. "They must be insane."

No one wanted this roadway, Jacobs went on to explain. No one but a few out-of-touch bureaucrats, she added. It would kill lively neighborhoods that had been standing almost since the Dutch first put down roots in Manhattan four centuries before. The crowd stood rapt, drinking in every word.

"The expressway would Los Angelize New York," she declared, pausing for a moment. This was the sound bite she hoped would make the evening news. This proposed highway is a "monstrous and useless folly," she added. "The arguments for it," she continued, "amount to piffle."

Applause rippled through the crowd, whistles and hoots of agreement, more camera flashes. True to form, Jane Jacobs was once again making waves, poking holes in official cant about the efficacy of urban highways, just as she had the city's earlier arguments for razing the West Village. She was tapping into a current of quiet discontent that lay slumbering just below the surface of the culture, demonstrating that ordinary citizens, if they were organized enough, could push back, even defeat the swaggering bureaucrats who for the last decade had been calling the shots, riding roughshod over the greater good, with no sense of consequence or need for accountability. This was her message tonight, and, like Rachel Carson's, it was hitting a nerve.

FOR MANY PEOPLE, THE SUDDEN APPEARANCE OF CARSON'S AND JA-cobs's brilliant and prescient books was one of those moments that seem, in retrospect, to have changed the very order of things. Both *Silent Spring* and *The Death and Life of Great American Cities* had an almost immediate effect on public sentiment. *Silent Spring* was not only a runaway bestseller but is credited with having led, in the short run, to the creation of the EPA; in the long run, it spawned America's environmental consciousness. Jacobs's book is said to have changed urban renewal policies across America and dethroned Robert Moses. It became the bible, ultimately, for the preservation movement and for the larger idea of self-emerging systems in cities rather than centrally imposed plans.

Like another wildly popular and transformative book that came out at almost the same time, Betty Friedan's *The Feminine Mystique*, which also unleashed a tsunami of change—and *also* had the effrontery to question the culture's most enshrined assumptions—Jacobs's and Carson's books were articulating for ordinary readers what many were beginning to feel. They were connecting with people on a visceral level, calling into question the nation's exuberant and self-assured path, suggesting that perhaps it wasn't as rosy as it had been cracked up to be. Both were pointedly addressing one of the central paradoxes of progress: in the blind embrace of technology there had been a loss of human scale and a degradation of the physical and social environment. What each stirred was the slumbering fear that technological innovation held the power to transform life as it was known into something not just alien, but inherently threatening. Change, the great hope and mantra of modernity, was perhaps not so benign.

Though they didn't yet have a name for it, Carson and Jacobs were sowing the first seeds of the green movement. They were putting a personal face on complex questions about the country's collective future, drawing upon values that were deeply rooted in America's earliest, Edenic idea of itself. Both were openly indicting 1950s corporate culture and its misplaced priorities, its self-interest and shortsightedness in plundering the commons; they were putting forward a set of values different from those of the big boys and their big business.

The fifties had been an era of Cold War fears and conformity, of getting and spending, setting up house and keeping up with the Joneses. It was a decade driven by powerful strains of hypermasculinity, its poster boys the Marlboro Man and the rakish playboy bachelor, the cool corporate executive and the clever ad man. The culture was awash in a glut of consumer goods and spanking-new, high-tech weaponry. It was enthralled by its own power and reach. (Despite the terrifying specter of Armageddon that loomed.) It went without saying that men were the breadwinners in this brawny new world; women's place was in the home, or at the shopping center buying shiny new appliances. The popular understanding—even by educators at prestigious women's colleges like Radcliffe—was that the "proper goal" for an intelligent woman was marriage. To be a stay-at-home wife was an honor (or so the thinking went), a signal that one had reached the middle class. Family "experts" enjoined every housewife to help her husband "rise to his capacity." Everything from the legal system to the movies reinforced this idea. Women were not meant to compete with men, to act independently of men, or to have adventures or strong opinions of their own.

There were scores of women who *did* work, of course, whether out of economic necessity or because they wanted to. But they were relegated to underpaid, ancillary jobs. In the higher precincts of power, the presence of women was pretty much a nonevent. At the dawn of the 1960s, only 6 percent of doctors were women; only 3 percent of lawyers were women; less than 3 percent of U.S. senators, members of Congress, and ambassadors were female. It was assumed that women couldn't possibly be scientists, TV news anchors, movie directors, or CEOs. In some states, a woman couldn't get credit without a male cosigner. In others, they were barred from serving on a jury, as it might involve neglecting their domestic duties. When Carson and Jacobs published their books, the idea of gender equality wasn't part of anyone's vocabulary.

And so the question: What was it about the insights of two uncredentialed women working outside the mainstream that carried such power? And why at that moment? Was it a coincidence that also in 1962, half a globe away in a remote corner of Africa, a third woman, a wild-hearted young Brit named Jane Goodall, was working a parallel vein, having witnessed chimps using tools the year before, a discovery that in a single stroke would reposition mankind's place in the natural world, no longer the exception in the cosmos, but one among myriad creatures? Or that around that same time, in Berkeley, California, incubating amidst the first stirrings of the free speech movement and the be-here-nowness of the embryonic counterculture, an elfin, twenty-one-year-old Berkeley student named Alice Waters, the soon-to-be mother of the organic-fresh-locally-grown food movement, was thinking along similar lines, dreaming of starting a restaurant that would change the way people thought about food and eating and the farming practices behind what they consumed?

Maybe it was a coincidence. But the parallels in these women's stories, for all the differences in their lives and careers and personae, beg the question. Each of these four great women, in intriguingly analogous ways, profoundly transformed an important dimension of the way we think about the world; all were preservationists of one sort or another, humanists intent on the longer view; all were antiestablishment outsiders who took on idealistic eco missions before anyone had heard of "eco" or "green"; none were academic theorists—Jacobs and Goodall didn't even earn undergraduate degrees—but rather people who waded into the fray in their respective fields and got their hands literally and figuratively dirty. Each faced down powerful (mostly male) adversaries and against all odds prevailed; and all brought a fresh, much-needed scrutiny to their particular spheres. Although members of two or three different generations (born between 1907 and 1944), all, interestingly, had their personal breakthrough moments in the 1960s, just before and as the second feminist wave was cresting, becoming models of what a woman could do and be, even if that hadn't been their original intent. And their influence is still huge and inspiring and more relevant than ever.

Today we take it as gospel that marvelous older buildings rich with architectural detail are worthy of saving, that a masterwork like McKim, Mead & White's Pennsylvania Station in New York, with its soaring, vaulted, steel-and-glass ceilings, wouldn't meet with the wrecking ball, as it did with little debate in 1963. We understand the value of preserving older cities; of supporting local farms and sustainable fisheries; of eating fresh and wholesome food, embracing the satisfactions of sharing our table with family and friends. We grow wiser about the perils of chemical contamination and rue the proliferation of tasteless, prefab convenience food bereft of nutritional value; we bridle at the sight of a zoo animal deprived of the light or space or physical stimulation necessary to its well-being. Our familiarity with these wise and salutary ideas is testament to how deeply they have embedded themselves into our collective consciousness and national debates.

But in 1961, these concerns weren't even blips on the radar screen. The future belonged to our astonishing technological know-how. Bugs would be eliminated with pesticides; weeds and food pathogens eradicated through genetic engineering and herbicides; slums torn down and replaced by gleaming towers or faceless cookie-cutter housing projects like Levittown, where everyone could own a little slice of the American dream. Big and standardized was better, mass-produced more efficient, more profitable. Architecture, as Le Corbusier once said, was a machine for living. Animals, insentient and unfeeling, were ours to use as we pleased, to entertain us in zoos, or engineer to make our industrialized food production more efficient. The natural world was ours to dominate and, if need be, despoil. Nature existed to serve humankind's needs.

Jane Goodall, Jane Jacobs, Rachel Carson, and Alice Waters changed all that, shifting the cultural conversation in profound and enduring ways. They saw and helped articulate the unconscious currents at play within the culture, and in so doing sparked the imaginations of ordinary people, offering them a new, more holistic way to think about the world and a more benign way of living in it.

In each case, their work underscored the fragile interconnectedness of the living world, in this way spurring a new public awareness of the links, respectively, between chemicals and the contamination of the biosphere; the preservation of neighborhoods and the sustainability of cities; the protection of habitats and the survival of animal species; the nurturance of farmland and the wholesomeness of food—or to put it another way, between man and his environment. In each instance, they argued for a radical shift in the way we collectively inhabit the planet, wary of the culture's blinkered enthusiasm for science and technology unmoored from ethical responsibility. And they all found their audiences during a decade that saw enormous changes and upheaval.

Writers have described the 1960s variously as a decade of carnivalesque "spectacle"; "an experiment in political theater"; a struggle for the nation's soul; a time of political and moral radicalization; a moment of buoyant change; a time of taboo-shattering sexual antics and instantaneous freedoms; and an era of strife and polarization riven with explosive, sometimes violent conflicts. Depending on one's vantage point, all of these descriptions hold true.

But what interests me here is another legacy that hasn't been much explored: the deep-seated shifts in the values and sensibilities of the culture, beyond the politics of feminism, that these four remarkable women catalyzed, and the ways in which their ideas simultaneously fed and expanded upon the larger issues shaping the era and its aftermath. What was it about the cultural soil of the 1960s that proved such fertile ground for their visions, allowing them to take root and flourish? Why was the culture so primed to hear what they were saying? For these women's dissident positions, their resistance to America's love affair with technological novelty, their embrace of the intuitive and the local, their contempt for the "machine," whether it was of war, or entrenched urban politics, or agribusiness, or the juggernaut of the chemical industry, turned out to be critically *of* their age. And I think the fact that they were female, interlopers almost by definition at that moment, was germane to their respective achievements—that is, their outsider status led them toward fresh ways of seeing their particular fields, and inspiring each was a kind of nurturing, arguably maternal instinct.

This is a group portrait of four accidental revolutionaries who made an indelible imprint on the world and the intriguing connections between them: a shy government science editor who wrote freelance articles to augment her pay; a silver-tongued editor-turned-activist-writer enamored of the urban environment; a plucky Englishwoman with a passion for animals who dreamed of going to Africa; and a soft-spoken idealist who loved cooking and the sensuous good food of southern France. That they came from disparate worlds and didn't know each other, that they grew up in different decades and stumbled into their respective fields in odd and utterly singular ways, that their temperaments were dissimilar and the degrees of their influence vary, only add to the unlikelihood that they should share so much and be so inextricably tied. But they *were* tied, it turns out, not only by the myriad and surprising common strands in their respective stories, but also by the striking similarities in the philosophical underpinnings of their thoughts, despite the differences in their origins. That they happened to be in the right places during an era when the culture was primed to have its moral compass reset is also part of their collective story, for their ideas both shaped and grew out of their moment. Each of these women was a visionary in her own right, a pioneer who spurred a powerful social movement that would change the course of history. Each made real the sometimes far-fetched notion that an individual, armed with grit and colossal courage and an abiding, outsized love for what she does, can make a difference. For these four women made all the difference in the world. Their ideas still have the power to touch and inspire and illuminate the way forward.

chapter ONE

....

RACHEL CARSON

As the winter of 1938 limped into spring, the news from Europe grew increasingly grim. On March 12, Nazi soldiers stormed into Austria, annexing the country in a single day, while the world looked helplessly on. That September, as the annual Nuremberg Rally opened with an ominous display of militaristic fervor—goose-stepping marches, human swastika formations, booming Wagnerian overtures—Hitler announced a spate of new anti-Jewish racial laws. Jews were by now barred from holding passports or practicing in most professions. There were limits on where they could live and work, restrictions on who they could marry, worrisome new policies of forced deportation. On November 9, Kristallnacht, "the night of the broken glass," Nazi thugs looted Jewish businesses and religious sites throughout Germany, torching more than one thousand synagogues and shattering the shop windows of thousands of Jewish stores. The violence sent shock waves across the world, as word of Hitler's anti-Semitic excesses spread. By September 1939, German storm troopers had invaded Poland and Czechoslovakia, and on September 2, Great Britain and France responded with a declaration of war on Germany. To many, it seemed the world was coming unhinged.

But not everyone's attention was on the bellicose Third Reich. That same cheerless September, a lone chemist named Paul Müller had a stunning breakthrough in his laboratory in nearby Basel, Switzerland. He was forty years old and employed by J. R. Geigy. Spurred by a severe food shortage that had nearly starved his country, for four long years he had been searching for a synthetic compound that would kill crop-destroying insects. One of his experiments involved a compound known as DDT (dichlorodiphenyltrichloroethane), which had first been synthesized in 1874. But no one had yet found a practical use for it. Now Müller saw one. Coating the inside of a glass box with the odorless white powder, he filled it with houseflies and waited. At first nothing seemed amiss. But by the next morning, all the insects were dead. A new batch of flies was added and they too suffered the same end. Even after he scoured the box with a solvent, the flies continued to expire, killed by invisible flecks of the substance.

By the following year, DDT was being hailed as one of the great triumphs of science, an illustration of mankind's increasing mastery over the natural world; it was a panacea that would rid the earth of insects forever. Tests showed it was a highly effective weapon against mosquitoes, fleas, lice, and ticks, all hosts for deadly human diseases. DDT was cheap and long lasting; it could be mass-produced, and at low doses didn't seem toxic to humans or warm-blooded animals. While no one really understood how it worked, it seemed to be some sort of "nerve poison." Whatever its properties, it annihilated insects almost instantly. Soon it was being sped to war zones to combat typhus and other insect-borne scourges. Closer to home, people began spraying their bedsheets with the stuff to kill bedbugs, amazed that just one application worked "for months." As refugees poured out of Nazi-occupied regions, DDT proved an effective delousing agent. By treating the interior walls of a house just twice a year, it appeared to stop the spread of malaria. As the war ground on, entire islands were sprayed aerially in advance of invasions. DDT could be dispersed as an emulsion with water, or mixed with chalk power and dusted on large target areas. When in 1944 a typhus epidemic in Naples was averted after the U.S. Army sprayed DDT on hundreds of thousands of civilians, its spectacular success made international headlines. Four years later Müller was awarded the Nobel Prize in physiology for medicine, his miracle compound celebrated as one of the "greatest medical discoveries in history." All but a few lone wolves agreed.

IN JULY 1938, AN OCEAN AWAY, AND ONLY MONTHS BEFORE MÜLLER'S breakthrough, a shy young science writer with deep-set blue eyes and an intent, slightly preoccupied expression, sat behind the wheel of her father's car. Her wavy, chin-length auburn hair was pinned behind her ears, to help with the withering heat, and she wore a sensible skirt that fell below her knees and a modest white button-down blouse. She might well have been taken for a demure small-town librarian, or a prim schoolteacher, except for the pile of technical volumes stacked beside her on the seat and the spiral notebook and serious binoculars, which spoke of something more single-minded and ambitious, more intense and sharply focused. She stole a glance in the mirror, aware that her attention had wandered. In the back seat, limp with exhaustion, drooped her sister's two daughters, ages twelve and thirteen, their spindly legs akimbo. In the seat beside her, clutching the road map, sat her seventy-year-old mother, Maria Carson.

Though she had just turned thirty-one, Rachel Carson was already no stranger to financial hardship or personal responsibility. That Europe teetered on the brink of war seemed far away, difficult to register given the weight of her own worries, the pressures so much closer to home. Her father was dead now. Three years before he had stepped into the backyard and toppled face-first into the grass, dying moments later and taking with him any lingering illusions that he might somehow reverse the family's financial woes. At the time, there wasn't even money for the Carsons to accompany his body to the burial plot in Canonsburg, Pennsylvania, where he was quietly interred by a brother. Rachel and her mother had done what they could, which wasn't much. In some ways his abrupt passing must have come as a relief, though neither Rachel nor her mother openly expressed it. Her father had been a distant, ineffectual figure, never a reliable breadwinner. At least now there was no longer a need for guilt or pretense for not having believed in him more.

Since then Carson's responsibilities had only increased. Her older sister, Marian, once wild-hearted and attractive, had been unwell for several years, often too sick to hold down even a menial part-time position. Deserted by her feckless husband early in their marriage, Marian and her two daughters had moved in with Carson and her mother, leaving it to Rachel to make up the financial shortfall. By all logic, it should have been her brother, Robert, who stepped into the breach. But Robert's income was erratic, when he had work at all. It was the depths of the Depression and the country was on its knees. The Carson family finances, precarious even in the best of times, were in desperate straits. Though Rachel was teaching part-time at the University of Maryland, to help support the family, while also enrolled at Johns Hopkins as a doctoral candidate in zoology—one of just five female students—she had seen no option but to drop out of Hopkins and look for a full-time teaching position. But in 1935, with millions still languishing in bread lines, such jobs, she discovered, were nonexistent, especially for a young woman.

What she *had* found was an odd position that no one seemed to want: a part-time government job writing brief, upbeat seven-minute radio scripts for an educational series on marine life, which her new boss at the U.S. Bureau of Fisheries described as a "problem assignment," explaining that it would involve both an intimate knowledge of science and literary skills, the latter evidently in short supply in the department. Carson hadn't seen it the same way. The job seemed to her a lifeline, except that it was only two days a week, and paid just $6.50 per day. But even its irregularity, she soon realized, came with a silver lining. As the year progressed and she interviewed shrimpers and oystermen, marine biologists and crabbers, sea captains and the owners of fish shacks, she found her head swimming with ideas for other, more in-depth stories, which she began to write on the days when she wasn't at the office. The first feature she did was about shad fishing, which she sold to the *Baltimore Sun*. Since then she had sold the *Sun* other pieces—on oyster farming; the tuna catch off Nova Scotia; the mysterious, circular migration of the eel. At $20.00 a story, it seemed like a miracle. For the first time she could recall, things were looking up.

But then Marian died suddenly too. A year earlier, she had come down with pneumonia a few days after her fortieth birthday; she'd grown gaunt and pale and within days was gone. There was no one else to raise the two girls but Carson and her aging mother. At twenty-eight, Rachel had become the sole provider for a family of four. Any privacy or open time she had known was over.

Her college mentor urged her to take the civil service exam. And finally, in 1936, she was offered a full-time position with the U.S. Fish and Wildlife Service, which carried the misleading title of junior aquatic biologist. In truth, she was allowed to do neither fieldwork nor anything even vaguely aquatic. Mostly she was charged with "women's work": editing and rewriting the reports of scientists who *were* out in the field—all of them predictably male. But she was managing to keep up her own writing, working on weekends and in the evenings when she got home from her job, often pushing deep into the night. It was a grueling schedule that ruled out any social life beyond the company of her mother and two nieces. But her perseverance was paying off. She was chipping away at a writing project she hoped would change her circumstances. It was to be a book about the life of the sea. She was driving to the small seaside town of Beaufort, North Carolina, at the southern end of the Outer Banks, to do research for it, using her long-awaited ten-day vacation from work to collect visual impressions for the story that was beginning to incubate in her head.

CARSON GRIPPED THE STEERING WHEEL, WATCHING THE ROAD AHEAD. It was a stifling day and the windows were down. Hot air blew through the car like a blast furnace. But as with so much in her young life, if Carson felt burdened by the heat, or the weight of her enormous financial responsibilities, if she resented the imposition of her sister's children, or worried that she would never reconcile the division she felt between the needs of her family and the pull of her work, she chose not to dignify it. Carson seemed to accept her lot in a way peculiarly her own; it was not so much an absence of self-pity, but rather a swallowing and acceptance of troubles before pain or resentment could even arise. Though she was quiet and reserved, and there seemed a fragility about her, she was also uncommonly stoic. Fiercely ambitious for herself, Rachel Carson possessed a strength of mind and a determination that set her apart from most people. She had a rare doggedness, a capacity for sustained focus that even those who knew her found remarkable. She could be working into the early hours of the morning, almost asleep in her chair, yet still driving herself, as if fatigue were an indication of weakness or lack of will. It was not so much escape, or even freedom that Carson yearned for, but time and a measure of security. And family obligation was all she knew.

They reached the outskirts of Beaufort late that afternoon. Carson helped her mother and the girls get settled in their rented cabin, then climbed back into the car, keeping her eye out for landmarks so she could find her way back—the tidy white church on the corner where she turned, the pocked blue door of a shop. She was headed for the fisheries lab on nearby Pivers Island, just west of the historic town; from there she would decide where to commence work the next morning. Beaufort was an old and stately town, a sleepy port that pulsed with the ebb and flow of the tides, the flux of the sea winds. She piloted the car past tidy, white clapboard houses, some dating back to the 1700s, and on to the harbor, her eyes on the broad sound stippled with islands unspooling before her, the grassy salt marshes wading boldly into the shallows. At the fishing piers she slowed, taking in the gray-shingled fishermen's shacks huddled at the water's edge, the vacationers milling about, buying ice cream and buckets of fried seafood at the fish stands by the wharfs. She didn't stop until she reached the fisheries station, where she finally got out and stood for a long time on the empty beach, breathing in the pungent salt air, staring intently at the view. From the low sand-and-scrub island, she could see the placid blue sheet of the inner sound, and in the distance, the checkered spire of the Cape Lookout Lighthouse. Farther out, she could just make out the long, nine-mile strip of beach and dunes called Shackleford Banks. It was here, Carson decided, she wanted to go.

Carson must have found someone from the fisheries station with a boat to ferry her over, because every morning for the next ten days, she chugged across the busy channel to the lovely uninhabited barrier island that separated the inner sound from the sea. Shackleford Banks was long and exceptionally narrow. Even at its highest point, it rose no more than five or six feet above the high-tide mark. But the variety of marine life and terrain she found there was perfect, she felt, for her purposes. On the ocean side, the thrashing sea hurled wave after wave against the strand; along the island's inner shore, the water was almost still. There were tide pools and low-lying marsh ponds in the flats where the sand dunes descended to the sea. And she discovered that by walking just thirty minutes along the wild ocean side she lost sight of all civilization. Alone on the beach, with only the sea and sky and the long white ribbon of sand unfurling in either direction, time fell away.

Day after day Carson walked the beach, watching the surge and fall of the tides, the shorebirds patter across the wet sand, the puffs of blown sea froth rolling like "thistledown" along the strand. She stood at the ocean's edge and listened to the waves, smelled the salt air, felt the sun on her back. She waded into the tide pools, spending hours studying the tiny sea creatures there: "little transparent worms with sharp biting jaws;" barnacles that tentatively opened their shells, rhythmically sifting the waters; clams stirring in the mud. Standing in the shallows, she noted how the crabholes "honeycombed" the beach with their "burrowings"; how the glittering light danced gently across the ribbed bottom of the sand. Swishing a hand into the salt ponds, she tasted their bitter tang.

Sometimes she lay in the dunes, arms behind her head, and watched the gulls wheel and glide across the sky, or closed her eyes and listened to the sounds—"the quick sharp sibilance" of a gust of sand blown over the dune; the "soft tinkling" of the ocean turning the shells on the wet sand; the harsh bark of the gulls.

Nothing escaped her hungry gaze. Carson studied the flora and fauna at Shackleford Banks with a kind of blinkered intensity, drawing upon all her senses, alert to the shifting light and hues of shore and sea, the cycles of the day, the smell of the marsh, the sounds of wind and sand and shorebird. Now and then, stopping to sit on the beach and record her thoughts, her small spiral notebook balanced on her lap, her earnest, down-turned face tensed in concentration, she would scribble out bits of broken narrative, ideas and sensory impressions, filling page after page with meticulous descriptions, intent on capturing all she saw.

Tireless, passionate, often she went out at dawn, when the sky was powdery and the dark waves flashed silver against the glassy sea; often she didn't leave until dusk, when bird forms became "dark silhouettes" and the only light that remained was that mirrored in the pools of water left on the beach. Sometimes she returned after dark, having gone home briefly to eat and check on her family, and there, flashlight in hand, walked the night strand.

Many times before, in the midst of researching a story, Carson had stared out at the open sea and contemplated the long sweep of time it represented. She found it calming to think of the timelessness and circularity of its processes, the endless cycling of life and death and rebirth that was continuously reenacted there. To stand at the ocean's edge, to feel "the breath of a mist drifting over a great salt marsh, to watch the flight of shore birds that have swept up and down the surf lines of the continents" for untold generations, was to bear witness, she thought, to things that had been going on for "countless thousands of years," to see patterns that were as "ageless as sun and rain, or as the sea itself." The sea was a place where one got a sense of the "great antiquity" of life, she felt. At its glacial depths, years gave way to centuries, and centuries, as she understood it, into ages of geologic time. It was a way of looking that she found helpful, a focus on the living flow of things, the eternal rhythms and relationships that had endured for eons. Viewing the sea from this vantage point provided "a little better perspective on human problems."

Now, as her days at Shackleford Banks drew to a close, besides the details of the marine life she collected, it was this larger picture of the sea that began to preoccupy her. Before coming to Beaufort, she had conceived of the book as a shore guide of sorts, a descriptive account of the daily life of a handful of sea creatures. But now she saw that the book needed to be something more, a view of the sea that was bigger and more viscerally charged, that would dramatize the fragile web of connections that bound each sea creature to its particular home, as well as to its place in the savage pecking order that held sway among the other sea-dwelling residents there. Observing the splendid variety and strangeness of sea life at Shackleford had underscored what she already knew, which was that no single marine creature or its habitat could be understood in isolation. All were part of a greater system. Every living organism, no matter how huge or infinitesimally small, belonged to a "larger diverse community," all "sustained by interdependence." It was this she needed to animate for the reader—the elaborate interconnectedness of this delicate web, the mysterious and complicated "interplay" between the ecosystem and all its inhabitants. All life, she saw, "is connected."

Carson's holistic view of the earth and the ocean, her sense of the living world as a web of relationships and interconnections—what we now call ecology—was not an approach to science much practiced or honored at that moment. Most biologists at the time tended to count and categorize species, each according to its physical attributes, to see the world in atomized parts. Carson, by contrast, was focused on mapping the connections between living things, in understanding the links and interactions that sustained the entire web of life. At a time when the culture gave priority to specialists, when scientists were learning to break down the world to its molecular elements, she was seeing the natural world as a single, integrated system, as more a *process* than an inert place. The natural world was a balance of live and ever-evolving forces.

It was a way of thinking that would become Carson's signature contribution, distinguishing her writing from others working at the time. Carson would effectively begin the modern environmental movement by popularizing the principles of ecology, writes Linda Lear, author of Carson's definitive biography, *Witness for Nature*. She would become one of the most beloved and widely read nature writers of her day, the author of three bestselling books about the sea and a stunning, history-changing book that would break through the Cold War paranoia of her time to discuss the environment, thus winning the hearts and minds of everyday people. She would also become one of the first modern thinkers to get people wondering if "all technology was progress," notes environmental historian Elizabeth Blum. Her position as someone outside the mainstream would allow her to challenge the status quo of 1950s culture and industry in a way that no one before her had. But all this remained in the future.

CARSON RETURNED TO HER HUMDRUM JOB AT THE U.S. FISH AND WILD-life Service the day after driving home to the modest split-level rental in suburban Maryland where she lived. Her days at Shackleford behind her, she fell back into her old schedule. Every morning, before leaving for her tiny, light-starved office, she handed her mother several pages of government stationery, on the back of which, drafted in longhand, was her previous night's work. A slow, meticulous writer, on a good night she might log fifteen hundred words; but usually it was less, sometimes just five hundred. When she returned home each evening, she shared a quiet dinner with her mother, reviewed her nieces' homework, and then padded off to her room to begin the night's toil, finding the pages her mother had typed for her on the desk beside her Olivetti. It was a routine that

Rachel and her mother would follow until the last draft was done, and that would continue for every book she wrote, a pact between mother and daughter that reached back into childhood. For in a sense, Rachel's budding writing life was the joint realization of both mother and daughter. Every milestone Rachel achieved drew Maria closer to her long-cherished dream that Rachel have all the chances she herself had sacrificed on the day she met Robert Carson.

MARIA CARSON HAD MARRIED BELOW HER SOCIAL STATION. THE DAUGH-ter of a stalwart Presbyterian minister who died early of consumption, she had come from some means. Delicate-boned, with deep-set eyes and the high forehead and chiseled features of her Scotch-Irish ancestors, she had attended an all-female seminary school, enjoying a robust classical education, a source of considerable pride. Cool and cerebral, a touch brusque, she'd been known as a driven student and a gifted musician. Educated to take up the mantle of "civic responsibility" and modest "Christian motherhood," it had never occurred to her that she would find neither the time nor money to pursue her own musical or literary interests, or that marriage would narrow her prospects.

Maria was teaching school when she met Robert Carson. Reed thin and of medium height, with sympathetic blue eyes and prematurely thinning hair, at thirty he seemed worldly and exotic compared to the other young men she'd encountered. That he hadn't completed high school, or was from a more hardscrabble family than hers, didn't seem to her a problem. They both shared a love of music. Robert sang in his church choir. He had beautiful manners, and an appealing reserve. Quietly he courted the twenty-five-year-old Maria, and when he asked her to marry him less than a year later, she accepted, despite her mother's disapproval.

They had their first two children sooner than they'd hoped. Marian, the eldest, was born in 1897; a son followed two years later, whom they named Robert, after his father. They had been living uneasily with Maria's mother, and now they needed additional room.

Robert Carson went out in search of a home for his growing family. When he found a parcel he liked just eighteen miles up the Allegheny River from Pittsburgh, he secured an $11,000 mortgage, brashly confident that he could make the economics work. Sixty-four rolling acres, the land was just outside Springdale, Pennsylvania, a down-at-the-heels river community of twelve hundred that Robert was convinced was on the upswing. Graced with meadows and woodland, apple and pear orchards, the property had a two-story log house and a barn, a springhouse and a chicken coop, a honeysuckle-smothered porch, and a standing fireplace. If the house itself was meager—just four drafty rooms, with neither electricity nor indoor plumbing—the land around it was wild and untrammeled, the views expansive. It seemed full of potential. Maria set to work, and in the black dirt behind the barn, planted a large kitchen garden, which provided the family with fresh vegetables and flowers. Robert kept a few farm animals, though he showed no interest in farming, claiming to be a "city boy." To the bankers he dealt with, Robert presented himself as a "developer." His plan was to subdivide the lower part of his land into lots, which he hoped to market for $300 each. But until this scheme took off, he was working as a traveling insurance salesman, though this income too was erratic. Paid on commission according to what he sold, Robert often came home with pockets empty. To make up the shortfall, Maria gave music lessons and sold chickens, making do as best she could. But the family was still desperately poor.

Rachel entered the world on the morning of May 27, 1907. Her father, now forty-three, was often away, on the road for weeks at a stretch. Marian was by then in fifth grade, Robert Junior in first. To Maria's delight, she had her pudgy-cheeked newborn all to herself. Rachel was a solitary, sweet-natured child, quiet but noticeably determined. At eight months, she was talking; as soon as she could walk, Maria took her outside, where they spent long hours wandering the meadows beyond the house. Maria encouraged her to wade into streams, to peer closely at seed cask and thistle, to read the light and listen for the wind. She showed her how to be slow, how to watch with all her senses, how to navigate the woods and know the stars. She gave her the names of every weed and wildflower they saw. It was a kind of second knowledge that Rachel learned at the same moment that she was beginning to read, and its precision and pleasures were at least as vital to her as her books.

To Maria, such interests were anything but idle. Like many women of her ilk, she had embraced the nature-study movement, a popular metaphysics in fashion at the turn of the century. The movement was in some respects a theological one, its focus the mystery and uplift of the natural world. The idea was that by studying nature, the "intricate design of the creator" would become apparent. Conservation was therefore a "divine obligation." And so it followed that as a mother, one could do no greater good than to instill in one's children a kinship toward nature and a love of the out-of-doors. It was a cause to which Maria devoted herself.

How disappointing, then, that of her three offspring, it was only Rachel, her youngest, for whom this earth love took. Marian dropped out of high school at the end of tenth grade, and eventually found work as a stenographer. At eighteen, she married a local boy who was four years older, but no more mature. Unable to afford rooms of their own, the couple moved in with the Carsons, adding to what was already a burdened household. Then one day the boy skipped town. Amidst the turmoil, Robert Junior moved out to the backyard, where he lived in a tent. And soon he too dropped out of high school. He worked for a while in a radio repair shop, and then volunteered for the army air service. When he returned to Springdale in 1919, having served for a short time in France, he was rumored to be arrogant and opportunistic. Later, one of his Springdale friends would remember Robert as "the only man I knew who would steal chickens from his own mother."

All this lost potential only made Maria more determined to chart a different, brighter path for her youngest child. On inclement days, she read and sang to Rachel. She saw to it that she had art materials and paper, encouraged her to draw and write stories, to make her own little books. She surrounded her with games and puzzles, new books from the library, the latest nature magazines published for children. Another mother might have welcomed a moment of time to herself, a respite from the toil of housework and childcare. But Maria was not a woman who suffered defeat easily. Her answer to her older children's hollow and hopeless lives was a furious and compensatory energy, all of which she directed toward Rachel's enrichment. She wanted her last child to be independent and to excel, to escape the small, pinched world that she herself had been snared by.

Her efforts, fortunately, didn't go unanswered. If Rachel was the proxy for her mother's foiled ambitions, she was also the source of her greatest satisfaction. Maria couldn't have asked for a more fervent pupil. Rachel was an omnivorous reader, keenly observant and diligent at school. She loved books and stories. Almost as soon as she mastered the use of a pencil, she began to compose her own. At eleven, she already longed to be a writer.

Rachel's school in Springdale only went as far as tenth grade. At Maria's urging, Rachel traveled by trolley to a town on the other side of the river for her two last years of high school. As always, Rachel rose quickly to the top of her class. But her teachers noted that she was often absent. Maria was a compulsive mother and whenever she heard rumors of an outbreak of illness, she kept Rachel at home, choosing to tutor her instead. Rachel never fell behind. But the pattern was socially isolating. Rachel had few friends and she wasn't able to stay for after-school activities; the trolley schedule was irregular and she needed to get back to Springdale by the last train. On those rare occasions when she did ask someone back, it was awkward. Maria hovered and her manner frightened most of Rachel's potential friends. One girl later remembered that winning the approval of the "stern-looking Mrs. Carson was an achievement."

The Carson family's insolvency was by now a subject of town gossip, which only added to Rachel's troubles. Fiercely loyal to her family, she was mortified that even perfect strangers knew they couldn't pay their bills. It seemed better to hold herself apart, to bury herself in books and the beauty and nurture she found in the natural world. Outside and alone, she could lose herself in the green immensity about her. Nature made no judgments, after all. Indeed, whatever unspoken anguish Rachel suffered, and certainly there was some, her separateness would also bring strength, serving to kindle a deep inner fortitude that would stand her in good stead later, a resourcefulness and sense of self-worth she might not otherwise have had were her circumstances easier. Rachel's social isolation taught her to turn the other cheek in the face of human pettiness; it allowed her to follow her own heart.

Robert Carson's bet on real estate, like most everything else he did, had been woefully wrong. When he bought the land, he'd been certain that Springdale would be a beneficiary of the booming Industrial Age. But in less than a decade, the town had become a casualty, a landscape of last resort. Now, from places on their property, Rachel could see smoke belching from the stacks of the American Glue Factory, where horses were marched up a covered wooden ramp to be slaughtered. At the train station, disembarking passengers were assaulted with the stench of horse parts ground up for fertilizer and glue. Industrial flues sullied the skyline, pumping out an acrid soot that burned the eyes and blackened the snow. In the craggy hills nearby, bulldozers clawed at the ore-filled rock, scarring the once-green landscape, replacing it with open sores of rubble and industrial waste. Springdale wasn't a place where hopeful young families dreamed of moving. Robert Carson had misread the signs. No one wanted to buy his lots. None of this would be lost on Rachel, who would draw upon these memories thirty years later as she began *Silent Spring*.

IN THE SPRING OF 1925, A PARTIAL SCHOLARSHIP ALLOWED RACHEL TO go to college. As always, it was Maria who decided where, selecting Pennsylvania College for Women, an elite women's school on the edge of Pittsburgh. That summer, desperate to make up the shortfall in Rachel's tuition, Maria sold the family silver and china. Robert, who was now working part-time at a local power company, promised to raise the rest by borrowing against the Springdale lots. But his health was poor and he seemed increasingly irrelevant in the family.

Early that September, with a mix of uneasiness and anticipation, Rachel left for college. Her father had borrowed a car, a beat-up Model T Ford, so that he and Maria could drive her the sixteen miles to Pittsburgh. Rachel had just turned eighteen. Thin and fine-boned, with the same deep-set eyes as her mother, she pressed her face to the half-open window as the Ford passed through downtown Pittsburgh, her eyes widening as an electric streetcar rumbled by. The Ford entered a wealthier enclave then, and the road began to climb, narrowing to a ribbon as it ascended the leafy hillside to the ridge, where Rachel got her first glimpse of the campus: a cluster of elegant buildings dominated by an imposing three-story Gothic mansion crowned with medieval crenellations.

The family sat for a moment in the car. Well-dressed young women ambled along the walkways that traversed the lawns. Others gathered in small groups beneath the vine-covered arches of a second, equally imposing edifice. If Rachel felt a flutter of apprehension, if she felt suddenly conspicuous in her plain homemade clothes, the serviceable cotton dress her mother had sewn that summer, her sensible tie shoes, brown and badly scuffed, in spite of her mother's efforts to cover the wear with polish; if she wished, now, that she'd made the time to do her hair, or worried about how she'd endure the days away from her mother, even if it was to be only a week, as Maria had promised, she did her best not to show it. Whatever flickers of uncertainty that went through Rachel's mind, she swallowed them, as always. For even amidst these worries, even in the face of the manse before her, so different from her own home with its lack of indoor plumbing, she must also have been sharply aware of her elation, conscious that she had been working toward this moment for as long as she could remember.

Rachel settled in quickly. She chose rigorous classes and threw herself into the work. A passionate student, compulsive about being well prepared, in class she was always quick to raise her hand, the first to fire off an answer. It was a trait that didn't always endear her to classmates, many of whom kept their distance, put off by the intensity of her ambition. The few girls who did get to know her found that she actually had a gentle wit and could be "slyly observant." But to most of her peers, she seemed an overzealous grind, awkward and drearily earnest, a girl too freakishly studious to be much fun. And her apparent apathy about appearances didn't help. Rachel was prone to bouts of acne, which hid her prettiness. Her thick oily hair required daily shampooing, which she often didn't make the time to do. Though outwardly she claimed not to care about her looks, one friend felt she suffered inwardly, and for that reason skipped out on social events, preferring to spend her time studying in the library.

Maria Carson's regular visits also complicated matters. Every Saturday after classes, Maria arrived at Rachel's dorm and stayed until she went to bed. At first the girls with rooms near Rachel were sympathetic, imagining Rachel to be suffering from homesickness. But when Maria continued to appear every weekend throughout the year, they amended their impressions: behind Rachel's back they mocked her.

It wasn't until her sophomore year that life began to look up. Though Rachel was an English major, like all students, she was required to take at least one course in science. And so, in the fall of 1926, she signed up for introductory biology. Taught by one of the most charismatic figures on campus (and in some quarters one of the most controversial), Mary Scott Skinker's beginning biology was a class the unambitious avoided. Skinker was a passionate and inspiring professor, but she was also a demanding one. Science, she believed, was the highest calling, and she expected nothing less than stellar work. Gentle but forthright, Skinker was known to periodically tangle with Cora Coolidge, the silver-haired president of the college.

Coolidge hailed from a wealthy Massachusetts family of patrician stock. She had grown up surrounded by politicians and men of letters, in drawing rooms where discussions of art and poetry were

as common as air. A "large bosomy woman" at home in the world
of men, she was a tireless champion of women's education. And yet,
despite her otherwise enlightened stance, she didn't believe women
should be encouraged to pursue advanced science. Women, in her
opinion, possessed neither the "intellect" nor the "stamina" for such
high-pressure careers, echoing a gender bias common in the 1920s.
Though Coolidge herself was unmarried—and happily so—she felt
that the foremost goal for all educated women was to be literate and
capable wives and mothers. Skinker, who had a master's degree in
zoology and hoped to go on for a Ph.D., strenuously disagreed.

Skinker's allure on campus went beyond her teaching. Tall and
willowy, her chestnut hair cinched in a loose chignon, she had the
swanlike grace of a dancer. Every evening, in the great chandeliered
hall, the dining tables laid with silver and crystal, she arrived in
formal dress, drawing the eyes of all assembled. She always wore
a fresh flower pinned to her shoulder or waist. (It was rumored
that the flowers were a gift from an amorous suitor waiting in the
wings.) But then one day the flowers stopped arriving, leading to
whispered speculations about a breakup. Skinker later revealed to a
former student that she had abandoned the idea of getting married
while at PCW.

Rachel felt an almost immediate affinity for the popular professor. Within weeks of joining her class, she had formed a deep and
abiding attachment. Rachel not only admired what Miss Skinker
taught, but also what she stood for: her high seriousness, the
breadth of her scholarship, her absolute commitment to science. If
the elegant and impassioned professor was an object of interest to
many students, to Rachel she was something more: a role model, a
woman in whose image she might remake herself. In class, seated
up front, Rachel felt her imagination roar to life. Skinker taught that
"all life was interconnected"; that time past and time present were
eternally linked in the long, slow process of evolution. This "holistic" view of the living world—and humankind's place in it—was to
become the foundation of Rachel's worldview. Under Skinker's guiding hand, Rachel was learning the principles of ecology, although

the word wasn't yet in use. Rachel too started to wear a flower on her shoulder at dinner in an act of homage.

By junior year, Rachel was feeling increasingly unsettled. She was spending more and more time in the drafty, top-floor biology lab of Dilworth Hall, often returning after dinner to dissect specimens. A literature major, she had always assumed she would be a writer, as she and her mother planned. Writing wasn't an easy career choice, certainly, but it was an acceptable profession for a woman. And it didn't necessarily require that she become a teacher, which was where most women landed. Science, by contrast, was a nearly impossible path. Few women, no matter how exceptional, were accepted into graduate science programs. And those who were rarely found work afterward in full-time research or business. Instead they were shunted off to teach in women's colleges, where, like Miss Skinker, they were often marginalized. Even the brilliant and well-qualified Miss Skinker had been unable to get beyond her master's degree. Rachel, who was usually pragmatic, was for the first time finding it hard to do her writing assignments. "I have gone dead," she scrawled to her adviser in a worried note. That February, after much agonizing, she finally changed her major to science and the conflict lifted.

Rachel had by now shed much of her social awkwardness. One winter night, after a fresh snowfall, she joined a group of girls on an impromptu sledding party. Flying downhill on trays temporarily lifted from the dining hall, they horsed around in the snow. Afterward, their sweaters and underthings soaked, they peeled off their clothes and showered, donned pajamas, and sat in front of the enormous fireplace in the dorm, eating potato salad and sandwiches. Then, turning off the lights, they sat in the firelight and sang until well past midnight. For Carson it was a rare moment of lighthearted abandon.

That same winter, Rachel's roommate helped fix her up with a date for the yearly prom. The young man, Bob Frye, was a junior at a nearby college. Rachel purchased "silver slippers a size too small"—as was the fashion—and spent the few days before the dance working to break them in. Afterward, she wrote a friend that she'd had a "glorious time," although oddly she made no further mention of Bob. Instead, Rachel went on at length about how glamorous Miss Skinker had appeared that evening: "Miss Skinker was a perfect knockout at the Prom . . . She wore a peach colored chiffon-velvet, with the skirt shirred just about 8 inches in front and a rhinestone pin at the waist." According to Rachel's roommate, Bob had looked like he was having difficulty making conversation with Rachel. Bob showed up the next day to take Rachel to a basketball game, and she saw him once more that semester. After that, Rachel never dated again. Perhaps the collapse of her sister's marriage stood as a cautionary tale, perhaps the quiet disappointments of her mother's. Or perhaps she simply wasn't interested.

IN JUNE 1928, RACHEL GRADUATED MAGNA CUM LAUDE FROM PCW. MAria made sure the news appeared in the local Springdale paper. The attractive young woman staring out from her yearbook picture bore little resemblance to the awkward girl of four years before. Early that spring, at Skinker's urging, Rachel had applied to Johns Hopkins and been awarded a full scholarship. She also learned she had been chosen for a fellowship at the Marine Biological Laboratory in Woods Hole, Massachusetts, for August. Excited, if a bit apprehensive, Rachel left for Baltimore in late July, stopping briefly at Johns Hopkins. Then, limp with the heat, she boarded a bus for Luray, Virginia, where she was joining Miss Skinker at her family's cabin in the Blue Ridge Mountains.

The two women spent the next three days lost in conversation. They hiked and rode horses, bird-watched and played tennis. But mostly they talked, incessantly, sitting together each night before the crackling fire, warmed by the mutual kinship they felt. Skinker had become the most significant figure in Rachel's life, both intellectually and emotionally, her mentor as well as a treasured friend. It was a pattern she would repeat throughout her life. Rachel always had one woman to whom she was deeply attached, and on whom she leaned for support and emotional ballast. Independent in so many important ways, Rachel was also often needy. And yet, fiercely protective of her privacy, she wasn't easy to know. Few were able to get beyond her austere reserve.

The next few years were emotionally fraught. Rachel loved her time at Woods Hole, and, as always, she worked with all her heart. But she was not entirely settled. She struggled in the lab, unsure of what she wanted to study. She felt self-conscious about her training, convinced she was unprepared. Unlike most scientific institutions, WH was welcoming to women, even if they were woefully under-represented. Though Rachel had never worked in a lab with men, she got on well with them. But seeing their work, she was struck by the superiority of their skills. She felt less clever, a bit lost. It was her first experience of the sea, and she often walked the shoreline, fascinated by the curious marine life she found there. She did her best to be socially outgoing, joining in the impromptu picnics on the beach; the excursions to the scrub islands in the sound; the collecting trips aboard the lab's dredging boat, which chugged up and down Buzzards Bay. But she was still distressingly aware that she felt more at home in the library than in the lab. It was the first time Rachel had encountered her own limits.

When she got to Hopkins that fall, her sense of being overwhelmed only grew. Her lab research was moving slowly; after months, she still hadn't produced enough for an original thesis. Though she was working long hours, often fifty a week, she was beginning to slip behind. Since the collapse of the economy, her financial pressures had deepened. The Depression was affecting everyone, but for her own family, all still in Pittsburgh, the hardship was acute. In early 1930, hoping job prospects in Baltimore would be better, she made arrangements for them to join her, renting a house in a rural backwater outside Baltimore. Big and empty, with no central heating, the house did, at least, have indoor plumbing and a fireplace. One by one her family straggled in, her parents first; then in June, when school was out, Marian and her two daughters, Virginia and Marjorie. A friend who visited Rachel during this

time recalled that Marian's two girls "clung to Rachel, talking incessantly until she gave in and read them a story." Her father sat by the
fire, and "looked ill." Rachel's reasons for wanting her family closer
were, as always, complicated. She didn't have the money to travel
home, or even to pay for a long-distance call. And she missed her
mother terribly. It was the longest the two had ever been separated.

In her second year at Hopkins, Rachel cut back her schedule,
becoming a half-time student. She found a part-time job as a lab
assistant, but it was an extremely tense period. She was now the
only one in the family who was steadily employed. A neighbor remembered dropping by the Carsons' one night and seeing them
"seated at the table with only a bowl of apples for dinner." Weary
and stretched thin, her progress in the lab stalled, there seemed
little chance that she'd be able to complete the kind of ambitious
study she'd envisioned. "I don't have time to think any more," she
wrote worriedly to a friend. Finally, in June 1932, a year behind
schedule, she delivered her master's thesis in zoology. Barely one
hundred pages, it was neither brilliant nor groundbreaking work.
But it was enough to secure her degree; in the recommendations
that followed, her professors suggested Rachel would make a fine
teacher, but they expressed reservations about her ability to do pioneering scientific research.

That summer, Rachel took on two more part-time jobs, both
teaching; when fall arrived, determined to move forward, she returned to Hopkins to begin work toward her doctorate. But her
father had fallen ill by now, and then her sister became sick too.
After a year and a half, with five family members to support, and
the economy still in free fall, Rachel, at twenty-eight, dropped out
of Hopkins to look for full-time work. At Skinker's urging, she
took the civil service exam; it was soon after that she was hired to
write radio scripts. "I've never seen a word of yours, but I'm going
to take a sporting chance," her new boss, Elmer Higgins, told her.
Later, another boss would remember Rachel as "extremely shy:
almost unable to get the words out when she came to him to ask
for a writing job."

IN 1935, THE YEAR CARSON JOINED THE U.S. FISH AND WILDLIFE SER-
vice, "conservation" was still a relative term. For the first time in
America's history, however, misuse of the nation's resources had
become a subject of terrible relevance. That summer, for the fifth
straight year, severe drought descended upon the Great Plains,
scorching the once-loamy topsoil to a powder that the winds
hoovered up into howling dust storms that blew steadily east, de-
nuding the land and uprooting hundreds of thousands of people.
Some storms were so dense they blotted out the sun; others dumped
dust and black rain on cities as far east as Buffalo and Boston. In the
winter of 1934–1935, red snow fell on New England. The following
spring, on April 14, 1935—Black Sunday—one of the worst storms
of the decade roared east across the Plains, carrying away three hundred thousand tons of topsoil, "more dirt than had been dug out
to build the Panama Canal." Five days later, the "swirling murk"
arrived in Washington, D.C., blackening the sky and interrupting a
Senate hearing on soil erosion. By the close of the month, Congress
had passed the Soil Conservation Act. Twenty thousand workers
were dispatched to the Great Plains to replant sod. President Roosevelt, convinced that nature could be reengineered on a massive
scale, hastily put through a scheme to plant trees from the Canadian
border to Texas. The trees, it was hoped, would help anchor the soil
and trap rainfall. But the prairie soil and climate proved ill suited
to trees. While some 220 million were planted, most either died
or were uprooted by farmers once the rains returned and the land
could again be cultivated for crops.

Yet even with the absolute and admitted failure of Roosevelt's
grand plan, the hubris behind it lived on, which was that nature
could be remade to mankind's specifications, subdued and then domesticated to serve his ends. It was an idea that hadn't much changed
since the first God-haunted settlers set foot in the new world, and
which informed the mission at the Bureau of Fisheries where Rachel now worked, albeit in a slightly more benign form. ("A howling wilderness" would, the Puritans vowed, be reduced to "fruitful
subjection.") In his annual summary, Elmer Higgins pointed out

that even when practical applications for marine research weren't immediately clear, such knowledge made "permanent contributions to social progress." Knowledge, he wrote, permits nature to be "harnessed, controlled and directed to economic advantage." It was a polite way of saying that behind the agency's push to unlock the ocean's secrets lay the perennial hope that it might be harvested more profitably. Even so, in practice it was an approach to conservation deployed mostly to gentle ends: fish numbers were watched, marine food monitored, hatcheries recommended for restocking species, all in the name of sustaining commercial fishing. If occasionally a marine creature like the starfish, which interferes with oyster harvests, was deemed an antisocial neighbor, and thereby slated for eviction (usually by poison), most of the bureau's efforts were focused on conducting in-depth research on the high seas rather than on eliminating species.

This was not so for the sister agency to the Bureau of Fisheries, the Bureau of Biological Survey, whose sole charge was to "control" agricultural pests that diminished crop yields, as well as larger predators that preyed on livestock. Here the mission was to actively "rebalance" the natural order, to make nature "more friendly to modern man" by selectively culling certain species, writes William Souder, author of *On a Farther Shore*, an elegant recent biography of Carson's life and legacy. For the long list of so-called pests—everything from birds and rodents to wolves and coyotes—it was open season; any means available—shooting, trapping, poisoning with lethal gases—was fair game.

By the beginning of the 1930s, the Bureau of Biological Survey could enthusiastically claim that they had substantially thinned, if not decimated, a multitude of animals in "stock-raising" regions of the country. "For wolves," they crowed, "the end is in sight." And great progress was being made on "cougars, lynx and bobcat." Full eradication—the tacit goal for most of these species—was believed to be "only a matter of time."

Complicating this mix were the sometimes-competing interests of sportsmen and hunters. As early as 1820, the pioneer and bird artist Audubon, an avid hunter, recognized that the clearing of forests for cropland was decimating the habitats of the many magnificent bird and wildlife species he loved to hunt and then paint. "The greedy mills told the sad tale," Audubon lamented, "that in a century the noble forests . . . should exist no more." It was Audubon and other well-born sportsmen, in fact, who were the first to press for some form of wildlife conservation. When Theodore Roosevelt organized the Boone and Crocket Club in 1888, the idea was to "promote manly sport with a rifle." But it was also to protect America's untamed frontier. Roosevelt felt that untrammeled wilderness was necessary to sustain the "vigorous manliness" at the core of the national character. In time, Roosevelt would go on to establish the first federal game management areas in America, sowing the seeds of what would eventually become the national park system. But not before the "wanton destruction" of the American bison and other species by market hunters in combination with the government was complete.

It all came down to competing definitions of nature, one from the Enlightenment, the other a preservationist ethos. To the former, nature was the realm of "rational laws and exploitable resources" that could be marshaled for monetary gain; the latter, in the spirit of the great naturalist John Muir and fellow travelers at the Sierra Club, was romantic: nature shouldn't be subject to human intervention or manipulation. Carson, for whom killing anything was abhorrent, was firmly in the second, conservationist camp, out of step, in many respects, with the government's positions. But she kept her head down, acutely aware of her need to keep the job.

IN THE SPRING OF 1936, THE RADIO SCRIPTS NEARLY FINISHED, HIGGINS asked Carson to write something "of a general sort," for a brochure about the sea. Carson would later recall losing herself in the project, claiming "the material rather took charge of the situation." A few weeks later, she delivered the eleven-page essay to Higgins's office, sitting with him as he read it.

"I don't think it will do," he said flatly, handing it back to her. "Better try again."

He looked her in the eye, seeming to relish the pause. "But send this one to *The Atlantic*," he added, winking now. The essay, he went on to say, was far too eloquent for a government publication; it belonged in a top literary magazine. This was shortly after her sister's death and Carson didn't immediately follow up. It would be a full year before she finally sent off the essay. But when at last she did, *The Atlantic* editors took it immediately, calling the essay "uncommonly eloquent." The piece, they added, would surely "fire the imagination of the layman." They asked for a few minor changes, and then sent her a $100 check—to Carson a princely sum.

Carson's essay, which appeared in the September 1937 issue under the title "Undersea," would prove life changing. Waiting in her mailbox one morning was a letter from the celebrated author and illustrator Hendrik Willem van Loon, who urged her to come up with a book proposal along the lines of her article. He invited her to visit him in Old Greenwich, Connecticut, so he could introduce her to his editor, Quincy Howe, at Simon and Schuster. From those first four *Atlantic* pages, Carson would later reflect, "everything else followed."

Carson's vision of the sea, from the shoreline to its abysmal depths, was a departure from her earlier work, probing deeper into the unseen machinery of primordial realms than anything she'd ever written. It was also stranger and more experimental in style. In it, she imagined for her readers the look and feel of life underwater, as if they were trekkers on a journey there.

> If the underwater traveler might continue to explore the ocean floor, he would traverse miles of level prairie lands; he would ascend the sloping sides of hills; and he would skirt deep and ragged crevasses yawning suddenly at his feet. Through the gathering darkness, he would come at last to the edge of the continental shelf. The ceiling of the ocean would lie a hundred

fathoms above him, and his feet would rest upon the brink of a
slope that drops precipitously another mile, and then descends
more gently into an inky void that is the abyss.

The sea as Carson described it was a savage and storm-tossed place at once beautiful and menacing, a realm teeming with eerie examples of adaptation and survival in which every creature was both predator and prey. It was a place at once perpetually in flux and eternally changeless, a world in which time past and time present were continuously linked in an endless cycling of elements from one generation to the next.

Individual elements are lost to view, only to reappear again
and again in different incarnations in a kind of material
immortality. Kindred forces to those which, in some period
inconceivably remote, gave birth to that primeval bit of
protoplasm tossing on the ancient seas continue their mighty
and incomprehensible work. Against this cosmic background
the life span of a particular plant or animal appears, not as
a drama complete in itself, but only as a brief interlude in a
panorama of endless change.

These were not the sorts of ideas readers were accustomed to contemplating, and they were riveted. Carson's picture of a dark and watery world churning with evolutionary wonders, a place at once inconceivably ancient and eternally present, was something new: a different measure by which to think of time. The life of the sea, as Carson explained it, was a "continuum" in which all living organisms were bound.

That January, a nervous but excited Carson took the train to Old Greenwich, Connecticut, where she met Van Loon. True to his word, he took her to see Quincy Howe at Simon and Schuster that same day. The following day, as Van Loon had predicted, she left with a publishing commitment.

RACHEL RETURNED HOME AND BEGAN THE SLOW, PAINSTAKING WORK
of making the book real, shoehorning the research into her crowded
days, chipping away at the writing as she could: on weekends,
nights, vacations. Three years later, the manuscript at last done, she
sent it off to Simon and Schuster. It was New Year's Eve, 1940.

Like all first authors, Rachel had high hopes for the book, which
she had titled *Under the Sea-Wind*. And the early reviews indeed
boded well. The Scientific Book Club picked it up as its November
selection. "There is poetry here," the reviewer reflected, "but no
false sentimentality." The *New York Times* pronounced it "so skillfully written as to read like fiction, but in fact a scientifically accu-
rate account." Other critics commented on its "lyrical beauty" and
"faultless science." Even the scientific world was respectful, an unusual turn given how dismissive they were of most popularizations
of science. But timing in life is everything. Less than a month after publication, on December 7, 1941, Japanese warplanes bombed
Pearl Harbor. The world's eyes turned inexorably to war. Speaking
later of the book's reception, Carson wryly observed that "the world
received the event with superb indifference." Barely two thousand
copies of the book were sold.

ON JUNE 5, 1945, A FEDERAL AIRPLANE BUZZED BACK AND FORTH OVER A
117-acre expanse of leafy forest in the Patuxent Research Refuge, a
wildlife sanctuary tucked between Baltimore, Maryland, and Washington, D.C. The crop duster swooped low over the treetops, spraying them with a mix of DDT dissolved in fuel oil. As it stitched
across the sky, it left a misty trail that wafted down through the
trees unevenly. Although the pilot tried to be methodical, the winds
were up that day; it was hard for him to see where, exactly, the toxic
murk was landing. He made multiple passes over some places; others he missed altogether. Some of the vaporous cloud landed on a
mile-long stretch of the Patuxent River, a meandering stream "that
was home to twenty odd species of fish."

In the months that followed, a clutch of federal researchers kept watch on what befell the mammals, birds, frogs, and fish exposed to the DDT. At first glance, the airborne DDT hadn't appeared to cause appreciable damage. But then ten hours in, dead fish began to bob to the surface on the Patuxent River. Follow-up tests in well-maintained artificial ponds showed that, even in more diluted concentrations than those used in the spray campaign, DDT still caused massive fish kills. In their first report, the researchers cautioned that the initial excitement about DDT should be "tempered by grave concern."

Further lab studies ensued. It was unclear if the animals that escaped poisoning in the aerial spraying were simply "lucky," or if somehow they'd avoided exposure to high doses. In the lab, when wildlife was fed DDT directly, every species tested became gravely ill; many of the animals died outright. No one yet understood precisely how DDT worked. But its symptoms—convulsions, twitching, and rigidity—were leading some researchers to liken it to the nerve poison sarin, a chemical-warfare compound that disrupts nerve impulses in the same way. When ingested, sarin causes asphyxiation by disabling the muscles needed in breathing. In their follow-up report, the researchers noted that the symptoms of DDT poisoning, no matter which animal was tested, seemed to be the same: "excessive nervousness, loss of appetite, tremors, muscular twitching, and persistent rigidity of the leg muscles, the last continuing through death." Though they didn't say it outright, the implications for humans weren't good. Quietly, the laboratory studies continued.

But the growing market for insecticides wasn't waiting for long-term studies. The chemical companies had a surfeit of product at the end of the war, which they were eager to turn to peacetime uses. They saw a robust new market for DDT among American consumers, a host of commercial possibilities. Soon DDT was being touted as a "must-have" for every American household, a low-cost, easy-to-use domestic product that would banish insect pests forever. Attractive and convenient, it was marketed in shelf paper, "white or tinted to match one's color scheme." It was added to soap products and lotions, floor wax and furniture polish, sprays for application to clothing and skin. One could purchase strips "impregnated" with the stuff to hang inside closets and clothing bags. It could be had in "pocket-sized" dispensers perfect for a lady's purse, or beach and golfing gear. The local hardware store sold special fogging gadgets that attached to the muffler of any lawn mower. While the lawn was being groomed, a mist of lethal poison spewed from the fogger, killing insects on contact. And its utility didn't stop there. DDT rapidly won acceptance as an effective agricultural and institutional fumigant as well, thanks to aggressive marketing efforts. It was sprayed from airplanes over giant tracts of woodland to control gypsy moths and other problem pests; it was used to douse the walls of hospitals, restaurants, and school kitchens; it was sprayed on suburban neighborhoods where insects were killing trees. In the South, children chased behind the DDT fogging trucks, playing in the white drifts of poison that trailed behind. Airplanes were deployed to spray football stadiums before concerts and big games, to rid them of mosquitoes. DDT could be dispensed as a powder or a liquid spray, through an "electric vaporizing device" or as an aerosol "bomb" that a housewife could release in her own kitchen.

DDT's status as a miracle compound had been cemented during the war, when it was mass-produced to fight deadly human diseases. In the furious push to prevail against the Axis powers, any questions about its long-term impact on health or the environment were ignored. Now, with the war over, the applications for the poison were expanding more rapidly than science could keep pace. Though tests on DDT's safety to humans had barely begun, few in government were paying much attention. America in the 1950s was a buoyant and triumphant nation, its citizens at home, if not in bed, with an advanced industrial society. The economy was booming, focused again on consumer goods. The "infallibility of material ingenuity," as the scientist E. O. Wilson has written, was all but assumed: "An ethic of limitless progress prevailed." Highways were being built to connect newly minted suburbs to jobs in the city. Farmland was being plowed under to make way for vast housing developments in places like Levittown on Long Island, and the San Fernando Valley outside L.A., where "acres of tract houses . . . almost compensated for the absence of individual character." It was the dawn of the organization man, the "team player," the bland GM conservative; the age of household convenience. Science and big business were king, and the spinmeisters of Madison Avenue their spokesmen. The new front, according to the popular press (and the Mad Men with their come-hither sells), was the ongoing battle against insect pests, which, thanks to the almighty reach of technology, could now be won. One ad even went so far as to "place Adolf Hitler's head onto the body of a beetle." In July 1945, *Time* magazine had showcased pictures of the first atomic bomb explosions in Alamogordo, New Mexico, alongside an article extolling the benefits of DDT as the equivalent weapon in the "war on insects." DDT, like the bomb, was a new technology for a new age, yet another exterminating agent born of the miraculous powers of science. It too could eliminate the enemy almost instantly. But DDT's powers would be a benefit to mankind, the press suggested, rather than a terrifying agent of its destruction like the bomb.

The war had changed women's status as well, albeit temporarily. In the heat of the conflict, women working in heavy industry had been a patriotic necessity. Females became pipe fitters and mechanics, welders and carpenters—jobs "previously unthinkable for their sex"—enjoying a dose of economic independence that would be difficult to give up. *Ladies' Home Journal* even went so far as to put a woman combat pilot on its cover. Eight million women had joined the workforce in the course of the war, many of them going into relatively skilled factory positions. Two months after the conflict subsided, eight hundred thousand of them were fired from the aircraft industry, and the same went for jobs in the auto industry. J. Edgar Hoover, the director of the FBI, pronounced paid work inappropriate for mothers in any instance. "A mother," he proclaimed, "already has her war job . . . Her patriotic duty is not on the factory front. It is on the home front!"

Madison Avenue's postwar message to women, if slightly more

covert, was no less constraining. Their pitch was crafted to glamor-ize consumption. A perfect homemaker was a savvy shopper, the message went, discriminating in her choice of washing machine and toaster, floor wax and tweezers. Her satisfaction would be found in perfectly polished floors and artfully applied makeup, an immacu-late house and well-mannered kids. Some "experts" even suggested that a woman's most important role was to "rebuild her husband's self-esteem," which was surely damaged by the discovery that she had successfully held down the fort while he was away.

Carson, for whom work was an economic necessity, was free of some of these pressures. But as an unmarried female scientist, her future was equally limited by the cultural biases and shifts taking place in the wake of the war. High-tech was the way of the future. Obsessed with the remarkable triumphs of the molecular revolu-tion, the scientific community held chemistry and physics in the highest esteem. Conservation biology, by contrast, was barely a blip on the cultural agenda. "To a populace whose forebears had within living memory colonized the interior of a vast continent and whose country had never lost a war, arguments for limit and constraint seemed almost unpatriotic," writes E. O. Wilson. Though Carson had been steadily promoted, rising seamlessly through the govern-ment bureaucracy, she was feeling stuck. Sleep deprived and fre-quently ill, worn out by her nocturnal writing schedule, she was at a crossroads, trapped in a job to which she was well suited, but which kept her from greater things. "I'm definitely in the mood to make a change of some sort, preferably to something that will give me more time for my own writing," she confided to a friend. "At this stage that seems the prime necessity." She wanted to give every-thing to her own writing, she added, but she knew she couldn't risk it. Money was still a grinding pressure, and her job paid the bills. She admitted that her life wasn't "well ordered" and that she didn't "know where she was going."

A close, discerning reader, Carson was always on the lookout for ideas she could turn into salable magazine pieces, which by now she was producing at least once a month. Discouragingly, most of what crossed her desk for editing seemed ill suited to the popular press: dry, technical, and achingly dull. Many reports required a complete rewrite, their syntax brutally scrambled, at times almost unintelligible. An organized and purposeful editor, Carson could turn even the most intractable prose into crisp, clean sentences. This was not to say, however, that such travesties against the English language went unremarked. Though tactful with clumsy writers, Carson's private views "were often more pungent," a friend, Shirley Briggs, recalled. Briggs and another woman, Kay Howe, were both graphic designers and recent hires, the only other females in the agency. They shared the office next to Rachel's, and though a decade younger, they soon became her friends, nicknaming her "Ray." At lunchtime and over "illicit" tea brewed on a hot plate Rachel kept hidden in her closet, the troika of women would huddle over brown bag lunches in Carson's office, poking fun at the day's editorial fare. Examples of particularly atrocious prose were read aloud and gleefully dissected, helping to relieve the monotony. "Nothing could pass the wry scrutiny of that gathering and still seem insurmountable . . . ," Briggs would later recall. "Intransigent official ways, small stupidities, and inept pronouncements were changed from annoyances into sources of merriment." Briggs said that Rachel always made the best of what was often a bore: "her qualities of zest and humor made even dull stretches of bureaucratic procedure a matter of quiet fun," she added. It was a side of Carson that few people saw.

Despite the tedium of her duties, occasionally a report crossed Rachel's desk that snapped her to attention. This was her response the day she began reading the Patuxent reports, which like so many important studies, came to her office for editing. Troubled by what she was reading, Carson immediately issued several press releases warning of DDT's potential health hazards, the first to the operators of fish-processing facilities, where the pesticide was routinely used for insect control. Recent studies, she warned ominously, suggest that DDT is toxic to humans when ingested, and could contaminate food products with "serious consequences." A few weeks later, as more findings from Patuxent trickled in, she followed up with a second, more detailed alert. New experiments indicated that DDT killed birds as well as fish, even in diluted concentrations, she reported. DDT "could conceivably do more damage than good."

Unable to stop thinking about the Patuxent studies, that July Carson proposed a story to *Reader's Digest*, explaining that she was in a position to cover the progress of the tests "first hand" and write a "timely" story. "Practically at my backdoor here in Maryland, an experiment of more than ordinary interest and importance is going on," she wrote. "We have all heard a lot about what DDT will soon do for us by wiping out insect pests. The experiments at Patuxent have been planned to show what other effect DDT may have if applied to wide areas."

At issue, as she saw it, was not just the elimination of a few species; it was the possibility of disrupting the entire web of life. But in July 1945, Carson was a voice crying in the wilderness. *Reader's Digest*, never keen on downbeat stories, politely declined. Though Carson continued to worry about DDT, her attention turned elsewhere.

IN THE SPRING OF 1946, ITCHING TO GET OUT IN THE FIELD, CARSON initiated a series of conservation booklets. The idea was to describe the federal wildlife preserves—and to make a compelling case for their importance. At the time, natural resource conservation was still anathema to many Americans. In some quarters, local residents were actively hostile to the idea of preserves, which they saw as impinging upon their hunting rights. It was Rachel's hope to change that sentiment. While others would contribute to the series, Rachel did the bulk of the writing, Shirley Briggs and Kay Howe the illustrations, which meant that one or the other traveled with her for the fieldwork. It was the beginning of an interlude Rachel would later remember as more carefree and satisfying than any she had known.

The conservation booklets were extremely popular, their beauty

and narrative sweep an anomaly in the world of government publications. The fifth in the series, titled "Guarding Our Wildlife
Resources," was different from the others, however; rather than
spotlighting a single sanctuary, as Rachel had done in the earlier
booklets, it was an appraisal of the overall state of wildlife conservation in the nation. Rachel framed it as a "serial tragedy," writes
Souder. It was presented as a tale of plenitude "repeatedly squandered," habitat and species losses differing only in the degree and
magnitude of their devastation. Yet even with all this, she assured
readers, small gains were beginning to appear: species on the verge
of extinction were coming back. These hopeful developments, she
added, were due to a rising awareness that problems in nature
couldn't be tackled "in isolation." Whether this was actually so was
debatable; certainly it was Carson's hope. But it fed into her deeper
purpose, which was to emphasize the interconnectedness that un-
dergirded the living world. Conservation, she gently explained, was
not only a safety net for endangered species, it served the entirety
of existence: "Wildlife, water, forests, grasslands—all are parts of
man's essential environment; the conservation and effective use of
one is impossible except as the others are also concerned." Already
prophetic in her ecological approach, Rachel's thinking was a subtle
departure from the guiding tenets of conservation up until then,
which had been forged to serve the interests of commerce and the
hunting fraternity. Rachel was already speaking to the future, to a
movement that was still to coalesce.

IT HAS BEEN SAID THAT REVOLUTIONS DON'T TAKE PLACE IN A VACUUM,
that certain ideas in the air seem to surface in multiple places during times of critical change. Coincidental or not, in 1948, the same
year Carson published "Guarding Our Wildlife Resources," another
seminal figure in the still-embryonic environmental movement, a
writer named Aldo Leopold, learned that the book he'd been chipping away at for seven long years had finally found a publisher.

A former U.S. Forest Service employee, Leopold was a gentle, Thoreau-like character, a naturalist and sometime hermit who had been teaching at the University of Wisconsin for some years. His book, penned while he lived alone for stretches in a shack he kept in the Wisconsin woods, was a record of his reflections on the natural world. Though Leopold died barely a week after receiving the good news, he would live on in legend as one of the great heroes of the sixties counterculture, his book, later titled *A Sand County Almanac*, becoming a bible of the environmental movement when it was reissued in 1966.

In one of the book's essays, Leopold introduced an idea that echoed Carson's own. Humankind's place in the world was not as a "conqueror of the land community," he suggested, but as a "plain member and citizen of it." All organisms in nature were part of "a biotic community," which depended on the totality of the earth's creatures for its stability. It was a mistake, he believed, to reduce conservation to a tool that served economic ends; it should be undertaken to serve the entirety of existence. "A thing is right," he wrote, "when it tends to preserve the integrity, stability and beauty of the biotic community. It is wrong when it tends otherwise." Leopold's point, like Carson's, was that nature "was in charge of humanity and not the other way around." "We fancy that industry supports us, forgetting what supports industry." Both were arguing for a broader and more holistic vision of preservation. Conservation, they insisted, couldn't be confined to a narrow swath of the natural world, but was "of necessity" about the preservation of the elaborate relationships that made up the larger community. Protecting nature was protecting our own interests, since humankind was not apart from nature but *of it*. It was a markedly different view of where humankind stood in nature's pecking order. Leopold, like Carson, was seeing the world from an ecological point of view. "We are only fellow-voyageurs with other creatures in the odyssey of evolution," Leopold observed, "but for one difference—technology had given man the whip hand over nature." Like Carson, he felt the whole idea of undesirable species was "entirely synthetic."

BY 1948, CARSON WAS STARTING TO THINK ABOUT A NEW BOOK, ONE that would probe mankind's dependence on the ocean, which she felt was growing in urgency as modern civilization plundered more of the land. Her regret over the commercial failure of her first book was fading, and she was intrigued by the oceanographic knowledge that had been surfacing in the years since the war, eager to be inside a big project again. Though she revealed little to friends, many at the office suspected she was working on something. The librarian at the Department of the Interior noted her requests each morning for a bewildering array of books and articles. Another colleague remembers seeing piles of technical volumes burying the back seat of her car.

Rachel's perseverance was by now legendary. Bob Hines, who illustrated many of her books, recalls Rachel working in Maine, standing in the tide pools through an entire tidal cycle. She would grow so cold and numb in the icy water that she couldn't feel her feet or walk back over the rocks. Bob would have to wade in and carry her to the shore, where her mother would be waiting with a blanket to wrap her in. Once warm again, however, she'd wade back into the water and resume her work.

Rachel plowed through exceedingly technical papers with the same kind of obsessive determination, corresponding with scientists and oceanographers throughout the world. For all her social reticence, she was never shy about approaching scientific experts, no matter how eminent, if in the service of her work. Graciously, unobtrusively, she solicited the information she needed, often asking those same specialists to read and review her work later for its accuracy. Wise about cultivating alliances, gentle in her approach, she came off as earnest and unthreatening, exceedingly respectful of others' expertise, which in the competitive world of science was no doubt disarming. Rachel asked each new scientist she interviewed for an introduction to the next, in this way building a vast and loyal network of people to whom she could turn. Unafraid to express her gratitude, secure enough to let herself be small, she always gave credit where credit was due. In some respects, hers was a very "female" style. Rather than competing, she worked to foster connections with those to whom she spoke. She was unashamed to be, in some senses, reliant.

Sometime during this period, during a birding expedition, Rachel and Shirley Briggs came upon the celebrated nature writer Louis Halle. Overcoming her normal reticence, Rachel approached Halle, expressing her appreciation for his work. Halle, who had no idea whom he was talking to, later remembered her as quiet and diffident, wholly without affectation. There was "something about her of the nineteenth century," he said. "She had dignity; she was serious." She gave him no hint that she was considering another book, but "put herself in the role of the pupil," not a fellow practitioner. He remembered her as being "always attentive, always listening, always wanting to know." She could not have "got away from being a simple human being," he felt. It was not until the enormous success of her next book, *The Sea Around Us*, that Halle realized, with some embarrassment, that he been giving advice to one of the most famous nature writers of their time.

BY THE FALL OF 1948, RACHEL HAD COMPLETED WHAT SHE HOPED WAS A solid first chapter for the new book. A friend recommended she find a literary agent, and after interviewing several, she settled on a former editor and mystery writer named Marie Rodell, who was opening her own agency. Although they couldn't have been more different, the two women took to each other immediately when they met for lunch at the Algonquin Hotel in New York. Rodell was a shameless extrovert—urbane, quick-witted, flamboyant, a chain-smoker with a raucous laugh and a penchant for strappy heels. She was a sharp negotiator, but had a reputation for integrity. Rachel signed on as her first client.

That November Rachel received shattering news. Mary Scott Skinker was dying of cancer. Skinker, who had been teaching in Chicago, was found collapsed in her apartment and rushed to the hospital. She gave Rachel's name as the person to contact. Rachel left immediately for Chicago, where she stayed with her friend until she lost consciousness. Three weeks later, Skinker was dead. Rachel was devastated. Mary Scott Skinker had fired her imagination and seeded her "ecological consciousness," Lear notes, encouraged her ambitions and been a model for what an unmarried female scientist could achieve. Beyond her mother, Rachel had loved her former teacher more profoundly than anyone else she knew, remaining in touch despite the geographical distance between them. With Skinker's death, she lost her most beloved friend.

By July, Rachel needed a break. She and her mother drove to Boothbay Harbor, Maine, where they rented a cottage on the shore of a saltwater estuary of the Sheepscot River. Surrounded by birch and spruce, the house was so close to the water, Rachel told Briggs, "that if you jumped out the windows on one side, you would fall in." There were no other cottages in sight, she added, the only sounds the "sharp, staccato cries of the gulls" and the sluice of water against the rocks. "The gulls go so high," they looked like "stars." Rachel spent the week bird-watching and wading through the tide pools near the shore. Every evening she watched the herrings' twilight arrival in the cove: "Suddenly the silken sheet would be dimpled by a thousand little noses pushing against the water film. It would be streaked by a thousand little ripples moving eagerly toward the shore . . . Then the herring would begin flipping into the air . . . They looked like silvery coins skipped along the surface." Overwhelmed by the beauty, Rachel wrote that if she could only figure out a way to do it, she would gladly spend the rest of her life in Maine.

RACHEL HAD BEEN WORKING ON THE NEW BOOK FOR ALMOST A YEAR when Rodell called to say that Oxford was interested. Rachel was elated, as she had been writing without any guarantee of a publisher. The advance was small, but Oxford was highly reputable. Within weeks a contract was signed, and by July, she had made plans for a helmet dive in Florida, asking Shirley Briggs to come along. The trip was a bust, though it would become a legendary feature of Rachel's biography. The winds were up that week, turning the seas choppy; the skies were brooding. Every time Rachel ventured out, she had to turn back. To kill time, she and Shirley walked the beach and lunched at the local Howard Johnson's. Finally, near the end of the trip, Rachel suited up in her helmet diving gear—a glass-fronted metal helmet hitched to an air hose attached to a pump—and in a protected area that was barely eight feet deep, climbed unsteadily down a ladder that almost touched the sand. A weak swimmer anyway, she clung to the ladder in a state of semiterror, and for a few minutes peered out of her faceplate, discombobulated by the "whooshing sounds of the pump" pushing air down the hose and into her helmet. She sighted a few colorful fish and then, heart pounding, struggled back up the ladder. Writing to an acquaintance later, she referred expansively to her "diving experiences," claiming that the difference between having dived and not dived was "tremendous." It was a rare instance of overstatement, amusing, no doubt, even to her.

All that winter Rachel pushed hard. She'd taken a month's leave from work, but was so behind now that she was bringing her office work home at night, defeating the purpose. Her niece Marjorie had been ill and the family had to move. Writing to Rodell, she tried to strike an optimistic tone about her progress. "Despite everything it begins to seem as though the book might some day be finished—that never has seemed possible to me until now."

By mid-February 1950, the pressure had become grueling: "None of the present or future is very favorable for the last desperate push but I am grimly determined to finish somehow. I feel now that I'd die if this went on much longer!"

A month later, still having to take office work home at night and on weekends: "Not a single walk, and spring almost gone! I am really upset about it, but don't seem to have the energy to tuck that in too." She mentioned the push she and others at the office were making to get the work there to the printers: "Then we shall all quietly relapse into a sanitarium, if I am not there already owing to the added strain of my own literary affairs."

Finally, in July 1950, she handed in the manuscript. She was surprised by how bereft she felt: "Oddly enough, I am less relieved at being delivered of my book than I expected," she told a friend.

Money worries, meanwhile, worsened. The second installment of her Oxford advance arrived, but it fell short of her expenses. She had hoped Rodell would be able to sell chapters as she completed them. But so far, of the twenty or so magazines Rodell approached, all had politely declined, even *The Atlantic*, who sat on the chapters Rodell sent for a full three months before finally saying no. "I don't like Miss Carson's writing at all," responded an editor at *Town & Country*. Rachel tried not to feel dejected. She was thinking about a new title, *The Sea Around Us,* still polishing sections she thought needed reworking.

Then something gave. *Science Digest* made an offer of $50 for a condensed version of one of Carson's chapters. Rodell was on the verge of saying yes when word arrived from a more unexpected quarter. She heard from Edith Oliver, a young, up-and-coming editor at *The New Yorker*. Oliver had written radio quiz shows before joining the ranks of *The New Yorker,* where she read submissions and, when she could, contributed droll pieces for the "Talk of the Town." A high-spirited character with an original voice, she was a discerning reader; eventually she would serve as the magazine's drama critic for thirty years. Oliver told Rodell that she was impressed with Carson's excerpt from *The Sea Around Us*; she wondered if she could see more.

Rachel tried to keep her hopes in check. She was aware that the smart and sophisticated magazine held the power to change a writer's life. Oliver continued to express interest in seeing more chapters. By the middle of the summer, she had read five and asked for another eight, each time sending them on to William Shawn, *The New Yorker's* editor-in-chief, with her recommendation. Rachel, who assumed the magazine was deliberating over which chapter to take, tried to be patient. Finally, in mid-August Rodell got word that *The New Yorker* was not interested in publishing a chapter from *The Sea Around Us*; they wanted to excerpt most of the book, a highly unusual turn, even then. Carson was stunned and, of course, elated. Shawn would do the condensation, as well as the editing.

More good news followed. Rachel had already begun planning her next project, a seashore guide, which Paul Brooks at Houghton Mifflin now signed on to publish. She applied for a Guggenheim grant and got it. Up until then she had been broke, unable to come up with the $150 she needed to buy back the rights to her first book, which was now out of print. She hoped to find a publisher to reissue it, preferably as soon as *The Sea Around Us* came out. In May, when *The New Yorker* check for $5,200 arrived, Rachel couldn't believe it; it was equal to a year of her government salary. Feeling the wind at her back, she decided to apply for a year's leave without pay from the FWS. She told Rodell she hoped the book would do well enough to tide her over for a few years. "If I'm not solidly established as a full time writer by that time I ought to be shot anyway," she added.

The only ripple in all this bright news was a small matter. Four weeks after hearing from *The New Yorker*, Rachel told Rodell she would be in the hospital for a few days. She explained she was going to have a "small cyst or tumor" removed from her left breast. The surgery was minor, she assured Rodell. "The operation will probably turn out to be so trivial that any dope could do it," she added. "But of course there is, in such cases, always the possibility that a much more drastic procedure will prove necessary." She would be in excellent hands, she said. She was going to "get it over with next week."

The mass removed from Carson's breast was benign and she quickly put the event behind her.

THE SERIALIZATION OF *THE SEA AROUND US* APPEARED IN *THE NEW Yorker* over three consecutive weeks in June 1951. It ran as a profile, a legendary corner of the magazine that had always been reserved for portraits of people. The first piece in the series ran for an astonishing fifty-nine pages, and readers were enthralled, writing in at once to express their admiration for its author, who had made the sea come beguilingly to life. Alice Longworth, President Theodore Roosevelt's daughter and a woman of considerable influence in her own right, phoned Rachel in a state of breathless excitement. She had spent the night reading the profile—finished at 5 A.M.—then read it a second time. She said it was "the most marvelous thing she had ever read!"

The critics were equally enchanted. When the full book arrived two weeks later, they were unstinting in their praise, lavishing it with superlatives. *Newsweek* called its lyrical style "hypnotic." Others remarked on its "biblical sweep." There was consensus that it was "one of the most beautiful books of our time." A critic for the *Atlantic Monthly* marveled that a marine biologist could "write what is a first-rate scientific tract with the charm of an elegant novelist and the lyric persuasiveness of a poet." The *Buffalo Evening News* described it as "half-way between the Thoreau of the *Journal* and the Darwin of the *Beagle*," calling it "a superb book."

It was inevitable, perhaps, given the era's general view of women as second-class citizens, that some of the enormous praise heaped on Carson sounded vaguely patronizing. Male readers expressed surprise that a woman had written such a rigorous book, the suggestion being that the difficulties of science and the physical dangers of the sea should have made the subject inaccessible to her. One reader addressed his fan letter to "Miss RC," but began it with "Dear Sir," explaining that the salutation was because "he had always been convinced that males possess the supreme intellectual powers of the world, and he could not bring himself to reverse the conviction." The *Boston Post* described Carson as "both bold and feminine," as if such qualities were rarely paired. Perhaps, the *Post* conjectured, she was a mermaid. "Apparently there are few photographs of Miss Carson . . . but we have worked this out. Rachel is probably no lady scientist at all, but an enchantress who lives in a cave under the sea and there the light is awfully bad for pictures of authors." Even John Leonard, in his glowing review for the *New York Times*, allowed that it was a "pity that the book's publishers did not print on its jacket a photograph of Miss Carson. It would be pleasant to know what a woman looks like who can write about an exacting science with such beauty and precision."

Within weeks, to Rachel's confusion, she was a household name, the subject of profiles in scores of publications. Everyone wondered who this woman was who had written so eloquently about the wind and the waves and the birth of islands, the sweep of the tides and the origin of storms. Carson told interviewers that the "backbone" of her book was "just plain hard slogging," searching through the "exceedingly technical papers of scientists for the kernels of fact to weld into my profile of the sea." In truth, it went far beyond this modest explanation. Carson's genius was a rare sort of alchemy: the ability to transform dry, seemingly lifeless concepts about biology, physics, and geology into beautiful and animated stories—stories so enchanting that few could put her book down. What drew readers to Carson's writing was its gentle explanations of large and complex things. But also to something else: Carson's book, as Lear notes, touched a deeper yearning that many at that moment were feeling.

It was 1951 and Americans were terrified of nuclear Armageddon, worried about sending their sons to a war in Korea they didn't believe in, haunted by Joseph McCarthy's witch hunt for domestic Communists. As William Styron wrote, theirs was a generation traumatized not only by the "almost unimaginable presence of the bomb, but by the realization that the entire mess was not finished after all: there was now the Cold War to face, and its clammy presence oozed into our nights and days." Carson's book, by focusing on time's eternal cycles, the birth and death of continents and seas, gave readers another perspective on the pressures they were feeling, a longer yardstick by which to measure time and man-made problems. Carson was giving readers another way of seeing, a temporary lift from life's anxieties and dissonances, which was deeply reassuring.

The Sea Around Us made the New York Times bestseller list on July 22, 1951, where it remained for the next eighty-six weeks. By November 8, sales had reached one hundred thousand; by Christmas it was flying out of bookstores at a rate of four thousand copies a day, and had climbed to number 1. Carson was in fine company that season. Also on the bestseller list were Herman Wouk's *The Caine Mutiny,* James Jones's *From Here to Eternity,* and a powerful little novel by J. D. Salinger called *The Catcher in the Rye. The Sea* would eventually appear in thirty-two languages.

CARSON WAS FEELING INCREASINGLY AMBUSHED BY ALL THE ATTEN-tion. Wary of the public eye, unaccustomed to being a subject of scrutiny, she found the crush of publicity unnerving, the invasions of her privacy hard to understand. While thrilled by the glowing reviews, and cheered, of course, by book sales, she was less comfortable with the attention directed at her private life. The *Saturday Review* had run a long biographical profile of her in its July 7 issue, in addition to a review, putting her picture on the cover, which she found unsettling. "I'm pleased to have people say nice things about the book," she told a friend, "but all this stuff about me seems odd, to say the least." Shy at heart, uncomfortable speaking off the cuff, she dreaded the personal exposure that came with her sudden renown. In this, she was a curious combination. While she craved the rewards that attended her success, especially the financial security it brought, she would always shun the limelight, feeling it was enough to write her books and send them out into the world. No matter how often she had to appear in public, she never got over her wariness. When she gave speeches, and she would deliver many over the next several years, she never digressed from her prepared text, which she wrote out on three-by-five note cards. Nor did she modulate her voice, which was steady but flat, never imbued with the incandescence that lit up her books. Rodell was continually reprimanding her for refusing interview requests. She urged Rachel to put in more public appearances, to go to cocktail parties, agree to more book signings. Rachel politely declined, claiming her work couldn't go forward this way. What distressed her most, she said, was how easily people violated her privacy. Once, she told Paul Brooks, she was in a beauty parlor, in a strange town, when she was summoned from under the hair dryer by someone who wanted to meet her: "I admit I felt hardly at my best, with a towel around my neck and my hair in pin curls." Another time, a knock came at the door of her motel and a determined woman pushed past her mother to find a drowsy Rachel still in bed. Insistent nonetheless, the woman presented two books for her to autograph, which Rachel found unbelievable. She was barely able to hide her annoyance.

In early January, Carson was stunned to learn she'd won the prestigious National Book Award. The award ceremony took place at the Commodore Hotel in New York a few weeks later. Dressed in a demure silk dress and a fashionable feathered toque, an exceedingly nervous Carson shared the head table with the poet Marianne Moore, who was bedecked, as always, in her trademark tricornered black hat.

Her speech that evening was pithy. Rising to speak to the overflow audience, Carson sounded a theme she had been thinking about for some time: the growing elitism and insularity of science, a worrisome trend in the culture, she felt. "We live in a scientific age; yet we assume that knowledge of science is a prerogative of only a small number of human beings, isolated and priestlike in their laboratories," she said. This is not true, she insisted. "The materials of science were the materials of life itself." It is impossible to understand man "without understanding his environment and the forces that had molded him physically and mentally"—forces that had been at work for billions of years. Viewed this way, she suggested, human follies appear in a far different light. "Perhaps if we reversed the telescope and looked at man down these long vistas, we should find less time and inclination to plan for our own destruction," she added, in a barely veiled reference to the horrors of the nuclear age.

It was a subject Rachel would return to again and again, mankind's increasing disconnection from the natural world. "Mankind has gone very far into an artificial world of his own creation," she observed in a later speech to more than a thousand women journalists. "He has sought to insulate himself," with "steel and concrete, from the realities of earth and water" and the growing seed. "Intoxi-cated with his own power, he seems to be going farther and farther into experiments for the destruction of himself and his world," she added, referring again to the nuclear peril. The answer, she sug-gested, was that "the more clearly we can focus our attention on the wonders and realities of the universe about us, the less taste we shall have for destruction."

IN THE SUMMER OF 1952, FINANCIALLY SECURE FOR THE FIRST TIME IN her life, Carson quit her job and bought some land in Maine. It was on Southport Island, near Boothbay Harbor, a tiny lot just 350 feet deep. But it had everything she wanted: shoreline frontage, a spruce and fir wood, tide pools teeming with sea creatures, what was al-most a beach. Rachel ordered a kit house that could be assembled by the following summer. It would have a writing studio and a long deck with white railings, a red brick fireplace and walls of knotty pine.

Southport was a slow-moving, old-fashioned place. Many of its long-standing residents, including Dorothy and Stan Freeman, had been summering there for years. The summer before, the Freemans' children had given them a copy of *The Sea Around Us*, which they'd both loved. When Dorothy learned that its celebrated author had bought land so close to them on Southport, she decided to send a welcome note, mailing it to Rachel's publisher, with no expecta-tion that she would hear back. Rachel, however, was delighted. She wrote back immediately, thanking Dorothy for her "charming and thoughtful" greeting. She added that she hoped Dorothy and Stan would stop by and introduce themselves in June.

They didn't make it until July 12, 1953, calling on Rachel just after supper. Rachel found them warm and genial. Stan was tall and lanky, an avid sailor and a passionate photographer. Dorothy had a spirit and vitality Rachel instantly liked. They were older than Carson, but they shared her love of the Maine coast. Rachel insisted they come back in a few weeks for a collecting expedition, when the tides would be lower on her beach. That visit was convivial too. The little party collected shreds of algae and clots of mud in small specimen bottles. Then, back at the cottage, after tea by the fire, they looked through Carson's microscope at what they'd collected. Dorothy expressed surprise that Rachel seemed so unimposing, so natural. She seemed "tiny" and wore a "wistful expression." She sensed in Rachel a deep, buried sadness. She wondered if she'd been overwhelmed by her sudden celebrity. On Rachel's end, she was sorry to learn the Freemans were leaving so soon. She wrote to Dorothy to say good-bye, suggesting that they henceforth call each other by their first names. Dorothy wrote back immediately, expressing her worry that she might be interfering with Rachel's work. Rachel assured her she was not, that she welcomed Dorothy's letters. Writing, she confided, was often a difficult and lonely endeavor. She urged Dorothy to keep writing letters and not to worry about interrupting her work.

A flurry of letters now passed between them, multiple mailings a week, becoming more open and descriptive with each exchange. Rachel said she was having trouble getting started on a new chapter; perhaps, she mused, it might be easier if she typed the words "Dear Dorothy" on the first page. She hadn't realized she was lonely, she said. In passing she mentioned she would be in Boston at the end of the year for a scientific meeting. She wondered if Dorothy would meet her for lunch. Dorothy wrote back to suggest that rather than just meeting for lunch, Rachel should come visit them in West Bridgewater, so they could spend the whole afternoon and evening together before she caught the train back to Maryland. Rachel's return letter barely hid her "disappointment" at not having time alone with Dorothy. She told Dorothy she liked to imagine arriving in Boston and "stepping off the train into your arms," even though she knew this was impossible. She said she was struggling with the book, "going mad." She wrote at length about her favorite authors. She wanted Dorothy to know them, she explained, to share everything she loved. She had never been so unguarded, so magnetically drawn to another person. It was the beginning of what for both women would be a profound and startling connection.

Rachel's lecture in Boston went well. She filled it with emerging science, with thoughts about evolution and ecology, the delicate balance between living creatures and the fragile ecosystem that sustained all life. As she approached the auditorium door to leave, she was amazed to see Dorothy standing there. Rachel impulsively kissed her. "We didn't plan it this way did we?" she whispered. They went back to Rachel's hotel room and sat together on the bed, smiling shyly at each other, unsure of what to do. Driving back to Dorothy's house in West Bridgewater, they stopped the car a moment, both women aware they wanted to say something, though neither daring. Whatever it was that was happening, they couldn't stop it, nor did either want to. On the train home that evening, Rachel referred in a letter to being able to feel "the sweet tenderness" of Dorothy's presence. She felt sure "Dorothy had sensed the same thing about her after she was gone." Later Rachel and Dorothy would refer to this time as "the thirteen hours" when a "little oasis of peace" entered their lives. Rachel felt it had been "truly perfect." There was "not a single thing" she would change about Dorothy, she said. From that moment forward, they began to call each other "Darling" and, in their letters, to openly express their love.

Both women now worried that Stan and Rachel's mother might be hurt by things they shared with each other in their letters, that the "craziness" between them might be misconstrued. Given the number of letters flying between them, and the length of some, it seemed awkward to read aloud just one paragraph. Rachel said they needn't discuss it further, that they both understood. What they should do, she suggested, was to write two letters, one "general and newsy," the other for their eyes only, which could be folded inside the general letter. They called these private notes "apples." Whenever one or the other felt something she'd written should remain confidential, she said it should be "put in the strongbox." Months now passed and the letters grew longer and more demonstrative, sometimes arriving one a day. At one point Dorothy suggested they stop until the seashore book was finished, imagining it would help Rachel's concentration. Rachel wouldn't hear of it. In a letter unlike any she had ever written, she told Dorothy she'd been in love with her even before she and Stan had left Maine the summer before. She said she needed to have Dorothy in her life, comparing their love to the parable in which a man said that if he had only two pennies, he would spend one on bread and the other to buy a "white hyacinth for his soul." Dorothy was her "white hyacinth," Rachel wrote. Without her, life was now unimaginable. From that moment forward, the two always referred to Carson's avowal as "the Hyacinth Letter." The flower itself became an emblem of all they felt for each other.

The seashore book, which Carson and Brooks had envisioned as a "field guide," was now years behind schedule. Carson traveled to the Florida Keys, exploring its reefs and mangrove swamps; to Myrtle Beach, South Carolina, to collect sensory impressions; to St. Simons Island, Georgia, to wade the vast tidal flats. She returned to Woods Hole, to use their world-class science library. When she updated Brooks, she claimed to be making progress. But privately she knew she was stuck. She wrote to tell him she was taking a short break, that she thought she was "over-concentrating." But it was more than that: Rachel was blocked. How ironic, she told herself, that with nothing to impede her travel or her work, no full-time job requiring her to limit her writing time only to nights, that she should find herself at such a standstill. This had never happened before—her time had always been too pressed for such a luxury—and eventually, once she got some distance, she realized the problem: the book wanted to be something else, not a simple field guide, with sketches of individual creatures, but a story with an "overarching narrative." What she needed, in effect, was to start over: to describe the shoreline as a web of living communities, played upon by a collision of shifting forces, much as she'd done for the ocean's depths in *The Sea Around Us*. In short, she needed to take an ecological point of view, to approach her subject more holistically. "As I write of it," she told Brooks, "it sounds so very easy; why is it such agony to put on paper?"

Rachel returned to work, but her involvement with Dorothy continued to deepen. She regretted "having taken so long to put into words" what she felt for Dorothy, she wrote. "How blind I was not to realize sooner that I should say it!" What they were experiencing, she said, was a process of "discovery" in which "each progressive stage of getting to know each other led to still more urgent feelings." She found herself continuously fighting the temptation to drop her work to go and see Dorothy. "But, oh darling, I want to be with you so terribly that it hurts!" she confessed. They had discovered that sometimes there was an eerie and marvelous synchronicity between them: the same thought would occur to each at the same moment. They called these inexplicable occurrences "stardust."

Rachel began her rewrite, determined to resist all distractions. But in mid-May 1954, a moment presented itself that she couldn't let pass. Stan Freeman was away on a business trip. She and Dorothy decided to travel up to Maine together, eager to spend the time in the place they most loved. Their brief idyll passed as if in a dream. They filled their days with drowsy breakfasts, with bird sightings and walks in the pine-scented woods. Sometimes they read aloud from the books they loved: *The Wind in the Willows*, E. B. White's *Charlotte's Web*. They stayed at Rachel's mostly, taking their dinners late, lounging before the fire long into the night, reveling in their hours together, the tenderness and the incessant talk, never wanting it to end. Afterward, Dorothy likened their visit to a "symphony." Rachel dubbed it "the Hundred Hours." Both would refer to it later as their "Maytime."

Rachel and Dorothy gathered their memories, slowly and carefully, like sea glass from the beach, quietly building their private world. They returned to them in reverie, savoring the comforts and pleasures of their shared affinities. But such interludes, though they would slowly accumulate as the years passed, were in fact rare. Much of what passed between them took place not in person, but from afar, in the day-to-day moments of reflection they exchanged in their letters. At bottom, theirs would always be a long-distance love, with all the longings and excitements and idealizations this entailed. In the long months that followed each summer's leave-taking, it was their mutual passion for the natural world that bound them, a love that ran deep as an underground river. Rachel wrote to Dorothy of lilac skies and the sleighlike chorus of spring peepers, of the thrill she felt at the arrival of the spring tides. Dorothy answered with her own impressions of the wild places they loved, the beauty of nature, and the rugged Maine coastline.

Dorothy's devotion to Stan was unwavering, which she knew Rachel understood. There was nothing between them they felt they couldn't say. She once wrote Rachel a beautiful and melancholic letter explaining the meaning she attached to her marriage. She felt lucky, she said, to have lived such a rich and feeling life. And yet, there was a part of her that belonged to Rachel alone. "Darling," she wrote, "you and I on our Island are looking at a light so bright— invisible to others—a glorious, miraculous light that has brought to me . . . untold happiness."

But their circumstances would never be comparable, which on some level they must have known. For Rachel, Dorothy was the "one great love of her life," although her world would always be her writing. Dorothy was happily married with a large and close-knit family. She cherished the time she shared with Rachel, but her life would have been rich and full even without it.

RACHEL AND DOROTHY REVELED IN THE RITUALS OF CHRISTMAS. IT WAS always a time of taking stock: looking back at the shared moments they treasured, looking forward in anticipation of the times ahead. For the Christmas holidays of 1954, they were planning a few days alone in New York. They would stay at either the St. Moritz or the Barbizon-Plaza, somewhere, said Rachel, "out of range." Rachel worried that were they both to arrive at the hotel at the same moment, they would be unable to contain their feelings until they reached their room. Dorothy felt sure they could feign an offhand air.

Rachel and Dorothy passed two nights together in New York. Afterward, Dorothy wrote to say that she had "no regrets" about any of it "thus far." Oddly, Rachel's response was unusually reticent. She said she could now return to her writing with new energy, and that it had been "a lovely interlude." Had intimacies passed between them that they now wished to disclaim? "Sex seems not to have been part of their relationship, or at least not an essential feature of it," Souder ventures. "Their surviving correspondence describes a transcendent, romantic friendship that existed in a realm above ordinary physical love and desire." As they were rarely together, he adds, their relationship seems to have "existed mainly on paper and in their own hearts and minds."

CARSON DELIVERED THE FINISHED MANUSCRIPT OF *THE EDGE OF THE Sea* on March 15, 1955. It was three years behind schedule, but everyone at Houghton Mifflin was pleased, convinced that they had another bestseller on their hands. This book was by far Rachel's most personal, the only one written in the first person, and the only one drawn almost entirely from her own fieldwork. Like its predecessor, it was filled with uncommon beauty, a lyric sweep even more seductive than the earlier book, which had now sold more than a million copies. Once again it was serialized in *The New Yorker*, and widely lauded by the critics, praised for its "lucid yet poetic force and simplicity," its "direct crystal clear prose." Four weeks after its publication on October 26, 1955, *The Edge of the Sea* climbed onto the bestseller list. By early December it had stepped up to number 4. Carson was elated. She was now the most popular nature writer in America.

All that remaining winter, Rachel felt happy and fulfilled. She and Dorothy were bursting with plans and the book continued to do well. And then quite abruptly family troubles intruded. One morning in October 1956, Maria Carson "toppled over like a felled tree" on the kitchen floor; it was clear to Rachel that she could no longer leave her mother untended. Three months later, her young niece Marjorie fell ill and was hospitalized for pneumonia and severe anemia; she seemed for a while to be getting better. But something went terribly wrong during her convalescence. Within weeks of returning home, Marjorie died at age thirty-one. Maria Carson was now eighty-eight; Rachel was just shy of her fiftieth birthday. There was no choice but to adopt Marjorie's five-year-old son, Roger, who was now Rachel's charge. Marjorie's sister, Virginia, wanted no part of raising a child; Rachel's brother, Robert, was "openly hostile" to the boy.

IN THE SUMMER OF 1957, A WOMAN NAMED OLGA HUCKINS LOOKED ON helplessly as a federal crop duster flew overhead, dousing her Duxbury, Massachusetts, property with a rain of DDT mixed with fuel oil. The aerial spraying, which went on intermittently all summer, was part of a massive, three-state government campaign to rid the Northeast of mosquitoes, tent caterpillars, and gypsy moths. The first time it occurred, Huckins watched as the supposed "harmless shower" of poison killed seven songbirds in her yard outright. By the next morning, there were three more corpses at her back door, and others scattered around her birdbath. She saw a robin drop suddenly from a branch. When she padded over to investigate, she noted that the beaks of the lifeless birds were open and "gaping"; their "splayed claws were drawn up to their breasts in agony." Yet all summer, she observed, the mosquitoes were more "voracious" than ever, while grasshoppers, bees, and other harmless insects were all but gone.

In January, infuriated by a glib letter in the *Boston Herald* claiming there had been no wildlife loss linked to the spraying, Huckins, who was a former writer for the *Boston Post,* responded with a letter of her own, enumerating all she had seen. "The testers must have used black glasses," she remarked acidly. "And the trout that did not feel the poison were super-fish." Calling the government spraying "undemocratic and probably unconstitutional," she closed with a plea to halt the aerial spraying program immediately. Then, pulling the letter from her typewriter, she sent it to the *Herald*, where it appeared a few days later. She also mailed a copy to Rachel Carson, a casual friend, whose book she had once reviewed, hoping she might know people in Washington who could help.

Huckins's attempt to enlist Carson was perfectly directed. Rachel had never stopped thinking about the dangers of DDT since 1945, when she had tried to interest *Reader's Digest* in an article. Now, with Huckins's letter—and with two massive federal spray campaigns in progress, each with the unconditional support of the pesticide industry—she was returned again to the urgency of the problem.

Recently, and perhaps not so coincidentally, Rachel had learned of a group of fourteen Long Island residents who were suing the government over just this issue. Led by an energetic woman named Marjorie Spock, the plaintiffs were seeking a permanent halt to all government spraying of DDT over private land. The trial was coming up soon. Spock was the eccentric younger sister of the celebrated pediatrician and later antiwar activist Benjamin Spock. Buoyant and audacious, seemingly tireless, she had forsaken Smith College to study abroad with Rudolf Steiner, a quasi-mystical philosopher who advocated a variant of organic agriculture called "biodynamic gardening." Upon her return, Spock and her live-in partner, Mary Richards, "a digestive invalid," had put in an organic garden to help with Mary's health. It was onto this garden that federal crop dusters had repeatedly showered DDT throughout the summer of 1957, as part of the same spray program that Huckins described to Rachel in her letter.

Rachel didn't see herself as an investigative reporter. But she did feel it was critical that a seasoned journalist cover the suit, so she wrote *The New Yorker*'s E. B. White, urging him to take it on. White, also the author of the beloved children's classic *Charlotte's Web*, had been one of the first to sound the alarm about the potential dangers of DDT, writing a piece for the magazine soon after the war. Like Rachel, he owned a house in Maine, a state rumored to be included in the summer's spray campaign; the subject, she reasoned, would touch him personally. White wrote back immediately to say he couldn't do it. But he agreed the issue was of "utmost concern." He would pass on her letter to Shawn, he promised, with a suggestion that he find a reporter for the story. Perhaps, he added gently, Carson should consider writing the piece herself.

RACHEL'S INITIAL RESERVATIONS ABOUT TAKING ON THE PROJECT WERE complicated. With Marjorie's death the year before, her life was changed. At six Roger was a handful. He had a short attention span and could be distressingly energetic. In a letter to Rodell, Rachel described him as "lively as 17 crickets." At moments, she said, she was the only person who could "hold him down." He was also physically fragile, vulnerable to respiratory problems. Rachel was attached to her grandnephew, but she was also painfully aware of the burden he represented. Dorothy had offered to take him soon after Marjorie's death, but Rachel had declined. Better to keep him close, part of a household that now included one aging mother, one active child, and a beleaguered author who felt blocked and entangled in domestic commitments beyond her control.

Carson was also suffering an existential crisis of sorts. Troubled by the coming of the Atomic Age, she suspected the reason she was having difficulty getting down to her work was the idea that humankind had discovered the means to irrevocably alter the world. She had become increasingly alarmed that the dumping of radioactive waste in the seas endangered not only the oceans, but also life itself. She told Dorothy it was hard to think about and "harder still" to put into words. "But I have been mentally blocked for a long time . . . for a reason difficult to explain . . . Some of the thoughts that came were so unattractive to me that I rejected them completely, for the old ideas die hard, especially when they are emotionally as well as intellectually dear to one. It was pleasant to believe, for example, that much of Nature was forever beyond the tampering reach of man . . . that the stream of life would flow on through time . . . without interference by one of the drops of the stream—man. And to suppose that, however the physical environment might mold Life, that Life could never assume the power to change drastically—or even destroy—the physical world."

Ultimately, Rachel overrode her reservations. By the time White's letter reached her, she'd already committed herself to doing some sort of book project on the pesticide issue for Paul Brooks at Houghton Mifflin. She'd also conferred with William Shawn at *The New Yorker*, who indicated he wanted to run a two-part piece by her on the pesticide problem. However grim the subject—in private Rachel was calling it "the poison book"—she knew *The New Yorker* would ensure her a large block of readers, even if the book itself bombed.

The project quickly ballooned. Rachel worked feverishly throughout the summer and fall of 1958, sifting through a mountain of technical reports. All that summer, Spock had been forwarding Carson materials from the Long Island trial, which that February the group lost on a technicality. Though disappointing—the judge suggested they didn't trust Spock's "experts" because they were into "organics"—the damning evidence she was sending was proving enormously helpful. "You are my chief clipping service," Rachel wrote Spock. A "surprise witness" from the Mayo Clinic in Minnesota, Dr. Malcolm Hargraves, had testified on behalf of Spock's group. Hargraves, who specialized in blood diseases, told the assembled he was "convinced" there was a link between DDT and the development of leukemia and lymphoma. While it was a position the Mayo Clinic didn't yet formally espouse, most of the doctors there believed it was the case. The more they were learning about the effects of DDT and other chlorinated hydrocarbon pesticides on human health, he added, the more perilous they appeared; the situation was analogous to the conclusions they were drawing about human exposure to atomic radiation.

Equally jarring was another set of findings Spock sent—this by an ornithologist at Michigan State University named George J.

Wallace. Since 1954, the campus had been sprayed with DDT every spring to kill off Dutch elm disease. Wallace and his team had been monitoring the effects of the DDT on bird populations, with a special eye on robins. Every year, after the spraying, all the robins in the treated area died. But the picture was actually much worse: the number of dead or dying robins they retrieved far exceeded the number that had been present in the spring, well *before* the spraying began. This meant that robins living *outside* the sprayed areas who happened to swoop in after the treatments were being poisoned by DDT residues lingering in the environment.

More ominous still, there followed a long-term decline in the overall robin population. This, Wallace surmised, was because of reproductive issues. The robins built nests, but they didn't lay eggs in them; and those few eggs that were laid didn't hatch. Ongoing tissue analysis indicated DDT loads in the birds' "testes and ovaries."

The conclusions Wallace was coming to were grim: DDT had entered the food chain. Heavy spraying had killed off the bark beetles that transmit Dutch elm disease. But the tree foliage was still coated with DDT. In the fall, when the leaves drifted to the ground, they were eaten by earthworms that feed on leaf litter. Some of the worms died outright, but those that survived took in "a heavy body burden of DDT," which they stored through the winter. In the spring the robins returned and ate the toxic worms, dying of "poison used in the *previous year's* spray program."

The most insidious property of DDT, then, was that it persisted in the environment, where it became concentrated in food sources far beyond its intended target. DDT was both lethal and long lasting, and it was hidden from view, a chilling combination. For Carson the nagging question was, if this "rain of death" produced so disastrous an effect on birds, "what of other lives, including our own?"

Rachel's research was going well. By the time she, her mother, and Roger returned from Maine, she was corresponding with an army of specialists—geneticists, entomologists, doctors, cellular biologists, botanists, agronomists—each of whom was providing her with new perspectives on the problem. Carson still had a number of friends from her government days, some who had been promoted and now held key federal posts, others who had joined private research institutes. Everyone she approached expressed an eagerness to help. Many were willing to pass on confidential information, trusting her discretion, putting their reputations, and even their jobs, at risk. It was an enormous vote of confidence.

Rachel's deepest worry now was her mother. Marie was eighty-nine and her health was failing. She had slept through much of the summer, often in a wheelchair just outside Rachel's study so Rachel could keep an eye on her. Then, just before Thanksgiving, Maria suffered a minor stroke, which turned into pneumonia. Rachel had an oxygen tent delivered to the house in Silver Spring, hiring a full-time nurse to help. But Maria continued to grow weaker, and finally, on November 30, she lost consciousness. Rachel sat up all night beside her mother's bed, her hand tucked under the border of the oxygen tent. At 6:05 A.M. Maria Carson slipped away, Rachel's hand still clutching hers.

Privately devastated, Rachel, as was her wont, hid her grief from all but a few. Writing to a friend, she spoke of her concern for Roger: "Poor little fellow, this is a new blow for him . . . it is obviously recalling to him all the memories of the loss of his mother, less than two years ago. It is good for me, I'm sure, that I am forced to think of him." In truth, it was the saddest Christmas she had ever spent.

By the middle of January, though still grieving, Rachel had returned to the Library of Medicine at the National Institutes of Health. She was resolved to resume work, if only for her mother, knowing how fiercely Maria had believed in the book. She wrote Brooks, telling him it was becoming increasingly clear to her that the book's heart was really how pesticides threatened human health. What she found curious, she added, was how contradictory many of the official recommendations seemed. "It is an amusing fact that although the AMA, when asked to take a stand, are rather on the fence, their various published statements constitute quite an indictment."

DDT, as Carson well knew, had now been available to consumers for nearly fifteen years. Not only had sales of the popular pesticide skyrocketed, there had been a deluge of new and even deadlier chemicals flooding the market in its wake. There were now two hundred registered synthetic pesticides—not just insecticides, but herbicides, rodent killers, fungicides. In addition, there were at least six hundred *other* products formulated with the same active ingredients. Annual production of pesticides had increased 700 percent since before the war; sales had ballooned to nearly "a quarter of a billion dollars" a year. Pesticides were now big business.

Yet even with these staggering numbers, Rachel was finding it oddly challenging to get definitive information about how these poisons affected human health. What little hard data she could dig up was scattered and highly technical, siloed in separate fields. And a lot of it was still circumstantial.

Typical was the case of Detroit, Michigan. After planes hovered low over the city, dropping pesticide pellets on bugs, animals, and residents alike—showers of supposedly "harmless" poison descending on "people shopping or going to work and on children out from school"—city authorities were barraged with calls from worried citizens, at one point more than eight hundred in a single hour. Though the callers were assured that the rain of pesticide dust was "harmless to humans and will not hurt plants or pets," they already suspected otherwise. Their house pets were suddenly stricken with illness. Local veterinarians reported their offices filling up with dogs and cats with "severe diarrhea, vomiting and convulsions." The local health department was besieged with citizens' calls complaining of severe throat and chest irritations. One Detroit internist reported that an hour after the planes circled, he was summoned to treat four patients with similar symptoms: "nausea, vomiting, chills, fever, extreme fatigue and coughing." Yet city officials insisted that these problems must have been due "to something else."

The American Medical Association, meanwhile, had been quietly collecting cases of pesticide poisoning since the early fifties. An illiterate, thirty-eight-year-old farmer had sprayed his tobacco crop with parathion. Unaware of instructions regarding its proper use, he stood so close to the sprayer that he was "soaked from head to toe."

He died fifteen hours later. A thirty-one-year-old university ento-mologist, working with parathion and other pesticides for months, one day forgot to wear his mask and protective clothing. After sev-eral hours, he felt nauseated. He went home, and died soon after. A child found a whiskey bottle in the crotch of a tree. Curious, he took a few swigs. The liquid wasn't whiskey, it was TEPP, an insec-ticide similar to parathion. The child started to foam at the mouth and died fifteen minutes later.

The stories went on and on. But these, the AMA insisted, were instances of acute poisonings, shocking of course, but not typical. To Carson they demonstrated how little those who handled these compounds understood their hazards. Equally troubling, she felt, was the lack of consensus about the day-to-day risks of small-scale exposures to these poisons. The real threat, she told Brooks, was the "slow, cumulative and hard-to-identify long-term effects." While no one was sure what a lifetime exposure for humankind might mean, it was known that "every child born today carried his load of poi-son even at birth, for studies proved that these chemicals passed through the placenta." Both a mother's breast milk and dairy prod-ucts "showed some content of DDT or other synthetic pesticides." There is "also scattered evidence indicating that some of these chemicals may interfere with normal cell division and may actu-ally disturb the hereditary pattern," she told Brooks. Still, like that of the officials in Detroit, Michigan, the government's stated policy remained that DDT was safe.

Rachel was by now convinced, she told Brooks, that there was a "psychological angle to all this: that people, especially professional men, are uncomfortable about coming out against something, es-pecially if they haven't absolute proof the 'something' is wrong, but only a good suspicion. So they will go along with a program about which they privately have acute misgivings. So I think it is most important to build up the positive alternatives." In this, Carson, as a lowly female scientist, was up against more than entrenched federal policies about pesticide use; she faced a male-dominated culture still steeped in militaristic values, a top-down corporate order accustomed to strict hierarchies. In the competitive, hypermasculine playbook of the 1950s, to show uncertainty was a sign of weakness. It was never a good idea to stick out one's neck, especially when so many people were drinking from the corporate trough.

RACHEL WAS WORKING EXTREMELY HARD, BUT SHE HAD LAPSED INTO A kind of melancholy. There were certain realms that she felt were sacrosanct. In 1957, when the Soviets launched Sputnik (which was followed shortly after by a second Russian satellite), she was uneasy. She wrote to Dorothy, saying that everyone now faced a "strange future," admitting that it "made her feel ill." This was partly her concern about the uses of space for nuclear warfare. But partly something more that she couldn't completely put her finger on, although it had something to do with becoming disconnected from the biological cycles that shaped the living world. Despite her belief in science, she was feeling increasingly out of step with her own time, wary of modernity and resistant to change. While thrilled by the technologies that had opened the ocean's depths to the human eye, she was more circumspect about other efficiencies of the twentieth century: the gadgetry and time-saving conveniences, the push into space and the sprouting of highways. The spread of "man-made ugliness" and the "trend toward a perilously artificial world," she said, portended the "destruction of beauty and the suppression of human individuality in hundreds of suburban real estate developments where the first act is to cut down the trees and the next is to build an infinitude of little houses, each like its neighbor." There were worrisome signs everywhere, she felt: threats even to the restorative calm of urban sanctuaries such as Washington's Rock Creek Park, near her home, where planners were now proposing to build a six-lane highway; "commercial schemes" proposed for the national parks. "Beauty—and all the values that derive from beauty"—shouldn't be "measured and evaluated in terms of the dollar," she insisted. In 1953, she had written a letter to the *Wash-*

ington Post protesting the firing of the director of the U.S. Fish and Wildlife Service, a steadfast conservationist, to make way for a pro-business nonprofessional. She had used the occasion to undercut the anti-Communist rhetoric: "It is one of the ironies of our time," she observed, "that while concentrating on the defense of our country against enemies from without, we should be so heedless of those who would destroy it from within." Haunted by the misuse of science and technology, worried by mankind's tendency to go "farther and farther into experiments for the destruction of himself and his world," she wondered where humans would take these things. It was all, she wrote Dorothy, "deeply disturbing."

IN THE SPRING OF 1959, AS RACHEL WAS MIDWAY INTO THE BOOK, THE U.S. Department of Defense admitted that its calculations about how long radioactive debris lingered in the upper atmosphere after testing were wrong. Originally they had assured the public that the waste remained aloft for close to seven years, during which time it would decay, scatter, and gradually drop as "minimally radioactive fallout" spread evenly across the globe. Now officials admitted that it could be "as little as two years." This meant that the radioactive fallout was hotter, more lethal, fell sooner, and "over a more concentrated area." The public also learned that the U.S. showed higher concentrations of strontium 90—which is absorbed into bone tissue, and was being implicated in cases of leukemia—than anywhere else in the world.

Public alarm immediately spiked. Though nuclear testing had been temporarily suspended in 1958, after the Soviets promised they would halt testing for a year, provided the West did the same, radioactivity levels in the atmosphere remained high. Researchers at Washington University in St. Louis had lately discovered strontium 90 in babies' teeth. Then *Consumer Reports* disclosed there were detectable levels of strontium 90 in cow's milk. That September, the *Saturday Evening Post* ran a story called "Fallout: The Silent Killer," in which scientists warned of leukemia, bone cancer, and long-term genetic alterations. Doctors claimed that continuous low-level exposure to radiation further magnified the risk of cancer—and that over much longer periods, "subtle genetic mutations induced by radiation would cause steady increase in birth defects." The government insisted there was little reason for worry. The radiation from fallout, it reported, "was far below the normal background level of radiation from natural sources," which included cosmic radiation and radioactive elements in the earth's crust. But the public was unconvinced. By the close of 1959, the nation was in the throes of a massive fallout scare.

UNSEEN BUT OMNIPRESENT, UNTRACEABLE EXCEPT WITH SPECIAL INstruments, radioactive fallout was "a strange and chilling thing," writes Souder, a poison that carried acute hazards to human health that might not be seen or felt for years. Though still an abstraction for most Americans, it had been swirling unseen about the globe since the close of World War II, the eerie spawn of an escalating arms race that seemed without end. As the Cold War ramped up during the 1950s, a handful of countries, but most particularly the U.S. and the Soviets, had been conducting nuclear tests of one kind or another almost continually. Between 1951 and 1955, the U.S. carried out forty-nine aboveground nuclear tests in Nevada alone. Some were small, dispersing little radiation. But others were two to three times as powerful as the atomic bomb dropped on Hiroshima in 1945 (which had incinerated tens of thousands of people in a single horrifying blast). Wherever the site, whether in Britain or France, Nevada or New Mexico, USSR or the South Pacific, the mushroom clouds rose to the sky with breathtaking regularity, steadily pumping radioactive debris into the upper atmosphere, where it joined similar menacing clouds of radioactive material coming from other places. All would ultimately fall to the earth, though no one quite understood the gravity of this yet.

For some time, the U.S. had been working secretly on a weapon—

a hydrogen device—that would be even more devastating than the atomic bombs periodically exploded at Bikini, a fragile atoll in the Pacific Ocean's Marshall Islands chain, beginning with the first in 1946. Finally, in 1954, again choosing as its site Bikini, the government decided to explode one. The hydrogen bomb test, which was officially called "Castle Bravo," didn't go as anticipated. Not only was the blast twice as powerful as expected, flattening several islands in the atoll, a sudden shift in the wind sent the huge cloud of radioactive "fallout" catapulting through the upper atmosphere in an unexpected direction, where, a few hours later, it began raining down on a hapless Japanese fishing trawler called the *Lucky Dragon*.

The *Lucky Dragon* had been decidedly unlucky since the start of her trip. The captain was ill, there was persistent engine trouble, and the catch was consistently lousy. Hoping to salvage what was beginning to look like a five-week rout, the captain had switched course and moved into waters he didn't normally trawl.

On March 1, just before dawn, the *Dragon* was floating about eighty-seven miles east of Bikini. Unable to sleep, a young seaman arose and walked the deck. Suddenly a blinding white light flashed at the horizon, turning ocher and then an eerie blood orange. Mesmerized, the disoriented seaman began shouting to his shipmates below, "The sun is rising in the west! The sun is rising in the west!" as the sleep-disheveled crew clambered up the stairs to see. One by one they froze. The unearthly light was racing up the western sky, a sickly flare of illumination unlike anything they had ever seen. And then there came a seismic shock that convulsed the sea, shaking the vessel in a momentary blur of chaos, followed by two enormous, concussive blasts. The crew was thrown to the deck. Almost immediately the radioman began to calculate the speed of the sound. He looked up, frowned. They were eighty-seven miles from the explosive event, he told the others. His face went white. "It's an atomic bomb," he said quietly. The crew fell silent.

The men started working immediately, hauling in the immense fishing line, fifty-odd miles of it. As soon as it was in, they could leave. The work was slow, abrading to the hands. It was extremely humid that day, and their skin burned. Then, two hours in, a dense fog descended, followed by a light drizzle. But it was not the kind of rain the seamen knew. It looked like snow, only it was gray and granular, gritty like sand. It landed in their hair, stung their eyes; it was sticky to the touch. Several of the men tasted it, wondering if it was salt. That afternoon no one had any appetite; several men were nauseated. The next morning many could barely open their eyes, which oozed a gluelike discharge. By the third day, many men looked like they had bad sunburns. The radioman was feverish, and his skin had turned black.

When the crew reached port, they were immediately hospitalized, first locally and then at the University of Tokyo, where they would remain for a year. By now most were losing their hair. Their charred skin had turned yellow and many were suffering from other symptoms: "bleeding gums, falling white blood cell counts, compromised bone marrow," sperm counts that bottomed out to zero. But none was as sick as the radioman. All that summer his health declined. Then, on September 23, he died of liver failure.

The papers initially downplayed the issue. American authorities told journalists that the cause was "hepatitis." But the story wouldn't go away. Two more Japanese fishing trawlers were found with critical levels of radioactive contamination soon after the return of the *Lucky Dragon*. Both had been fishing in areas considerably east of the Marshall Islands, one in waters more than 780 miles from the test site. The *New York Times* revealed that the catch from the *Lucky Dragon* was "sufficiently radioactive as to pose a threat to human life." Worried Japanese housewives stopped buying fish. The American ambassador to Japan offered a public apology to the radioman's family on behalf of the "American government and people." Quietly, he followed up with a check made out to the radioman's widow for a million yen.

In truth, as President Eisenhower acknowledged at a press conference a few weeks later, no one had anticipated an explosion of such force, not even the scientists involved. A decade earlier, during the first tests for the Manhattan Project, scientists had believed that the devastation from a nuclear device would be caused by the scorching heat and radiation close to the blast, followed by the accompanying shock wave that would rumble outward, covering a larger area. Little thought was given to the "secondary contamination" of far-flung places from radioactive waste carried high in the stratosphere. But such "fallout," as it would be called, "was to become the great fear of the atomic age," writes Souder. It turned out that the mushroom clouds of radioactive dust and debris produced by a nuclear explosion didn't drop straight down, as was initially believed. Instead, the radioactive fallout rode the "upper-level wind currents" for years, where it sailed over land and seas, eventually merging with larger weather patterns and falling back to earth as rain or snow wherever it had blown.

Americans were rightly alarmed to learn that strontium 90 was now in their milk and even their children's teeth. But few understood how or why this had happened. They read about incidents like that of the *Lucky Dragon,* which was reported in American papers, and they knew about the Nevada tests. Like it or not, they were accustomed to living uneasily with the specter of nuclear war, which they hoped, of course, would never happen. But it was harder to imagine this new, more insidious threat, which they couldn't see, but which now they were being told was everywhere. It was harder still to believe that what happened half a world away, in a remote pocket of the Pacific, could have an immediate impact on their lives in Kansas or Vermont. No one really understood the extent to which fallout could travel, though they were beginning to get a sense.

In late January 1951, tracking planes monitoring a nuclear test in Nevada had followed the giant mushroom cloud as it rose and blew east, where at some point it merged with an incoming storm. Three days after the blast, radioactive snow fell in New York City's Central Park.

THE PARALLELS BETWEEN PESTICIDES AND RADIOACTIVE FALLOUT were becoming increasingly clear to Carson. In both cases, the hazards were cumulative and carried the same risks of genetic damage.

And yet, the public had been left in the dark about these chemical poisons. It was a disquieting example of technology outpacing knowledge. These toxic agents had been developed and sold to the public with no oversight and no understanding of their long-term effects.

Carson was feeling disquiet for other reasons. The book was terribly behind schedule. Her summer on Southport Island hadn't gone as planned. Roger was hospitalized for a week with a respiratory infection, and then she herself was ill. In a letter to Brooks in December 1959, she apologized for having failed in her promise to deliver the manuscript by then. She thanked him for his forbearance, admitting that it was almost "unbearably frustrating" to be so far behind. The one thing she could promise, however, was that the book would have an "unshakable foundation," which she felt was especially important given the "violent controversies" that surrounded the subject. Those with a stake in pesticide sales were sure to attack the book, she added, so it was critical she have the "weight of evidence" on her side. She said she hoped she might be done by February, but that she couldn't promise. In closing, she mentioned that she and Roger had been feeling better, though she had developed "some sort of thyroid condition" that left her with sharp headaches, which sometimes robbed her of hours, or even a day, of work.

Rachel's health issues continued to interfere. Shortly after New Year's, she learned she had a duodenal ulcer. She told Rodell that until it healed she would be living on baby foods. Then she was stricken with the flu, which grew into pneumonia, followed soon after by a sinus infection. Her immune system was clearly overtaxed. As February, the deadline she'd set for herself, came and went, she fought off the hopelessness she was feeling. "Sometimes I wonder whether the Author even exists anymore," she wrote Dorothy. In March Rachel finally wrote to Brooks, sharing the details of her various illnesses, and admitting it had been a rough few months. She was still hard at work on the cancer chapters, she told him, but she had to carefully apportion her time, as only sleep and calm would aid the healing of her ulcer. She assured Brooks that while it might appear that the book had given her the ulcer, this was in no way the case. She found the entire subject of pesticides, however sobering, "quite fascinating." Any decent ulcer, she joked, might have waited to strike until the book was finished. In closing, she told Brooks that she thought that "by far the most difficult part [was] done."

What she didn't tell Brooks, perhaps for fear of worrying him, was that two cysts had developed in her left breast—the same breast from which she'd had the cyst removed a decade before. Rachel immediately scheduled surgery to investigate what they were. Writing Brooks again, she alluded to a new health concern, though she tried to downplay it. She was going to have surgery that she hoped would not be "too complicated," she said, although she did allow that she couldn't be sure. Somewhat ironically, in addition to her letter, she enclosed the two chapters on cancer she had just completed.

As Rachel braced for her upcoming surgery, she and Brooks exchanged a flurry of correspondence. The operation was now scheduled for the following Sunday, Rachel wrote. She expected to be back at work by Wednesday, adding gloomily, "otherwise at the end of the week."

But the operation didn't go as she'd hoped. The doctors found two tumors in her left breast, one that was benign, the other "suspicious enough to require a radical mastectomy." The operation was brutal and she was in a great deal of pain. Rachel asked her doctor pointedly if the pathology report showed any malignancy, and was led to believe that the mastectomy had been "a precautionary measure." He told her no further treatment was needed, and she let the matter lie.

Anxious about leaving Roger, Rachel "talked her way out" of the hospital barely six days after the operation. She wrote to Marjorie Spock, saying that her "hospital adventure" had ruined her work schedule for the spring, but that she was thankful the cancer had been found so early, and that her prognosis was good. "There need be no apprehension for the future," she said.

Her chief concern now was in guarding her privacy. She feared her critics would use her cancer diagnosis against her, claiming that her illness was the rationale for her conclusions about the links between pesticides and cancer. She told Spock she was only sharing the details of her ordeal with "special friends" like her. "I suppose it's a futile effort to keep one's private affairs private. Somehow I have no wish to read of my ailments in literary gossip columns. Too much comfort to the chemical companies."

Brooks, who didn't yet know the reason for Rachel's surgery, wrote to commend her on the cancer chapters, which he felt were excellent: clear and convincing and beautifully composed. He remarked that the recent news about fallout and cancer was working in their favor. "In a sense, all this publicity about fallout gives you a head start in awakening people to the dangers of chemicals," he said.

RACHEL'S PROGRESS CONTINUED TO BE DISAPPOINTING, JUST AS SHE'D feared. When she got to Southport that summer, she was still weak and in considerable pain; it was hard to concentrate on much beyond looking after Roger, however much she tried. Even so, she hid her distress. By December Brooks was anxious for a progress report. It was only then that he got the sobering news. Rachel was sick again. She had discovered "a curious, hard swelling" near her sternum, which X-rays confirmed was cancer. Rachel, clearly caught off guard, was now in the midst of radiation treatments. She told Dorothy that it was hard to face all this "after being so sure the previous spring that her surgery had solved everything." Now it appeared she couldn't be sure that she would ever get well. It was especially cruel given the time pressure she felt because of the book. Writing to Brooks, Rachel admitted she was angry that she'd been misled: "I know now that I was not told the truth last spring at the time of my operation. The tumor was malignant, and there was at the time evidence that it had metastasized . . . But I was told none of this, even though I asked directly." Rachel's situation was in fact not unusual. Medical conventions for female patients in the 1950s and 1960s were still extremely paternalistic. If a woman was married, she herself was never told she had a malignancy; only her husband, if he asked, was given a full account. For a single woman like Rachel, the same must have held true, only with no husband to deliver the news, she was left in the dark. In Rachel's case this was especially disturbing, as she was more than capable of understanding the science behind anything the doctor said. That she was not given the choice of being fully apprised of her treatment options was unconscionable, if not demeaning. She was still fifty-two, and had she known, the outcome might have been different.

Rachel was so sick after the first two radiation treatments that she couldn't leave her bed. When the nausea at last lifted, she decided she had been too hasty in her decisions the previous spring. It was time to seek a second opinion. She left for the Cleveland Clinic in the middle of a blinding snowstorm, waiting on a windy corner for an hour before she was able to flag down a cab to the airport. Dr. George Crile Jr. was one of the foremost cancer experts in the country. He confirmed Rachel's worst fears—that her cancer had metastasized to other lymph nodes—but he recommended a different course, saying they should target only the affected site with radiation. In her letter to Crile afterward, Rachel thanked him for his respect and his absolute candor. "I have a great deal more peace of mind when I feel I know the facts, even though I might wish they were different."

In January, Rachel's ulcer flared up again, triggered by her radiation treatments. But the good news was that the mass was shrinking. Dorothy, daring to imagine that she and Rachel might still have more summers, sat down on the bed where Rachel had napped on her first visit, and wrote Rachel a long letter, reflecting on the richness of the times they had spent together over the past seven years. "You spoke of the moonlight shining in your room—how many happy memories that evokes. If we had only moonlight, shared, to remember, our storehouse would be unusually rich. But there are the Sea, the Shore, the Woods, the Gardens, the Marshes, Phosphorescence, Wind, Sun, Sand, Scents—oh, my Darling . . . For me it has meant more than I can tell you to have your love enfold me— love that was your arms around me in dark hours."

By the time Dorothy's letter reached her, Rachel was bedridden again. In January of 1961, she had developed a staph infection that progressed to severe phlebitis, settling into her knees and ankles. Unable to walk or stand, she was now confined to either a wheelchair or her bed. She wrote to Brooks, telling him she had "never been sicker in her life," which must have shaken him, given what he already knew. Rachel was taken to the hospital, which she tried to make light of, though she admitted to Dorothy that she had been "devastated" seeing Roger, who was just eight, "slumped and sobbing" as the medics slid her into the ambulance for the short trip.

As April neared, Rachel wrote Brooks to say she was finally writing again, albeit cautiously. She apologized for having burdened him with so many details of her health, but wanted him to know so he could understand why progress on the book had been so agonizingly slow. The only positive side to all this, she added, was the fresh perspective she'd gained in being pulled away from it. She now wanted to make the narrative leaner, to free it of all excess detail.

Rachel pushed on throughout the spring, continuing to converse with a number of experts about technical questions, winning allies with each new exchange. By June 1961, a jubilant Rodell was able to tell Brooks that it looked like Carson would be finished in a few months. "I am working late at night most of the time now," Rachel wrote Brooks. "If I can fight off the desire to go to bed around 11:30, I seem to get my second wind and be able to go on." In early January 1962, though Carson still had one remaining chapter to complete, the rest of the manuscript was in the hands of both *The New Yorker* and Brooks. One night the phone rang. As Rachel picked it up, the quiet voice of the caller said, "This is William Shawn."

Shawn described what she had done as "a brilliant achievement." He told her she had made the difficult subject of pesticides into "literature, full of beauty and loveliness and depth of feeling." As she hung up the phone, Rachel felt a surge of happiness. For the first time, she allowed herself to believe she had succeeded in what she'd set out to do, that now the book would have a life of its own, no matter what happened to her. She padded into her study, put on a Beethoven violin concerto, and allowed herself a good cathartic cry.

ALL EYES NOW TURNED TO HOW HOUGHTON MIFFLIN SHOULD HANDLE PRE-publicity for the book. Brooks knew that many of Carson's assertions would be explosive. He feared—rightly, it turned out—that the chemical companies might try to interfere with its publication. Better not to send out advance copies until *after* the public had read *The New Yorker* serialization, which was set for June 1962 and would appear in three consecutive issues. This way, any perception that Carson was leading a crusade against big business would be blunted by the public's reception, which he judged would be good.

Rachel, though normally outspoken about these issues, was preoccupied with more private matters. She was still undergoing radiation treatments. On hospital days, though desperate to begin revisions, she was often too sick to work beyond midday. She had discovered a new mass in her armpit and was anxious, as it was outside the treatment area. The mass *was* cancer, she learned. But the pain near her neck was only a side effect of the radiation, which was some relief. The real agony of cancer, she told Dorothy, was the loss of any trust in one's own body. "Every perfectly ordinary little ailment looks like a hobgoblin," she said.

The state of the world was also on her mind. In the summer of 1961, the Soviets had announced they were resuming nuclear testing. Over the next three months, they exploded thirty-one nuclear devices, one more than three thousand times the size of the bomb that had irradiated Hiroshima. Fearful the Soviets were getting a leg up, President John F. Kennedy, who had campaigned on a promise to secure a permanent ban on testing, "reluctantly restarted" America's program as well. Tensions between the East and the West had escalated in response to the construction of the Berlin Wall, followed by the Bay of Pigs debacle. The botched invasion, an embarrassment for Kennedy, had led to a worrisome new trade pact between the USSR and Fidel Castro, Cuba's brash new Communist leader, fanning fears of a nuclear incident. There were rumors that the Soviets were shipping offensive weapons to the island, which was only ninety miles away. Meanwhile, tests of milk from Minnesota, a major dairy state, were showing high levels of iodine 131, a component of radioactive fallout linked to leukemia. No one should have been surprised: the amount of fallout swirling about the globe had "doubled" in a matter of months. Unwilling to shut down dairy suppliers whose milk surpassed radiation thresholds, the government decided instead to revise the guidelines upward. It was a response that only heightened public anxiety.

THE NEW YORKER RAN THE FIRST EXCERPT OF SILENT SPRING ON JUNE 16, 1962. It opened with a brief, eerie fable in which Carson described a nameless town "in the heart of America where all life seemed to live in harmony with its surroundings." Fertile farm fields encircled the town and the hillsides were stippled with orchards. Birds chirped amidst the trees, which blazed amber and gold in the fall. The trout streams ran cold and clear, the wildlife was abundant. And then a "strange blight" descended on the area. Inexplicable illnesses began to appear. Farm animals sickened and eggs didn't hatch. Pig litters were so stunted they died within days. The wildflowers by the road turned brown, "as though swept by fire"; the townspeople were stricken with diseases the doctors couldn't treat. No one understood what had happened. Though on the roofs and gathered in the gutters, a mysterious white power still lingered, which weeks before had "fallen like snow" upon the houses and lawns, the farmland and fields. Everywhere there was a ghostly quiet, a "strange stillness," as if all living things had leaked from the world.

In three short, chilling paragraphs, Carson had described an archetypal town that she was certain her readers would recognize, for it was a portrait of where they lived—and all they feared. Though her subject was the new and invisible threat of pesticides, Carson's images of death and devastation couldn't help but call up for readers all the stark terrors of the nuclear age, right down to the white flakes of poison falling from the sky, reminding them again of the *Lucky Dragon* incident, and, by extension, all their fears of nuclear fallout and the specter of annihilation. "No witchcraft, no enemy action had silenced the rebirth of new life in this stricken world," Carson wrote. "The people had done it to themselves."

THE THREE-PART SERIALIZATION CREATED AN IMMEDIATE FUROR, DRAWING more letters of response than any piece in the magazine's history. Scores of readers wrote in to share their outrage at the chemical industry's shocking indifference to human health and the environment; many expressed their gratitude to Carson for having brought these abuses to light. But unlike the nearly unanimous applause for her earlier books, a number of the letters were also dismissive. One peevish writer suggested that Carson's characterization of the pesticide manufacturers reflected "Communist sympathies," and that she was probably a "peace-nut too." Anyone could "live without birds and animals," he said, "but not without business," adding that as long as we had the H-bomb, everything would be okay. Executives from chemical companies accused Carson of one-sidedness, saying she had failed to consider the economic benefits of pesticide use in food production. One pesticide manufacturer, the Velsicol Chemical Corporation of Chicago, threatened to sue, insisting that Houghton Mifflin stop publication of the book or take out all references to its products. Houghton Mifflin refused, and the matter was left hanging, though no one was certain if a lawsuit was imminent. Perhaps the most unjust charge against Carson was that she was trying to derail the global campaign against malaria, which relied on DDT for its mission. It was a misrepresentation of Carson's position that she continued to face. The truth was, at no time did she advocate a complete ban on chemical insecticides. Even later, testifying before a Senate subcommittee, Carson allowed that the use of potent chemical poisons was sometimes necessary, particularly in crises to combat human disease. What she objected to was the indiscriminate and "heedless overuse" of these poisons by persons "wholly ignorant of their potentials for harm."

Carson and Roger left for Maine toward the end of June 1962, as the first reactions to *The New Yorker* pieces were still coming in. Despite the spate of cranky letters she'd received, the advance press so far was overwhelmingly positive, which she found heartening. The *Times* ran a strong editorial in support of her positions, predicting that she would be branded an alarmist, despite the strength and balance of her arguments. Within days, it was followed by an even more stunning endorsement by a young investigative reporter named Robert A. Caro, who wrote a five-part series, which began that August, for the Long Island daily *Newsday*. Caro, who years later would be celebrated for *The Power Broker*, his magnificent biography of New York urban planning czar Robert Moses, was a brilliant and hard-hitting reporter who didn't mince words.

Scientists, he reported, had long been aware of the links between pesticides and human diseases such as cancer, leukemia, and abnormal gene development, yet the government, in collusion with the chemical companies, had for years systematically suppressed the damning evidence, allowing corporate profits to take precedence over threats to the environment and human health. The truth about these chemicals, he wrote, was being whitewashed by an ongoing public relations campaign orchestrated by an eight-hundred-million-dollar industry that touted pesticide safety and effectiveness in order to keep the wheels of commerce churning. The problem was compounded by the USDA, he added, which "leaped aboard the pesticide bandwagon as soon as DDT was introduced," and continued to mount "vast" spraying programs of its own, despite evidence of grave damage to wildlife species. Now, thanks to "famed biologist and author" Rachel Carson's *New Yorker* series, a flood of angry letters was finally reaching the USDA, he reported. And so his argument continued, never letting up on his stinging indictment of the government and the chemical industry for their heedless indifference to the perils of pesticide use—nor his admiration for Carson and her still-unpublished book.

THE GOVERNMENT WAS SCRAMBLING, UNCERTAIN HOW TO HANDLE THE controversy. Shirley Briggs warned Carson of rumors that the Department of Agriculture was scouring *The New Yorker* pieces for evidence of libel, or even errors, anything they could use to discredit her claims. Others insisted the controversy would blow over; better to soft-pedal the problem, they counseled, than to overreact. Either way, staff members were told to start brainstorming for ways to weaken Carson's position, should fighting her become official policy. The secretary of agriculture, worn down by the recent milk scare in the Midwest, assured a reporter that there was no reason for "panic and hysteria," insisting that on balance, "pesticides provided more benefit than harm." Another official promised that his agency was fully on top of the pesticide issue. The government, he said, was engaged in "nationwide surveillance" of all "pollutants and contaminants in the environment." Behind the scenes, there was less consensus. Internally it was acknowledged that the FDA couldn't possibly "keep pace" with the pesticide industry in setting guidelines for residues in food when the number of different pesticide products had mushroomed to forty-five thousand. Officials were instructed to stop "blanket" denials that pesticides posed no perils. The chemical companies, meanwhile, were circling the wagons, amassing an enormous war chest in anticipation of the coming battle. Word went out to certain publications that a positive review of Carson's book might result in canceled ads. The FBI was quietly launching an investigation, looking into whom Carson had been speaking with on the phone, if and when she had been in contact with foreigners, searching, one can only suppose, for Communist sympathies. Not surprisingly, they found nothing.

RACHEL AWAITED THE BOOK'S PUBLICATION FROM HER PERCH IN MAINE, trying to keep a cool head. She was overwhelmed by the fan mail pouring in each day, spending hours trying to acknowledge what she could. To date, the letter she most treasured was from E. B. White, who said he believed her articles were "the most valuable" the magazine had ever published, expressing his admiration for the courage she showed in going after "this formidable opponent." He predicted her book would be "the sort that will help turn the tide," and the work that ultimately she would be proudest of. "I'm unable adequately to express my gratitude," he added.

Rachel had worried that the public would forget her book between its June appearance in *The New Yorker* and its September publication. But a telling incident that summer dispelled such notions. In mid-July, in the wake of *The New Yorker* series, a news story broke that jolted the nation, reminding it of the ways that science and business could ride roughshod over the public interest: Americans learned that U.S. drug companies had attempted to market the drug thalidomide, which had been linked to horrifying birth defects in western Europe. The drug, which caused severely deformed limbs, had been routinely prescribed to pregnant women as an antidote to morning sickness. Though the FDA had failed to approve thalidomide for use in the U.S., it had been widely sold abroad. The controversy, which shocked American consumers, was just breaking out as *Silent Spring* was heading to press. Approached by the press for her thoughts, Rachel didn't flinch. It was "all of a piece," she responded. "Thalidomide and pesticides—they represent our willingness to rush ahead and use something new without knowing what the results are going to be."

SILENT SPRING CAME OUT ON SEPTEMBER 27, 1962, AND WAS AN ALMOST immediate bestseller, in part because of its appearance in *The New Yorker* that summer. Within days, bookstores were struggling to keep it in stock. Newspapers everywhere reviewed it, and many published excerpts. Over seventy newspapers ran editorials, the majority of which were overwhelmingly positive.

Brooks Atkinson, writing for the *New York Times*, praised Carson's thoroughness, reporting that the book came with fifty-five pages of references and source citations. He called Carson "a realist as well as a biologist and writer." Her contention that chemical pest control was an act of arrogance, a misconception of mankind's place in the natural order, was well taken, he said. "The basic fallacy—or perhaps the original sin—is the assumption that man can control nature." Another reviewer compared *Silent Spring* to *Uncle Tom's Cabin*, calling it "timely," certain to stir the nation. The Book-of-the-Month Club chose it as its October selection, with an endorsement from Supreme Court Justice William O. Douglas, who called *Silent Spring* "the most important chronicle of this century for the human race." Carson was swamped with requests for appearances and speeches, most of which she declined. She did, though, agree to an exclusive interview with CBS News for an hour-long segment on pesticides hosted by Eric Sevareid. Perhaps most surprising was President Kennedy's mention of *Silent Spring* at an August press conference. After the president had fielded several tense questions about the recent escalation of Soviet shipping traffic to Cuba, a reporter asked about the "dangerous, long-term side effects" of pesticide use. He was aware of the problem, Kennedy said, as if he'd anticipated the question. He was directing his science adviser, Jerome Wiesner, to launch an investigation, adding that he knew of "Miss Carson's book."

But the attacks on Carson's book rained down just as quickly, a storm of vitriol more caustic and personal than anyone had anticipated. Carson was accused of being subversive and antibusiness, a crackpot and a Communist sympathizer, a "fanatic" determined to threaten the food supply. A letter from Velsicol's lawyer insinuated that she was under the influence of "sinister parties" who intended to reduce the West's use of agricultural chemicals so that food stocks would be depleted to "east-curtain parity." "America's food supply had never been safer," proclaimed a shill for a trade group called the Nutrition Foundation. *Silent Spring*, he added, was "obviously the rantings of a poorly informed and probably deranged person." (The group's members, Carson later learned, were some of the nation's most profitable food-processing concerns.) Carson was "a journalist cherry picking the facts," her critics alleged. Her writing was "unfair, one-sided and hysterically overemphatic." She was not "a professional scientist," but an overwrought female, a generalist without title or credentials. It was a charge leveled at Carson again and again: that her scientific credibility was compromised because she wrote for the general public; that she had only a master's degree in zoology, and had never worked as a scientist; that she wasn't "published in peer-reviewed journals" and had "no academic appointments." And then there was the problem of her sex: Carson was a woman who kept cats, a "bird and bunny lover," proof, certainly, that sentiment would always trump reason. She was a "spinster," as if this alone was evidence of an unsound mind. What was really going on was something else: with a poet's eye, a scientific mind, and a powerful sense of mission, Carson had invaded what had heretofore been a man's world, writes Linda Lear. "In short, [she] was a woman out of control." Carson had "overstepped the bounds of her gender and her science."

And so the campaign of misinformation roared on, driven by an army of well-paid, fast-talking PR men. "I don't know of a housewife today who will buy the type of wormy apples we had before pesticides," ventured one industry insider, who was quoted in the *Washington Daily News*. Another warned darkly that Carson's book would have "a catastrophic effect" on the nation's food supply and economy.

But the multimillion-dollar chemical industry and its allies in government and big agriculture misjudged the public mood. The readers to whom Carson appealed weren't troubled by corporate bottom lines or downticks in the food economy. They were everyday citizens, many of them housewives, whose first concern was the safety of their children, and they were finding Carson's arguments profoundly unsettling. Women and women's groups were a demographic that hadn't been much considered until then. They were consumers to appeal to, of course, but hardly a threat to the status quo. But this was changing, and Carson knew her audience. In 1961, women across the country had organized to protest the tainting of the milk supply by strontium 90. Few could forget the cranberry scare of late 1959, in which cranberries grown in the Pa-

cific Northwest were found to be tainted by the weed killer ami-
notriazole, a known cause of thyroid cancer in lab rats. Already
rattled by a host of contamination issues—from radioactive fallout;
to what *Time* magazine was calling "illegal quantities of penicillin
and hormones" in the food supply; to health problems associated
with artificial preservatives, colors, bleaches, coatings, thickening
agents, and emulsifiers used in food processing—women at that
moment were primed to hear Carson's message. It was no accident
that Carson's most political and hard-hitting answers to her critics
would come when she spoke before women's groups.

Carson's critique was at once simple and complex: pesticides,
she argued, posed a threat to the "shared biology of all living things."
Mankind, she insisted, was not the overlord of nature, but simply
one of its citizens. As such, humans were deluded in believing they
could control nature through chemistry, allegedly in the name of
progress. Technological meddling, she reflected, could "easily and
irrevocably" disrupt the entire web of life. Nature was a fabric on
the one hand "delicate and destructible," on the other "capable of
striking back in unexpected ways." To ignore this would be at humankind's own peril.

Carson built her case brick by brick: she explained how pesticides poisoned soil and water; how they accumulated in human and
animal tissues; how their residues made their way into cow's milk
and the milk of human mothers; and how their use often backfired
when the very insects that had been targeted became resistant. Science and technology, she argued, had become the pawns of a chemical industry indifferent to all but its own profits. Humans were now
subjected to dangerous chemicals "from the moment of conception
until death."

Using the era's fears about radiation to focus her readers' attention, Carson ticked off the "exact and inescapable" parallels between nuclear fallout and pesticides again and again. The effects of
exposure to synthetic pesticides—like exposure to radiation—were
especially burdensome to children, she warned, carrying the potential to alter our hereditary makeup. "We are rightly appalled by

the genetic effects of radiation," she asserted. "How then, can we be indifferent to the same effect in chemicals that we disseminate widely in our environment?" Our bodies had their own ecologies, she insisted, which could easily and irrevocably be knocked out of balance. At stake was nothing less than "our genetic heritage, our link with past and future."

Carson felt it was arrogant to believe that nature existed to serve mankind's needs. "The control of nature," she wrote, was a "phrase conceived in arrogance, born of the Neanderthal age of biology and philosophy." Human health, she asserted, would ultimately be a victim of the environment's degradation. "Can anyone believe it is possible to lay down such a barrage of poisons on the surface of the earth without making it unfit for all life?" "The question is whether any civilization can wage relentless war on life without destroying itself and without losing the right to be called civilized."

Carson repeatedly described a world in which "nothing must get in the way of the man with the spray gun." She deplored the deceptions routinely practiced by a government that refused to acknowledge evidence of pesticides' damage. The system as it stood was one in which government officials permitted the poisoning of the food supply and then "blithely claimed" to police the results. "Lulled by the soft sell and the hidden persuader," she said, the average citizen was "seldom aware of the lethal chemicals" with which he was surrounded. When the public tried to hold the government accountable, it was "fed little tranquillizing pills of half-truth."

"Man has lost the capacity to foresee and to forestall," she warned, quoting Albert Schweitzer. "He will end up destroying the earth."

Carson never let up, or softened, her message. Her arguments were sharp and insistent throughout. Yet it was her appeal to readers' hearts that gave *Silent Spring* its stunning power. Much of the incandescence of her book, which weighed in at a modest 294 pages, was Carson's ability to humanize a harrowing subject, to buttress the hard science with real-life stories. Sad and senseless, at moments horrifying, the tales ran through her narrative like a gallery of horrors, each a human face on an otherwise abstract problem, each a piece of the larger story.

Carson was a persuasive writer. She walked her readers through her subject as a scientist, but also as a poet, imbuing *Silent Spring* with a lyricism unusual to such a polemic. All of the elements of her earlier, softer books were here: her long view of time, her luminous observations of the natural world, her focus on systems and processes and the interconnectedness of the living world. But added to this mix was a new moral urgency, a sharpness to her warnings that was laced with fury. Describing a dying squirrel unintentionally caught in a spraying incident, she asked readers to consider the implications: "The head and neck were outstretched and the mouth often contained dirt, suggesting that the dying animal had been biting the ground," she wrote. "By acquiescing in an act that can cause such suffering to a living creature, who among us is not diminished as a human being?"

"And what of human beings?" she went on to ask, describing a series of similarly grisly accidental poisonings. In each of these situations, "one turns away to ponder the question: who has made the decision that sets in motion these chains of poisonings. Who has decided—who had the right to decide—for the countless legions of people who were not consulted . . . ?"

Carson was in effect making it "personal," a problem that touched "everyday people." She was giving her readers both more responsibility and more agency by saying, "You can understand the science. You can change things." She was also exposing the perils "of not speaking out," underscoring "how one kind of silence breeds another, how the secrecies of government and big business beget a weirdly quiet and lifeless world," writes ecologist Sandra Steingraber. This was something new in an era of intense social conformity, a departure for a culture unaccustomed to challenging authority. Carson was debunking the idea that only the "experts" could stand in judgment on such matters.

Much of the research Carson collected wasn't new; scientists and doctors had been aware of aspects of her findings for some time. But most knew only what was relevant to their particular expertise. Carson was the first to translate the science into laymen's language, to piece it all together into a story that ordinary readers could follow. In so doing, she was able to stir a population just emerging from the Cold War to action in a way that no writer before her had. Carson not only alerted people to the dangers of pesticides, she awakened them to their power to press for sweeping changes. Armed with the facts, she believed, the public would demand accountability from both the government and the chemical manufacturers; they would insist on a search for alternatives. In the end, the choice was between "working with or working against nature," she insisted. "Every once in a while in the history of mankind, a book has appeared which has substantially altered the course of history," said Senator Ernest Gruening. "*Silent Spring* was such a book." In sounding the call to arms, Carson, the mild-mannered nature writer, unwittingly sparked a revolution. She changed the way people saw the world. We are a part of nature, she showed readers, and not separate from it. As such, we are equally vulnerable.

Silent Spring was more than a polemic about the perils of synthetic pesticides; it was a critique of the values of the 1950s: its love affair with technology, its deference to big business, its scientific elitism, its mania for national security, its increasing disconnection from nature. Technology had spurred a "golden age" of economic growth, supplying the nation with a pleasure box of consumer products from washing machines to weed killers, Cadillacs to tract houses. Technology had won the war, made the world safe for democracy. It would keep the red menace at bay. Or so the official script went. "No man who owns his house and lot can be a Communist," quipped developer William J. Levitt. "He has too much to do." Levitt had brought Ford's idea of mass production to housing; McDonald's had followed suit with mechanized food production. Now everyone could have a home and a little square of turf free of crabgrass; everyone could dine out on fast, assembly-line food. That the experience was homogenized seemed a small price to pay. The only shadow to all this was the Cold War: omnipresent, terrifying, a chill to honest debate.

It was an odd time for Americans, who teetered uneasily between untold affluence and fears of annihilation, optimism and quiet despair; a risky time, many said, to be idiosyncratic or outspoken. Eager to blend in, people looked inward to home and family, diverted themselves with easy pleasures, turned a blind eye to social and racial injustices. When Adlai Stevenson voiced questions about nuclear testing in the 1956 presidential campaign, he was viewed as an extreme, beyond-the-fringe liberal, far from the safe center; obviously he wasn't a serious contender. The "legislative monument" of the era was the interstate highway system, its justification the speed with which it could whiz the new class of gray-suited businessmen from shiny new suburbs to jobs in the city. But also, more gravely, the ability to move weapons and material into the cities and people out, should nuclear war break out. "These have been years of conformity and depression," wrote Norman Mailer in 1957 in "The White Negro." "A stench of fear has come out of every pore of American life, and we suffer from a collective failure of nerve." It was a time, in Mailer's words, "when one could hardly maintain the courage to be an individual, to speak with one's own voice." But Rachel Carson was doing just that, in open defiance of cultural norms, especially as they applied to women, who were expected to serve up coffee and not social critiques, to be seen but not heard. Carson was calling attention to the shadow side of unchecked technological progress, the irresponsibility of both industry and science toward the "long-term health of the whole biota." She was questioning the supremacy of science when it was used to alter the natural processes of the living world.

Carson had always been an outsider, and never more so than in the scientific world of the 1950s, first because she was a woman, which by definition made her a second-class citizen, but also because her field of interest, biology, carried no weight or prestige in the nuclear age, where chemists in white lab coats were king. But if her lowly station as a woman and a popular science writer consigned her to a measure of obscurity within the scientific establishment, it also accorded her enormous freedom and intellectual latitude. Carson had no academic affiliation or institutional voice to defend. She could follow the evidence she found, wherever it led her, free of institutional bias or the prerogatives of cronyism. And by the time *Silent Spring* appeared, her outsider status had become a valuable asset. The culture was changing.

Silent Spring came out at a moment when trust in the government and the corporate status quo was eroding everywhere, but most notably among the young. A new generation of Americans was coming of age. They were more open and optimistic than their parents had been, more willing to question authority, less sure that an exploding economy was the only measure of happiness or human achievement. Wary of the expansion of what Eisenhower had called "the military industrial complex," skeptical of the "vain chase for satisfaction through mass consumption," uninterested in the hierarchies of wealth and status that preoccupied their parents, this emerging generation was not so sure but that affluence and corporate indifference were corrupting to the human spirit. Uneasy with a world in thrall to the artificial and the mechanistic, the hectic, striving, industrialized life, they could see that greater material well-being was being traded for freedom and individuality, that the system as it stood not only disenfranchised people, but also laid waste to the land—all in the name of mammon. This gathering demographic, the first wave of a youth culture just beginning to coalesce, dreamed of an America in which every individual got a fair shot, a society where nature and natural systems were respected, where community rights might trump business rights. Brash and impatient, eager to shed the manacles of convention, it was a cohort that felt a new willingness to stand against the tide, a fresh and quixotic sense of possibility. And in Carson they found an affirmation of much of what they felt.

One age was passing into another. In voicing her unease about the increasingly technocratic focus of American life, Carson was raising many of the questions that the younger generation was beginning to ask about the trade-offs between material progress and the more ephemeral, sensory, and aesthetic qualities of life; she was setting the stage for the moral call to arms that would coalesce in the counterculture sixties, when the health of the earth became a live and pressing issue.

Silent Spring sold more than one hundred thousand copies in the first two weeks it was out. By Christmas 1962, it was number 1 on the bestseller list. In February 1963, it appeared in England, followed soon after by editions in France, Germany, and Sweden, where it also proved stunningly successful. The English edition opened with an introduction by Lord Shackleton, the son of the famous Antarctic explorer Ernest, who wryly told the House of Lords that cannibals in the South Pacific now preferred the flesh of Englishmen over Americans—as Americans had higher body burdens of DDT. He added that his comments were strictly "in the interest of the export trade."

IN LATE OCTOBER 1962, CARSON HAD A PREMONITION THAT THINGS were not good. According to her latest X-rays, her cancer seemed to be under control, but she told Dorothy she felt "a menacing shadow." By now the chemical industry's smear campaign was in high gear. To blunt their attacks (which were becoming increasingly personal), Carson was doing more publicity than usual, speaking to garden clubs and conservation groups worried about the pesticide issue. In a speech to the Women's National Press Club in December, Carson told her audience that she found it breathtaking that so many who disparaged her book hadn't read it. Her critics had deployed "all the well-known devices" for weakening an argument, she said, including claims the book "said things it did not." One especially exasperating mode of attack, she added, was to undermine the credibility of the person behind the book. She allowed with some amusement that she had been labeled "a bird lover, a cat lover, and—heavens—'a high priestess of nature,'" although most charges were much sharper. Too often scientific truths were compromised to "serve the gods of profit," Carson went on to say. Many academic researchers were funded by the chemical industry for studies that reflected favorably on pesticides' safety, she reported. "When the scientific organization speaks," Carson asked, "whose voice do we hear, that of science or of the sustaining industry?"

Carson's speech went well. But by the next afternoon, she was doubled over with pain. She went to the doctor for X-rays of her back, and was relieved to hear that her spine showed no discernible masses. But when the pain hadn't lifted by Christmas, the doctor recommended another round of radiation treatments, as pain sometimes occurred in the vertebrae before cancer showed up on X-rays. However discouraged she felt, Rachel tried to feign normalcy for Roger's sake, reminding herself that no cancer had been found. Perhaps her back pain had nothing to do with a malignancy. And then a new worry arose. It was just before Christmas, and she was shopping in Chevy Chase for a gift for Roger. Without warning, she collapsed over a table of records. Terribly shaken, having never fainted before, Rachel thought she was okay. Her doctor wasn't so sure. Her symptoms indicated angina, he thought, which would mean considerable adjustments. A hospital bed was brought to the house and she was told to do minimal walking, no stairs, and to cancel any further speaking engagements until her chest pains eased up. Except for seeing Roger off to school, Rachel now stayed in bed. She felt terribly alone.

"It has been such a mixed year for us," Rachel admitted in a letter to Dorothy. "Joy and fulfillment," but also "the shadows of ill health. For me, either would have been a solitary experience without you." She told Dorothy that the past ten years—the time they had known each other—had been "crowded" with sorrows, tragedies, and problems, but also, she suspected, with everything she would be remembered for. She said she couldn't imagine those years without Dorothy: "Because of you there has been far more joy in the happy things, and the hard spots have been more bearable. And so it will be in the time to come."

But Rachel was feeling less sure than her letter indicated. Two weeks later, she got more dispiriting news. There were two new tumors, one above her collarbone, the other in her neck; the cancer had entered her bones. She told Dorothy she couldn't pretend to be "light-hearted" about this, but she was trying hard not to think about it. Her chest pains had sharpened, and she was fearful of doing even silly things, like pulling up the blinds. But she was determined not to be a burden, and, as always, concerned about her privacy. She admitted to Dorothy that she wasn't getting help because she'd told almost no one about her cancer. She was determined not to live "an invalid's life," she added. "The main things I want to say, dear, is that we are not going to get bogged down in unhappiness about all this. We are going to be happy, go on enjoying all the lovely things that give life meaning—sunrise and sunset, moonlight on the bay, music and good books . . . the wild cries of the geese."

HOUGHTON MIFFLIN HAD BEEN MAKING PLANS FOR RACHEL'S INTER-view on *CBS Reports*. The publicity people were filled with tips: she should wear no lipstick, as it would look "black on black-and-white film"; it was important that she not look "too stern," as her message itself was sobering; she should smile occasionally, as it would "relax" her face, keep it from looking too grim. They wanted her "gentle nature" to come through, her thoughtfulness and her modesty. Her mastery of the science would be obvious, they reminded her; they were confident her message would speak for itself.

But privately everyone was beginning to worry. Rachel's interview came and went and seemed to go well. But CBS continued to postpone the date for when it would air. Houghton Mifflin worried that the delays might mean that the program was tilting toward the pesticide industry, which they knew CBS was also interviewing. Jay McMullen, the producer, was known for being cautious and methodical. Paul Brooks hoped that "their friends in the chemical business" hadn't made too strong a case for pesticides. He had learned that the network had received more than a thousand pro-industry letters, all of them mimeographed, clearly a letter-writing campaign orchestrated by the chemical lobby. Then, two days before the broadcast, three of the five commercial sponsors pulled out. Everyone knew Carson's views were politically vulnerable.

By early March 1963, both Rachel and Dorothy were aware that the sand in the hourglass was slipping away. Rachel admitted to having excruciating pain in her back and ribs, and said her nausea made it nearly impossible to do anything. One night, convinced the end was near, she wrote out her last words to Dorothy, explaining that the pain that night had been terrible, but even more agonizing, she realized, was the thought that if she didn't make it past the night, she might not get to say good-bye. And so she was writing her something now, hoping it might make it "a little easier for you if there were some message."

She told Dorothy that she should have no regrets on her behalf. "I have had a rich life, full of rewards and satisfactions that come to few, and if it must end now, I can feel that I have achieved most of what I wished to do." She said this wouldn't have been so two years ago, when she first realized her time was short, "and I am so grateful to have had this extra time. My regrets, darling, are for your sadness, leaving Roger, when I so wanted to see him through to manhood." What she wanted to write, she said, was of the "joy and fun and gladness" they had shared, for these were the things she wanted Dorothy to remember. But then, having made it through the night, the moment passed, and she put the letter aside.

Carson's appearance on *CBS Reports* finally aired on April 3, 1963. It was the same evening that the popular astronaut Gordon Cooper was orbiting the earth in the second Project Mercury space capsule, vying for viewers' attentions. Dressed in a sage green suit, Carson sat in her office chair, answering every question with calm deliberation, never sounding anything but thoughtful throughout. It was a brilliant performance, made all the more so by its juxtaposition with her chief adversary, the wild-eyed and hyperbolic Robert H. White-Stevens, who spoke on behalf of the chemical industry.

Wearing a white lab coat and black glasses, and sounding more like a mad scientist than an evenhanded researcher, White-Stevens began with a prepared statement describing a world beset by hunger and pestilence, as hordes of rasping insects denuded forests and ravaged croplands, leaving humans prey to every scourge imaginable. "If man were to faithfully follow the teachings of Miss Carson, we would return to the Dark Ages and the insects and diseases and vermin would once again inherit the earth," he warned.

White-Stevens went on to claim that registered pesticides, used correctly, "posed no hazard" to humans or wildlife, a statement so obviously distorted that even the other pesticide proponents Sevareid questioned had to disagree, undermining the credibility of everything else White-Stevens said.

CBS had opened the program with shots of pesticides being sprayed over cropland and through neighborhoods where children scampered happily behind fogging trucks. Then it cut back to Sevareid, who proceeded to recite statistics about pesticide use, including a mind-numbing tally of new products on the market for home, farm, and gardeners' use—a sum that sounded so over-the-top scary that it must have given pause to even the most jaded viewer.

Between Carson, who read selected passages from Silent Spring, including many that highlighted her most worrisome findings, and the creepy White-Stevens, came a string of government experts, whose answers were generally vague, as if trying to gracefully sidestep the pesticide problem. By the end, viewers were left with the clear impression that Carson was correct: none of the people who were supposedly in charge knew all that much about the safety or long-term effects of pesticide use. It was a troubling thought.

Sevareid closed on a philosophical note, allowing Carson to share her view of mankind's place in the ecological balance. "We still talk in terms of conquest," she said. "We still haven't become mature enough to think of ourselves as only a very tiny part of a vast and incredible universe. Now I truly believe that we in this generation must come to terms with nature," she added. "And I think we're challenged as mankind has never been challenged before to prove our maturity and our mastery not of nature, but of ourselves."

The program was an unqualified success and both Carson and her publisher were pleased. The network estimated the audience that night numbered ten to fifteen million, a large portion of whom hadn't read Carson's book, but were roused by what they heard, writing in afterward to say how grateful they were to CBS for presenting such an "important" topic. Rachel's on-camera appearance was especially devastating to her critics. Any notions that she was a zealot or a Communist operative had been roundly dispelled by her manner that evening, which was intelligent, measured, and poised throughout. She came off as a responsible citizen-scientist armed with what sounded like the truth.

On May 15, *Silent Spring* got another boost. The president's pesticide committee released its report, which more than vindicated Carson's claims—this despite rumors that political pressure had been brought to bear to water down its findings. In a follow-up broadcast, Sevareid called *Silent Spring* "the most controversial book of the year," reporting that it had now sold more than five hundred thousand copies and started "a national quarrel." Dorothy wrote to Rachel telling her that May 15, 1963, would "go down in history as Rachel's triumph." She predicted her name would be remembered long after Gordon Cooper's—the Project Mercury astronaut still orbiting the earth.

Rachel was feeling optimistic enough to buy a new car. It had a radio and seat belts—the latter still a novelty. In early May 1963, she wrote Dorothy, telling her she was desperate to see her in person. There were "things I need to say to you, but they should be said with my arms around you." At the back of her mind was the worry that she wouldn't be able to make it to Maine that summer. Though she was feeling better, her angina was back and her spine was now riddled with compression fractures, making it more painful to walk. She had accepted an invitation to testify in Washington before Connecticut senator Abraham Ribicoff's subcommittee on pesticides on June 4. It was the one last thing she needed to do.

CAPITOL HILL WAS BUSTLING THAT DAY. THE WINDOWLESS HEARING room was already packed when she got there. Television cameras were trained on the dais. Soundmen and recording equipment crowded the aisles; cables snarled the floor, every square inch of available space taken. Rachel was dressed in a green suit. Her gait was unsteady and she walked with a cane as she hobbled to her seat. To conceal her baldness, she wore a brown wig.

She sat and waited calmly, speaking to no one, a still point amidst the tumult. On the table before her lay her note cards, a neat stack that was large and typewritten: her testimony for the upcoming hearing. She felt no nervousness, no reservations about being there. She had known for some time what she wanted to say.

Finally Senator Ribicoff stood, and the room fell silent as he introduced his lead witness. "You are the lady who started all this," he said, turning to Carson. "We welcome you here." Graciously she returned his thanks, taking a moment to acknowledge what he'd said. Then, putting on her black-rimmed glasses, she began to read, her voice calm and low, surprisingly strong.

Rachel's prepared statement went on for forty minutes. She called for more research, and less aerial spraying, especially of pesticides that left long-lived residues. She argued that citizens should have the "right" to be secure in their own homes against the imposition of poisons applied by other persons, and that greater restrictions on the sale and use of pesticides were needed. "Our heedless and destructive acts enter into the vast cycles of the earth and in time return to bring hazards to ourselves," she told the committee. "I speak not as a lawyer but as a biologist and a human being." Asked if she believed there should be a total ban on pesticide use, she said no, she did not, stressing that this had been a false charge leveled at *Silent Spring* from the beginning. Later Ribicoff would recall that one of the reasons Carson's testimony that day was so impressive was that she believed so strongly in her vision. No one who heard her presentation "could have questioned her integrity." She was that "rare person who was passionately committed when few others believed very much in anything." No one that day knew how close Carson was to death.

Rachel left immediately for Maine, despite her precarious health, arriving just in time for ten-year-old Roger to begin summer camp. Her doctors had warned that any kind of fall could have serious consequences; she needed to limit her walking. Still, she was sorely tempted. She sent a note to Dorothy: "Would you help me search for a fairy cave on an August moon and a low, low tide. I would love to try it once more, for the memories are precious."

They didn't go, but one day in early September 1963, they drove to the end of Southport Island, to a lovely inn, where they lunched together and then sat on a bench overlooking a field of goldenrod. It was a blue-skied day, clear and crystalline, the air still soft and warm. They listened to the wind in the pines and the lap of the sea against the rocks. Suddenly, flickering through the air, came a drift of monarch butterflies, all moving in the same direction, as if "drawn by some invisible force." As they watched, they spoke of the butterflies' mysterious life cycle, how they traveled thousands of miles in a single year, several generations living and dying in the course of that journey, both aware that none of the monarchs they were seeing on that slow, dreamy afternoon would return.

When Rachel returned to Maryland a few days later, she wrote to Dorothy, telling her what a joyful memory it was. "It occurred to me . . . that it had been a happy spectacle, that we had felt no sadness when we spoke of the fact that there would be no return." Carson offered that it was because they both knew that every living thing must come to the last of its days, which was as it should be. The monarch's life cycle is measured in months, she added. "For ourselves, the measure is something else, the span of which we cannot know . . . when that intangible cycle has run its course it is a natural and not unhappy thing that a life has come to its end."

Leaving Maine this time had been especially wrenching. On the actual morning, a friend had had to carry Rachel to the car. The cancer was advancing quickly now. There were new lesions on the left side of her pelvis, bringing new pain, more difficulty in walking. Dorothy tried to cheer her up with daily letters describing her days at Southport: the birds and the seals, the crisp autumn days, the walks through the piney woods. "I've been x-rayed practically from chin to ankles," Rachel wrote back. In October she began a cycle of testosterone and phosphorus treatments, which her doctors hoped would make walking less difficult. She hoped she could somehow make it to San Francisco for an upcoming speech at the Kaiser Medical Center. She was beginning to fear, however, that it wasn't to be. The pain was now moving from one part of her body to the next, with no apparent pattern. There were days, she told Dorothy, when she couldn't walk at all.

Somehow she managed to go, though for most of the trip she was confined to a wheelchair, even for the short visit she made to see the redwood forest at Muir Woods. A local newspaper, covering her lecture, which had packed the auditorium, described her as a "middle-aged, arthritis-crippled spinster." When she got home, Rachel admitted she was "as sick as she had ever been." Now, to get through the nights she was taking sleeping pills. Much of her time was spent in bed, though some days she felt strong enough to hobble through the house with a walker. She wasn't in constant pain, she assured Dorothy. And some days were better than others. What she didn't say was there were times when she had trouble writing because of numbness in her hands.

That November Rachel began going through her papers, which she had decided to donate to Yale. She told Dorothy she was surprised by how much comfort this brought. She remembered herself again as a young writer: tireless and filled with ideas, with still so much time ahead. She recalled her first days in Maine, meeting Dorothy, the amazement of having found a kindred soul. But even this interlude of peace was abruptly shattered when, on November 22, 1963, she learned that President Kennedy had been assassinated in Dallas. It was unbearably sad; she was unable to think of much else, she said. She wrote Dorothy she felt as if she'd lost a family member, that his killing had filled her with "shock, dismay, and revulsion at the black aspects of our national life—the bigotry, intolerance and hatred preached by so many."

The awards continued to pour in. Although Rachel now had unremitting pain, and the compression in her spine sometimes made it difficult to work her hands, she decided to go to New York in early December to accept the award from the American Geographical Society. Dorothy and Stan joined her in the city for the event, which went flawlessly. Afterward, Dorothy admitted in a letter that she had never expected she would see Rachel "out of bed again after she'd left Maine that fall." Seeing her up on the dais, looking so "lovely," had made her tear up.

When Rachel returned home she began another series of radiation treatments. Her doctor had found more cancer at the base of her skull, which he believed was the cause of her numb hands. Writing to Dorothy on the winter solstice, she alluded to the upcoming summer in Southport, saying she intended to plant roses. "I had not, until recently, allowed my thoughts to range so far into the future," she said. "Now I do." They would yet build more memories, she added.

Dorothy came to Silver Spring for a four-day visit after Christmas. After she left, Rachel wrote to apologize for having talked too much about her illness, and how little time she probably had left. There was stabbing pain in her head now, and she had lost all sense of smell and taste. Her doctor reminded her that it had already been three years since her cancer diagnosis. She was already "something of a miracle." What he didn't say, though Rachel inferred it, was that she shouldn't expect too much more. Even so, she hoped for one more reprieve so that she might make it back to Maine.

And then Dorothy's husband died quite suddenly of a heart attack. Rachel was devastated. She reproached herself for having burdened Dorothy with her own health issues. It was now her turn, she told Dorothy, to take care of her dear friend. Rachel flew to West Bridgewater for the funeral. It was a sad visit, but she was glad she could be there. Dorothy's son drove Rachel and her wheelchair to the airport, helping her to board.

Dorothy and Rachel spoke often in the months that followed, hoping to arrange a visit. By early spring Rachel's cancer had advanced to her liver. At the advice of her doctor, she returned to the Cleveland Clinic, where, near death, she was hospitalized for several weeks. At some point while there, she had an out-of-body experience, which she described as feeling suddenly surrounded by "a brilliant white light" and then being lifted up. Rachel told Dorothy it had been like being in the swirling fog near their houses in Maine. When she could, Rachel dictated letters to friends from her bed, but she hid the gravity of her situation from most, saying that she was in the hospital for arthritis, and would soon be leaving. In early April, though terribly weak, Rachel finally went home. Dorothy immediately flew to Maryland for a visit. Rachel was only partially aware during many of their hours together, but she was overjoyed to see her friend. When Dorothy got home a few days later, she told Rachel how happy she was to be able to now imagine her days. On April 14, she wrote again, saying how lovely it was that Rachel could hear the birds outside her window each morning. That same afternoon, Rachel's heart stopped. She died before sunset.

chapter TWO

....

JANE JACOBS

In the spring of 1955, a tall, self-effacing man in a rumpled suit paid a visit to a young editor at *Architectural Forum* magazine to ask her aid in writing about what he called "a bloodletting." William Kirk didn't know anything much about Jane Jacobs. He was the head of the Union Settlement, a community group that helped the indigent in East Harlem. But he hoped he would find a receptive ear. A handsome, bespectacled woman with a hawklike face and a thatch of pale, chin-length hair, Jacobs had just returned to work after a brief maternity leave, having recently had her third and last child. By her own admission she wasn't an architect, or an expert in city planning. She didn't even have a college degree. But she was a keen observer and a quick study, and with the help of her husband, who *was* an architect, she had learned to read architectural plans. She was becoming something of an expert on school and hospital design, in fact. Though lately, her editor, Douglas Haskell, had been assigning her to urban renewal pieces, which was one of the reasons he had shown Kirk in.

Kirk had an air of urgency about him. He had come, he told Jacobs, because he needed to talk about the massive "slum clearance" taking place in East Harlem; it was tearing the heart out of the neighborhood. Yet no one in city government seemed to care, or even to have the time to hear him out.

Three million dollars of federal money had been spent to erect new housing, he explained. But the problems in the community "were only getting worse." Under a new government program known as Title I, or urban renewal, block after block of the old neighborhood was being leveled. People were being bulldozed from their homes, small businesses driven out by the hundreds, whole streets wiped out. Yet the grim brick towers rising in their stead were proving more wretched than the tenements they replaced. Ugly and anonymous, they were dreary, soul-crushing places, petri dishes for crime. Teenage gangs prowled the grounds, terrorizing the tenants. What little public space there was seemed to invite vandalism. There were muggings in the halls, assaults in the elevators, no eyes anywhere to keep a watch on things. Residents reported they were frightened to go out at night. Mothers wouldn't let their children outdoors to play. From the upper floors, they couldn't see into the bleak no-man's-land between the towers. Fifteen hundred merchants had lost their stores; more than forty-five thousand workers their jobs. Those few commercial blocks still standing were cut off from their customers, the remaining shopkeepers, in their imposed isolation, increasingly susceptible to crime. But the problems went deeper still. When the old buildings fell, the locals not only lost their livelihoods and the services they needed, they were robbed of the intimate, time-worn places that connected the community. Without these social anchors, the fabric of the neighborhood was unraveling. Kirk paused now, temporarily out of words. He wondered if she'd like to come to East Harlem, so she could see for herself.

Jacobs was listening closely. Nodding, she said she would. There were things Kirk was saying that resonated. Only a month before, Haskell had sent her to write a story about a major slum-clearance project in Philadelphia. The ambitious plan was being overseen by one of the rising stars of the urban renewal movement, Edmund Bacon. Bacon, like Robert Moses in New York, had been systematically replacing Philadelphia's run-down neighborhoods with vast super-blocks of high-rise housing. The popular wisdom at that moment was that American cities were sinking into ruin. They were seen as crowded, crime-ridden places of poverty and blight, graveyards of substandard housing and festering slums. Much of this, in fact, was the unintended blowback of the government's own policies. At the end of WWII, in response to the flood of returning GIs looking for homes (and a baby boom of unprecedented size), Congress had rushed through a massive housing bill that guaranteed long-term, low-interest mortgages and a risk-free playing field for single-home developers. This, coupled with the passage of the multibillion-dollar federal highway act, had changed the face of American cities almost overnight, as money that had once funded public works, mass transit, and urban infrastructure (compliments of the New Deal), now poured into highways and suburban housing. What followed was hardly surprising: as the cash flowed out, so too did the people, at least those who could. By the mid-1950s, middle-class families were fleeing in droves, forsaking urban life for the freshly minted suburbs, discouraged by the dirt and the danger and the general air of neglect, the steady disinvestment that had left downtowns crumbling and public transit starved of funds.

Ideas about urban living, like those about nature, had long been contested territory of course. Faced with the perception of a gathering crisis, what many saw as the needs of a new and modern age, a succession of architects and thinkers had been charged with finding answers. Urban blight, they concluded, could only be tackled by thinking on a grand scale. "Make no little plans," proclaimed the Chicago architect Daniel Burnham. "They have no magic to stir men's blood." What was needed, it was hypothesized, was to level the old city and build anew, abandoning the outmoded urban grid for a fresh template that was sleek and streamlined as a machine. The new idea was to build high, to create a vertical city, "a whole city in the free air of the sky," in the modernist architect Le Corbusier's words. The metropolis of the future, according to such utopians, would be open and uncluttered—a sharp departure from the dense urban streetscapes of nineteenth-century European cities, with their short, crowded blocks and ornate classical buildings. It would be a place of unbroken perspectives, closely regimented, a marvel of order and geometry: tall, soaring towers surrounded by swaths of grass, or open plazas planted with trees, with living and work zones strictly separated, and aerial expressways looping around to connect the parts. The old chaotic city with its vivid street life, its turbulent clamor, its various modes of transport crawling between rows of low-slung buildings, was obsolete. "We must kill the street!" was Le Corbusier's war cry. "Cafés and places of recreation will no longer be the fungus that eats up the pavements of Paris," he wrote in *Towards a New Architecture*, his modernist manifesto of 1929. Technology was the "moving spirit of modernity," Le Corbusier and others insisted. It would solve all the "dissonances" of contemporary life, remake the urban environment in a more ideal form. The new man, he asserted, needed a "new type of street," that would be exclusively a "machine for traffic."

Edmund Bacon was an ardent believer in these grand themes. The wrecking ball approach to revitalization seemed the good and true way forward. A charming and cosmopolitan man, he had greeted Jacobs at the Philadelphia train station, and then escorted her to a crowded street in the downtown area.

"First he took me to a street where loads of people were hanging around on the street, on the stoops, having a good time of it," she told her colleagues when she got back to New York. "He said, well, this is the next street we're getting rid of. That was the 'before' street. Then he showed me the 'after' street, all fixed up, and there was just one person on it, a bored little boy kicking a tire in the gutter. It was so grim that I would have been kicking a tire, too. But Mr. Bacon thought it had a beautiful vista."

"Where are the people?" she had asked.

Bacon seemed not to hear her question. "They don't appreciate these things," he said, waving his hand airily at the view. Excitedly he explained the need for order in the crowded and unruly downtown, the importance of providing "a view corridor."

And then they moved on to the next block, where people were streaming down the sidewalk, popping in and out of stores, chatting amiably between errands, and Bacon explained that this was an example of what they were working to eliminate. Jacobs stared at him in disbelief. It was clear that the vitality she was seeing on this lively little block looked to Bacon like chaos and dysfunction.

Riding back later on the train to New York, Jacobs mulled over her day. By now, in preparation for the article, she was familiar with many of the theories leading up to the urban renewal movement: British-born Ebenezer Howard's ideas for what he called the "Garden City," towns of thirty thousand or so people built just outside the urban centers, each encircled by a green belt; Lewis Mumford's later, similar proposals for small, anti-urban settlements of people spread along a continuum radiating from the city's core; Frank Lloyd Wright's vision of the "horizontal city," low-slung houses built on separate plots dotting the prairie; Le Corbusier's "tower in the park" ideas. It was curious, she mused. All sought to spread people out more thinly and to separate the city into separate sectors—people here, work there, recreation somewhere else—trading the characteristics of cities for the characteristics of suburbs.

In principle, Jacobs wasn't necessarily against modern architecture per se. She admired the elegance and purity of some of Le Corbusier's white modernist cubes, with their open interior plans and praying mantis legs. She genuinely loved Mies van der Rohe's modernist tower, the Seagram Building on Park Avenue: the visual effect of the steel and glass "curtain wall" rising sleekly toward the sky. But examples of successful modernism were hard to find. There seemed to be a yawning chasm between the high-minded theories and the reality of the built environment. What had begun as an aesthetically pure and exacting movement had devolved into the dreary strip-mall architecture of postwar America, a landscape of boxy tinted-glass office complexes and sprawling shopping centers, each no different from the next. And now, in the cities, these drab public housing towers. Something seemed terribly amiss about the huge, project-scale demolition she'd just seen. Somehow, the human quotient had been lost in translation.

Back at the office, Jane's colleagues listened quietly as she shared her reservations about Philadelphia's urban renewal efforts. No one seemed comfortable refuting the accepted wisdom of a planner of Ed Bacon's stature, however. He was fighting the good fight, they argued, trying to staunch Philadelphia's spiraling economic decline, as was Robert Moses in New York. For several decades now, Moses had been remaking New York at a maniacal pace, razing old neighborhoods and building public works on an unprecedented scale. His signature was on bridges and tunnels, parkways and housing developments, swimming pools and parks, dams and power plants. He had built the Cross Bronx Expressway and the Central Park Zoo, Jones Beach and the Triborough Bridge, Idlewild (now Kennedy) Airport and the 1939–40 New York World's Fair, a full-throated paean to technology and industry for which he'd done everything from seizing the land to deciding on the exhibitions. Moses had "the energy and enthusiasm of a Haussmann," one of his acolytes had written in *Space, Time and Architecture,* the bible of the modern planning movement. He was the "one public figure" supremely "qualified to build the city of the future in our time." Indeed, Robert Moses at that moment enjoyed "quasi-mythological" status, and so, to a lesser degree, did Bacon. To a public hungry for the new, both men signified the "moving spirit of modernity." They were progress personified, heaven-sent agents of rebirth and renewal—or so the rhetoric went. It was only later that Moses, in particular, would be seen as one of the great destroyers. All this, however, was still in the future. To Jane's peers at *Architectural Forum,* Robert Moses and Ed Bacon were trying to save America's great cities. It seemed unwise, even ungenerous, to challenge either of them.

And yet now, barely a month after Jacobs's visit to Philadelphia, here was William Kirk standing in her office, expressing many of the same misgivings that she herself had felt.

She met him in East Harlem a week later, riding the subway up to 110th Street, a stop she hadn't yet explored. Jacobs's first impres-

sion, as she and Kirk started down the block, was of life and movement everywhere: people on the street strolling, shopping, talking,
lounging. She saw men playing cards on milk crates, teenagers loi-
tering on stoops, mothers leaning casually out of windows, women
in foam curlers chatting in groups, girls playing hopscotch and
jumping rope. The sidewalks were messy and congested, no question. Every few steps, little islands of commerce broke the pedestrian flow: mobile lunch stands manned by women in food-soiled
aprons, folding tables piled with men's working gloves and cheap
children's underwear, pushcarts stacked with pyramids of fruit. Everything looked a little shabby and makeshift, a bit disorganized.
But for all the seeming chaos, there was a palpable feeling of neighborhood here, a sense of solidarity and connection rising from the
communal life of the street. Everyone looked relaxed, comfortable,
as if they knew they belonged.

What amazed her was how abruptly the feeling changed as soon
as she and Kirk crossed into the projects. Where a moment before
there had been bars and bodegas, candy and cigar-rolling shops,
animated conversations everywhere, now there was only an empty
stretch of dun-colored grass with several dreary housing towers
in the middle. Except for a few menacing-looking teenagers, who
slumped against a broken slide in an otherwise empty playground,
there seemed to be no evidence of life anywhere, no connection
between the place and the people who lived there. It felt just as
Kirk described—like a "bloodletting," a place leached of all vitality
and cheer. Jacobs shook her head. Something wasn't working. She
wanted to go back and talk to the people they'd just passed, to get a
sense from them of what was happening here. "I can remember the
people in East Harlem hating a patch of green grass," Jacobs later
explained. "I couldn't understand why until one of them told me
that the tobacco store had been torn down, the corner newsstand
was gone, but someone had decided the people needed a patch of
green grass and put it there." The planners, in their self-importance,
had assumed they knew what was best for the people, with little
concern for what they actually wanted or needed. They were so

determined to change the way the neighborhood *looked*, they hadn't paid any attention to how these new places actually felt for those who lived there. It was arrogant and misguided, having nothing to do with what was going on on the ground—in real life.

Jacobs was keenly interested in what went on in real life. Inquisitive by nature, she liked to start from the bottom up, to look hard at the "hot, rich reality" of the world itself, rather than starting with dogma or abstractions. She had always been interested in puzzles and patterns, the repetitions as well as the intriguing irregularities, curious about how the world worked, how the pieces fit together. Nothing was too small or insignificant. The fine-grained details, she found, were often superbly revealing, the oddities in the pattern especially telling. There was no substitute for direct observation, she felt, using one's eyes and one's senses. Coming to *Architectural Forum*, she had had no official expertise or credentials, no preconceptions, which suited her just fine. She had made a point of learning with each project—looking carefully, collecting evidence, asking questions—and only then making up her mind. Which was what she'd done in Philadelphia.

Jacobs *did* go back and talk to the people in East Harlem, just as she'd vowed, returning with Kirk again, on other afternoons, lingering sometimes until well after dark. She stopped and spoke with people, listened to their stories, their worries about their own neighborhood. One summer night she noticed television sets outside, used publicly, on the pulsing sidewalks. "Each machine," she observed, "its extension cord run along the sidewalk from some store's electric outlet, is the informal headquarters spot of a dozen or so men who divide their attention among the machine, the children they are in charge of, their cans of beer, each other's comments and the greetings of passers-by." Strangers stopped as they wished, she added, to join in the viewing. "Nobody [was] concerned about peril to the machines." Who would or could steal a TV from the midst of such a crowd? This was the side of the neighborhood that was being systematically eliminated by the planners: the casual, communal life of the street.

At that moment, Jacobs had a quiet epiphany: there was another way of looking at the city, which was to see and appreciate its social fabric. Amidst the "moving chaos" of the old city, there was, in fact, a marvelously rich and complex human order that gave a neighborhood like East Harlem its life. Touring the neighborhood with Kirk, she would later write, "showed me a way of seeing other neighborhoods, and downtowns too. He opened my eyes."

JACOBS HAD MOVED TO NEW YORK FROM HER HOME IN SCRANTON, PENNsylvania, in 1935, joining her older sister, Betty, who arrived two years earlier, taking an apartment on the top floor of a six-story walk-up in Brooklyn Heights, a quiet neighborhood overlooking the East River. It was a tiny and disheveled place, but the sisters were thrilled to be together. Their father, a ruminative man, had counseled his daughters early on the need for having two career plans; one should be the vocation of their dreams, he said, the other a practical skill they could fall back on. Betty had aspirations to be an interior designer, but was happy to have landed a job as a salesgirl on the home furnishings floor of the Abraham & Straus department store. Jane's ambition was to be a newspaper reporter, but she soon learned that jobs in journalism were few, especially for a young woman with only a high school diploma. It was the depths of the Great Depression, the same year Rachel Carson had found work writing radio scripts for the U.S. Fish and Wildlife Service. Though the Empire State Building was going up and cars clogged the streets, there were signs of destitution everywhere: families living in vacant lots, in huts of cardboard and scavenged wood; men sleeping on steam grates; soup kitchen lines that wrapped whole blocks. Along the Lower East Side waterfront, there loomed a "mountainous open garbage dump," where "desperate men and women fought the sea gulls for scraps."

Jane felt well-armed, despite the hard times. Before leaving Scranton, she had completed a half-year course in typing and stenography at the Powell School of Business. Then for some months after, she had worked unpaid in the newsroom of a Scranton news-paper, where, shunted off to the women's pages, she'd covered wed-dings, church suppers, and civic events with names like the "Women of the Moose." Most edifying, however, had been her writing for the "Agony" column, where she composed letters filled with imaginary problems, for which she offered pithy answers, giving free rein to her prodigious imagination. The combined experience, she liked to say, constituted her journalism school.

Jane loved her new city, and within weeks her routine was es-tablished. Every morning, after combing the job listings for cleri-cal positions, she tromped across the Brooklyn Bridge to Lower Manhattan, showing up wherever there was an interview. Often the position she sought was already filled. Or the interview was over within minutes, leaving her with the rest of the day to do as she liked, which was to explore some new pocket of the city. Sometimes she set off on foot; other times, she sprang for the nickel it cost to ride the subway and disembarked at some new station—drawn by the name of the stop. It was in this way that she discovered Green-wich Village, the neighborhood that she and Betty would soon call home. One morning, after another dead-end interview, she alighted at a station called Christopher Street and decided to get off and ex-plore. As she emerged from the underground, she was immediately struck by the irregularity of the streets, some of which were curved, in giddy defiance of the grid, others narrow and cobbled, shooting off in multiple directions like bicycle spokes. The buildings were low and slouchy, many of them brick with small wooden dormers or four- and five-story tenements, with an occasional warehouse squeezed in between. There were cheese shops and corner grocers, coffee shops and ma-and-pa delis, a Laundromat tucked between two hole-in-the wall restaurants. Nothing was tall or imposing, ev-erything built to human scale. And wherever she looked she saw people and casual activity: merchants in shop doorways, banter-ing with passersby; young men with pushcarts clattering along the pavement; mothers talking in clusters, their children darting about their legs. Jacobs had been to New York once before, when she was twelve, but she hadn't visited this quirky little neighborhood that felt so vital and yet so low-key, so down-to-earth and alive.

When she returned to Brooklyn that evening, she told Betty that she had found the place where they had to live. Jane, of course, was hardly the first to feel the tidal pull of Greenwich Village. The Village had a long and storied history as a bohemian haunt to generations of artists, writers, drunkards, dreamers, and revolutionaries.

Soon the two sisters were ensconced in a tiny Village apartment on Morton Street, a narrow lane that ran four short blocks from east to west before it terminated at the Hudson River. The tatty, tree-lined street of disheveled town houses and four- and five-story brick tenements was home to an eclectic assortment of characters: truckers, longshoremen, railway workers, writers, artists, poets— among them Jackson Pollock, Willem de Kooning, E. E. Cummings, and Delmore Schwartz. The White Horse Tavern, one of bohemia's most famous watering holes, was just around the corner, on Hudson Street.

Unfortunately, Jane and Betty were rarely flush enough to go in. Sometimes they were so broke they were reduced to eating Pablum, a soupy but nutritious cereal for infants. Then one day that spring, Jane's persistence finally paid off. After months of pounding the pavement, she landed a job as a secretary for a candy manufacturing concern. It was the first of a variety of clerical jobs she would hold over the next several years, including positions at a clock maker's and a drapery hardware business. Though even these jobs, she soon learned, promised little security. Sometimes she was hired, only to discover that the business was about to go under, putting her out of work within weeks. "I think that's the hardest time I ever had," she later reflected. "But that gave me more notions of what was going on in the city and what business was like," what went on in the many trades. She was fascinated with the inner workings of the metropolis. In her off hours, or in between jobs, she had begun to carry a notebook so she could jot down her observations. Unwilling to let go of her dream vocation, she would head home after a day of exploring, toss off her shoes, and sit down before her typewriter, turning her notes and impressions into articles.

Over time, she had noticed that every few blocks of the city was home to a different specialty trade, each a little economic engine of its own. Stumbling one morning upon the wholesale flower district, centered at Twenty-Eighth Street and Sixth Avenue, she was riveted. For the rest of the morning, she wandered from block to block, poking her head into the "cool, sweet-smelling shops," making mental notes of the colors and the sounds, the tempo of the street. As the bustle began to ebb, she struck up conversations with the buyers and the sellers, the day workers and the florists, peppering them with questions.

She was intrigued by what they told her about the particulars of the trade, from the 5 A.M. arrival of hampers of lilacs, peonies, irises, and gardenias trucked in from as close as Connecticut and Long Island and as far away as South America, to the hundreds of thousands of cut flowers that changed hands on any day. One grower, she noted, sent in "twenty thousand dozen iris a day"; another, one hundred and fifty thousand roses. She learned how the market had started, fifty-five years earlier, and how certain rules had evolved as the system began to self-organize. In the beginning, most growers brought their market baskets to the ferry landing at Thirty-Fourth Street and the East River, where they were met by the retail florists. But as time passed, the buyers and sellers began to gather at a place near the docks called Dann's Restaurant, which stayed open all night, making it an ideal spot to congregate and "conclude their dickering." Gradually Dann's grew into a more organized market. But as the competition sharpened, the sellers began to arrive earlier and earlier, to get an edge on the day's sales. Finally, to make things fairer, they settled on some rules, the first being that no one could take the cover off his basket until a gong rang at 6 A.M. She was fascinated by this, interested not only in the way the system worked, but also in how it self-regulated. It was a world built upon a vast web of relationships and interactions, a system that was astonishingly complex and interconnected. No single entity could function independently of the whole. Yet taken together, the many parts of the system meshed. Jacobs was seeing in the city what Carson had seen in the sea: the city was a balance of live and ever-evolving forces, a fluid network of exchanges, as much a *process* as a place.

That night, when Jane sat down to write, titling the piece "Flowers Come to Town," she opened with a note of drama: "All the ingredients of a lavender-and-old-lace story, with a rip-roaring, contrasting background, are in New York's wholesale flower district . . . Under the melodramatic roar of the 'El,' encircled by hash-house and Turkish baths, are the shops of hard-boiled, stalwart men, who shyly admit they are doodles for love . . ."

She worked all that evening and into the next, chain-smoking as she banged out the piece. When at last she finished, she slid the crisp pages into an envelope and addressed it to *Vogue* magazine. To her surprise, several months earlier *Vogue* had taken an article that she wrote on the fur district, offering her a contract for four similar pieces over the next two years, for which they would pay her forty dollars each. It was a queenly sum, a fortune compared to the twelve dollars a week she was making as a secretary. Now, as she fingered the envelope, she hoped the editors would be interested in this piece too. When a few days later, a letter reached her indicating they were, she was elated. Her life as a writer was beginning to seem real.

Jacobs's interest in the systems that made up the city continued to deepen. In those first early years, she was working double time, holding down a secretarial job for forty hours a week while also searching out new corners of the city to spin into articles. In one story she sold to *Cue* magazine, she decoded the mysterious markings on manhole covers, mapping out the subterranean maze of gas, electricity, and steam arteries that surged unseen beneath the city streets, conveying heat and light to the buildings above. It was a system of astonishing complexity, an underground grid that stitched together the entire metropolis. In an article for *Vogue*, this one about the diamond district, she wrote about the arcane inner workings of the jewelry trade, which at the time was based in the Bowery, near the Manhattan Bridge on the Lower East Side. "Not a sound is heard at the auctions," she wrote. To an outsider, "the proceedings are baffling." She went on to describe how the dealers made secret notations as they examined the rings, broaches, and gems put up for auction each day, and how once the bidding commenced, they crowded around the auctioneer, squeezing his arm, nudging his ribs, or rubbing their elbows to raise their bids. It looked like "a cross between hocus-pocus and mind reading," she observed.

> *Upstairs, in small light rooms over the stores, diamonds are*
> *cut and polished and set or re-set, and silver is buffed. The*
> *doors and vestibules to the rooms are barred and there is*
> *no superfluous furniture, just the tools and tables where the*
> *skillful workmen sit with leather hammocks to catch the chips*
> *and dust of diamond and metal. Silver is polished against*
> *a cloth-covered revolving wheel . . . All the sweepings are*
> *carefully saved to be refined and the silver recovered. The*
> *walls and ceilings are brushed and the old oilcloth coverings*
> *and work clothes of the men are burned to extract the silver*
> *dust. Even the water in which the workmen wash their hands*
> *is saved. A small room where silver is polished may yield to a*
> *refiner hundreds of dollars' worth of metal a year.*

The piece, which closed with a quick, evocative sketch of the neighborhood, was at once quirky and colorful, plainspoken and theatrical: "Outside on the Bowery, the lusty, tumultuous life of the lower East Side converges," she wrote. The area pulsed with the "raucous chaos" that Jacobs was coming to see as the lifeblood of the metropolis.

Often, in those first years, she would go up to the rooftop of her apartment building and watch with wonder as the garbage trucks made their elaborate rounds, circling and zigzagging from street to street. "I would think, what a complicated, great place this is, and all these pieces of it that make it work," she recalled. Jacobs was trying to put those pieces together. The more she wandered and

explored the many working districts of the city, the more she began
to see the metropolis as an elaborate, self-sustaining human ecosystem, a marvelous, complicated mosaic that churned with life.

HOW DID A YOUNG, BESPECTACLED GIRL FROM SCRANTON, PENNSYLVA-
nia, become such a searcher? Born Jane Butzner on May 4, 1916, she
had grown up in a lively and genial household, the third of five
children in a solid, middle-class family of old Presbyterian stock. Her
father, John Butzner, was a respected doctor, well known in the community and much beloved; her mother, Bess Robinson, was trained to
be a teacher, though for most of her life she worked as a nurse. The
couple met in Philadelphia, at the hospital where John was doing his
internship. Bess was a night nurse at the time. Buoyant and resourceful,
with a sweet nature and a sly wit, she was drawn to the tall, avuncular
doctor in wire-rimmed glasses. He was kind and ruminative, attentive
to even the smallest details. She liked his quiet manner, his watchful-
ness and gentlemanly bearing. John was equally taken with Bess. Soon
the two were seeing each other regularly. After a short courtship, they
married, and in 1905, moved to Scranton, where John set up a private
practice, which was firmly established in less than a year, allowing
the couple to start a family. Their first child was a girl, Betty, followed
soon after by a son, who died of scarlet fever as a child. Jane was
the third and her two younger brothers, John and Jim, the last.

It was a happy and stimulating household, embracing, supportive, remarkably free of rancor. Dr. Butzner, though not wealthy, provided well for his children, whom he encouraged to be independent
and to think for themselves. An avid sports fan, he followed baseball
and boxing and urged his children, even the girls, to participate in
athletics. Beyond the fresh air and camaraderie, such pursuits, he
felt, were character building. Bess was a strong mother, able and
organized, who, like her husband, encouraged her children to be
self-reliant and venturesome. There were great tracts of time when
the children were left gloriously untended, free to play and explore
as they pleased. They were a wild-hearted brood, fellow conspira-

tors in a world mostly of their own making, given to pranks and small acts of mischief. They tore around their leafy neighborhood on bikes, carved ghoulish-faced pumpkins in the fall, rode the streetcar downtown on their own. The Butzner house was solid and stately, the most spacious on the block, a red brick Victorian with white balustraded porches and a backyard planted with flowering shrubs. Bess managed to keep a vegetable garden, even as she raised her four children and worked as a nurse. She made jams and pickles, chutneys and pies, rituals Jane would carry on in her own life in Greenwich Village.

Jane was an intensely curious child, a trait that set her apart, even at an early age. Once, when she was in grade school, she asked her father to explain the purpose of life. Her father, a reflective man to whom she would later ascribe much of her intellectual independence, answered her with a question: "Look at that oak tree. It's alive. What's its purpose?" he asked. Jane thought a moment. The answer, she decided, was, simply, to live. In one sense, the answer was a conundrum, but in another it was not: if the purpose of life was to live, then it made sense to try and understand the processes that made this possible.

Jane was extremely close to her mother, who lived to age 101, writing her amusing, news-filled letters nearly every day of her adult life. But it was with her father that she most identified, a feeling reinforced by their strong physical resemblance: Jane had the same owlish aspect as her father, owing to glasses; the same aquiline nose and height and sturdy frame. And she had a similar cast of mind, a habit of close observation and searching questions, a tendency to ruminate, to trust her eyes and intuition. She often spoke of her father as a deep and original thinker, recalling him hypothesizing, long before there was clinical proof, that mental illness was caused by chemical imbalances.

Both Jacobs's parents had come from rural backwaters, which they had longed to escape. Her father had been homeschooled by female cousins on the family's Virginia farm until a windfall from a wealthy uncle freed him to go to college. Her mother had grown up in a quiet little town in Pennsylvania, the daughter of a country lawyer. Their home had bordered an old canal, where coal barges passed by en route to Baltimore. Both felt urban life offered distinct cultural advantages. Cities, they impressed upon their children, were places of dynamism and interest.

Scranton in the 1920s, when Jane was growing up, was a booming city, having become rich on anthracite coal. It was a place of gracious old houses on tree-lined blocks, cultural activities, and a lively downtown, the first metropolis in the country to build an electric streetcar system, a point of pride that had earned it the nickname Electric City. Jane loved riding the streetcars, which she could catch on the corner of her own block.

Her father owned a car, a little red Ford, which he used for making house calls, as was customary for physicians in those days. But when he was going downtown, he always took the streetcar, which he considered a superior mode of transport.

Jane saw the advantages of the streetcar too. The streetcar could carry her to places well beyond the reach of her bike—to the science museum, the theater, the lovely old library downtown. The streetcar lines latticed the city like a giant spiderweb, connecting all corners of the metropolis. This idea pleased and interested her: the connectedness of things, the freedom and flexibility of choices it provided.

Jane was well liked and had no shortage of companions. A tall girl with a daredevil spirit, she liked to entertain her grade school friends with small acts of subversion. She was known for openly challenging her teachers, and often could be found sitting in the principal's office, serving time for some small insurgency or another. Once she amused a pack of friends by running up the down escalator at the Scranton Dry Goods store, only to meet the glaring store manager at the top. "She was a free spirit, clever, hilariously funny and fearless," a school companion remembered.

Jane also had several imaginary friends. The two she counted as her closest were Thomas Jefferson and Ben Franklin, who she considered regular companions. Franklin asked "very good questions," Jane claimed. "He was interested in everything . . . in lofty things, but also in nitty-gritty, down-to-earth details." She explained to him how the traffic lights worked, and the city's system of trash collection. She liked Jefferson too, but sometimes she found his questions "too theoretical."

Jane was interested in everything too, as long as it didn't have to do with school. Like Mark Twain, she felt that school interfered with her education. "I learned a great deal from the teachers in the first and second grades," she allowed. Thereafter, she took to reading in great quantities, with a book hidden under her desk, while the teachers, whom she considered ill-informed and hopelessly dull, droned on before the class. Her first official act of sabotage took place at age seven, in elementary school. Her father, who often mused aloud, had just told her "never to promise to do anything for the rest of your life while you're still a child." Promises, he added, "are serious." The next day, her third-grade teacher asked the class to raise their hands and promise to brush their teeth for the rest of their lives. Her father's words still fresh in her mind, Jane refused. Then she began "proselytizing" her classmates, warning them of the gravity of making such an absolute promise. The teacher was so incensed that she ordered Jane out of the classroom, telling her she was expelled. Left to her own devices, Jane drifted over to a forbidden area of rocks near some train tracks. She climbed the rocks, feeling like an "outlaw" for the first time in her life. After a time she went home for a quiet lunch with her mother, and then returned to school, unsure what to expect. The teacher never uttered another word about the episode. "It gave me the feeling I was independent," she later recalled. By fourth grade, her outlaw intellect was in full flower. When a teacher in geography class claimed that cities always formed where there was a waterfall to provide electric power, Jane took issue, arguing that while Scranton had a waterfall, which was lovely, it didn't have anything to do with powering the city or its economy. Mines were the thing in Scranton, she pointed out. "I was very suspicious," she later said.

Jane's acts of defiance were not without precedent. She hailed from a long line of sturdy and opinionated stock. Her maternal grandfather, a Union captain in the Civil War, had survived the Confederacy's most brutal prisoner-of-war camp, an infernal place called Libby Prison. Though he went on to become a successful lawyer, in 1872 he ran for Congress on the grassroots Greenback-Labor ticket, a point of pride for later generations, who applauded his progressive stance, his independence from the usual assumptions of his class. "I am proud that my grandfather stuck his neck out" for many of that party's planks, Jacobs would later write. Though "outlandish at the time," later those ideas became "respectable law and opinion." Her father's side displayed a streak of radicalism too. Although Virginians, some of her father's relatives refused to fight in the Civil War, asserting they didn't believe in either secession or slavery—a display of independence Jacobs thought well established the Butzner family spirit. "Perhaps it is partly because of such personal tradition that I feel our American tradition of freedom to deviate from the accepted viewpoint is not a cliché or of secondary value," she wrote. "I was brought up to believe that simple conformity results in stagnation for a society, and that American progress has been largely owing to the opportunity for experimentation, the leeway given initiative, and to a gusto and freedom for chewing over odd ideas." It pleased her that the women in the family were no less independent. A female cousin became the director of a sanatorium. Another relative, a Quaker who believed in "women's rights and women's brains," as Jane put it, refused to use a masculine nom de plume in her published writing, as custom then prescribed. Her great-aunt Hannah Breece, forsaking marriage in order to study anthropology, went to Alaska in 1901, where she taught native children, traveling to fishing camps and reindeer stations by dog sled or kayak, wearing a poncho made of bear intestines. Jane loved these stories, and if she had any doubts about what a woman could or could not do, Hannah Breece dispelled them.

Jane's parents, hewing to family tradition, were of a similar liberal bent, treating their daughters no differently from their sons. They saved money so that Jane could attend college, just as they did for her brothers. But Jane had decided by then that high school was quite enough, further schooling a distraction easily dispensed with. Her eyes set on New York, she signed up for the Powell secretarial course, decamping for the city soon after.

THREE YEARS HAD NOW PASSED, HOWEVER, AND JANE WAS BEGINNING TO reappraise things. It was 1937, the year *Vogue* published her stories, a positive turn certainly. But she still hadn't landed a full-time journalism job. Her usual optimism had been shaken by the unexpected death of her father that year at age fifty-nine. It was a shattering blow, as they had always been close. She began to reconsider the idea of college; perhaps further schooling would give her a leg up. The following September, she enrolled at Columbia University's School of General Studies, which had open admissions for "nontraditional" students and a program that allowed her to take classes in all departments, which pleased her. She signed up for a raft of challenging courses—chemistry, constitutional law, geography, political science, psychology, zoology—and was soon happily immersed, amazed to find how much she was enjoying herself. For the first time in her life, her grades were good; she was finding it bracing to bounce her ideas against those of her professors, to delve more deeply into how the world worked as part of a dialogue, rather than on her own. And then she hit an impasse. At the two-year mark, having now amassed a bundle of credits, she learned she was required to matriculate in order to complete her degree. This meant applying to Barnard, and taking some mandatory courses, which didn't sit well. Jane requested permission to waive the requirements, which she had already well surpassed in her course work, but was told that given her poor grades in high school, there could be no accommodation. It was just the sort of institutional intransigence she deplored, and it cemented her lifelong distain for academic credentials and the prestige of experts. Galled, she walked off campus and never returned.

"Fortunately, my [high school] grades were so bad they wouldn't have me and I could continue to get an education," Jacobs would later say.

Determined not to look back, Jane turned her attention to searching for a full-time writing position again. After several near misses, she did finally land something in publishing, a job at a trade rag called *Iron Age*, which catered to the metals industry. It was 1940 and *not* the job she'd imagined for herself, certainly. Besides knowing nothing about the steel or iron industry, she was hired as a secretary and not as a writer. But this would soon change. Her boss at *Iron Age* was lazy, and within weeks Jane was picking up the slack, coming up with story ideas and banging out articles that everyone liked. Soon she was promoted to associate editor, which allowed her to write full-time. Her pieces were distinctive, offbeat and unusually lively given the dullness of the subject matter; it was the side of the job she wrestled with most, finding stories that were interesting and relevant. But as the war deepened, and America's role in supplying men and material grew, she uncovered a metals-related story that didn't feel forced: the sharp economic downturn in once-booming Scranton and other steel-producing cities in Pennsylvania. The ensuing article, which emphasized the large number of unemployed steel workers in the Lackawanna Valley (and which she would later expand for the *New York Herald Tribune*), caught the eye of executives from a company that made parts for the B-29 bombers and other warplanes. Sufficiently moved, they decided to put a plant in Scranton, delivering an almost instant boost to the flagging economy. Jacobs was intrigued by the fact that a bit of journalism could spur such a change, and she quietly stored away the knowledge. (Her article would also become fodder for a group of labor leaders pressing government officials to locate more wartime manufacturing in the Scranton area.) Pleased that a local girl had done them proud, the editors of Jacobs's hometown newspaper, the *Scrantonian*, answered with a paean of their own. Shortly after the announcement of the incoming B-29 plant, they ran a piece with the attention-grabbing headline, "Ex-Scranton Girl Helps Home City. Miss Butzner's Story in *Iron Age* Brought Nationwide Publicity."

But life at *Iron Age* was not as rosy as it appeared. Almost from the moment she was hired, Jane's editor seemed to have it out for her. He hated the idea of a woman writing for the magazine, and certainly not one as smart and self-assured as the opinionated Miss Jane Butzner. Despite her position as an editor, in her file he continually referred to her as "only a typist." The harder she worked, and the better her pieces, the more determined he was to undermine her position. One night he sent her on a putative work assignment, which turned out to be a stag dinner that included, in Jane's words, "entertainment to which he was well aware no woman was invited." He paid her less than her male peers, and made "loose and untrue allegations about my morals," she recalled. When she learned of the salary gap, she launched a campaign for equal pay, and for the right of the rest of the company's employees to unionize. Though her boss's behavior was a clear example of sexual harassment, the syndrome didn't yet have a name, let alone any legal heft behind it. Even so, Jacobs was keenly aware of what such bias felt like, and she was unwilling to pretend it wasn't there. She never flinched in the face of a fight over issues of fairness.

It was time to leave *Iron Age*, writing job or no. Armed now with a portfolio of pieces bearing her byline, she began combing the want ads again. There was an opening as a features writer for the U.S. Office of War Information that sounded interesting. She rushed in her application, was summoned for an interview and immediately hired. Soon, she was pedaling her bike up to the State Department offices at Columbus Circle each morning, her tailored tweed blazer flapping in the wind, passing her days writing features for overseas distribution, often about American history, architecture, and cities. She enjoyed her colleagues and loved the work, grateful for the editorial latitude she was given. Her life felt buoyant and full.

Jane and Betty liked to socialize and on weekends they often entertained. By now they were living in a new apartment on the corner of Washington Place and Sixth Avenue in Greenwich Village, just a block from Washington Square Park. Betty had recently taken a job at Grumman, a defense plant on Long Island, where she was meeting lots of new people. One weekend in 1944, she invited a friend from work named Robert Hyde Jacobs back to the Village for a party. Jacobs was designing warplanes and doing engineering work at the time. Trim and bespectacled, with wavy hair and an open smile, dressed that evening in a natty tweed jacket, he exuded an air of bohemianism. As he walked through the door, Jane was standing on the other side of the room, chatting with a clutch of friends. She was wearing a beautiful green evening dress, and as she glanced up to see who had arrived, their eyes met. A few moments later Betty brought Bob over and introduced them. Soon they were lost in talk, seemingly oblivious of everyone else in the room. The rest of the evening passed in a blur. They sipped martinis and smoked, trading stories, leaning in toward each other occasionally, as if to listen more closely. Bob had taught art at City College, he said; he'd even tried his hand at acting for a while. But design was his true passion. He was a full-time architect now and loved the work. He had studied architecture at Columbia, and decided early on that good design was less about creating iconic forms than it was about being responsive to how people lived—and to what made them feel comfortable. He tried to approach every design project with this in mind. Jane watched him closely as he spoke. His eyes were bright behind his wire-rimmed glasses. He seemed warmhearted and sensitive, funny and alert, unusually open. He was great fun to be with, and she was excited. "Cupid really shot that arrow," she later remembered.

Bob was equally taken with Jane. She was earthy and intelligent, extremely articulate. She had a sharp wit and a wry, irreverent sense of humor. She told him she'd been hired to write for the metals magazine because she "could spell molybdenum." She had a mischievous glint in her eye, and he too was smitten. They made plans to see each other the following day. Later they would say it was already a foregone conclusion: over the next few weeks, tramping about the city together, they couldn't stop talking, lost in each other's company. They had discovered they shared a deep love of looking. Robert pointed out buildings he liked; Jane showed him parts of the Village he'd never seen. They went to galleries and tiny storefront theaters; coffeehouses and low-lit basement restaurants; they sat under the stars in Washington Square Park. A month later they married. Jane later claimed their engagement would have been even shorter, but before tying the knot, they needed to schedule a time for her to meet Robert's parents in Alpine, New Jersey.

The wedding, which took place in the spring of 1944, was small and intimate, the ceremony held in the living room of her childhood home in Scranton, with only close family in attendance and no best man or maid of honor. It was late May and Jane insisted on flowers from the garden for her bouquet. She and her mother filled the house with vases of lilacs, irises, and roses, all cut that morning. Jane didn't like fussy affairs. Even the white, street-length dress she wore was simple and unadorned. For their honeymoon, they bicycled through parts of northern Pennsylvania and upstate New York, exploring the quiet back roads, stopping for lazy picnics. They felt it a perfect idyll.

BY 1947, THE JACOBSES WERE BEGINNING TO THINK ABOUT PUTTING down more permanent roots.

Ambling through the Village one day, they noticed a run-down building that caught their eye. It was at 555 Hudson Street, between Eleventh and Perry Streets, considerably west of the more charming, residential blocks near Washington Square. Shoehorned between a Laundromat and a tailor, both with apartments directly above, the three-story brick building was in need of work. But it was an ideal size and looked empty. The ground floor had once been a candy store—a painted Canada Dry sign still hung out front—although it was vacant now. There was no central heating, and the tiny backyard was piled with garbage. But the place had character and the patina of age. And unlike so many couples their age, they were dead set against buying a house in one of the bland new suburbs outside the city, where every home was identical. The savings they had amassed since their marriage would cover the $7,000 purchase price—just about what a tract house in Levittown cost.

They decided to do the renovation themselves, an unusual turn for a professional couple, building a kitchen on the ground floor, where the store had been, living in the upper two floors as they worked. They scrubbed the bricks and hauled away the garbage, turning the rear yard into a blooming garden. In those days, few people equated city life with having a garden. A private yard was one of the seductions of the suburbs, not a part of the urban experience. Jane and Bob saw no reason to give up a place as lively and serendipitous as Greenwich Village just to have a lawn. What they liked about their neighborhood was the jumble of people and places within blocks of where they lived: the contrast between the gracious patrician homes overlooking Washington Square Park and the smoky jazz clubs on Hudson Street; the bars and coffee-houses on Bleecker Street, where the folkies and beatniks flocked, and the walk-up apartments close to the Hudson River, stomping grounds for a mix of Irish-Catholic longshoremen, Puerto Ricans, and working-class blacks.

Their home was similarly eclectic, if not bohemian. According to friends, there was a casual, slightly disheveled feel to 555 Hudson Street. Dishes teetered in the sink and ashtrays went unemptied, stuck absentmindedly on some makeshift shelf where they were promptly forgotten. Stacks of books buried tabletops and portions of the floor. There was always a jigsaw puzzle going—and a solemn pact between them that allowed only one peek at the cover image before it was put away. Thrifty and resourceful, Jane wore the same chunky costume-jewelry necklace whenever she dressed up. Bob trimmed her hair. Jane liked to support the local artisans in the neighborhood. The sandals she sported in summer were a particular source of pride, as they were handcrafted by a Village cobbler. Fall weekends were spent at Bob's uncle's apple farm in upstate New York. In summer, they camped with friends in Montauk, to be near the beach.

Jane enjoyed domestic life. And when they decided to start a family, little changed. They had their first son, James, in April 1948. Nicknamed Jim, he was followed two years later by a second son, Edward, whom they called Ned. A third child, a daughter named Mary, was born in 1952. "The click-clack of the typewriter was a sound of the house," Jim would remember. "When the house was quiet, Jane was reading." Jane baked cookies and made pies; she kept a vegetable garden and cooked feverishly in the cheerful buildup to each holiday. But beyond taking a brief maternity leave for each child, she didn't stop working, which made her an anomaly. "In my generation," she would later explain, "women were made to feel guilty if they didn't stay home and devote themselves to being wives and mothers. If we worked at jobs or at a profession, we had to struggle against regarding ourselves as irresponsible [and] selfish."

At war's end, the focus of Jane's State Department work had shifted. Now, as part of the reconstituted Overseas Information Agency, she was charged with producing upbeat articles that showcased American culture. Much of the writing she was doing was for a glossy publication called *Amerika*, which was aimed at a Soviet audience, with the hope that it would help bolster America's ties with its tentative new ally. But as relations with the Soviets began to unravel, and the Cold War grew chillier, the magazine recalibrated its focus again. More and more, it was Jane's job to promote American *values* as well as its culture. To this end, articles highlighting indelible American scenes—the Southwestern desert in bloom, New England's quaint village greens, Radio City Music Hall—were regularly featured, complete with seductive photos, the idea being that Soviet audiences would contrast these images of plenty with what were presumed to be meaner circumstances at home. Whether this was the case was an open question. But the magazine was a big hit in Moscow anyway.

Jacobs's articles for *Amerika* had made her curious about life in the Soviet Union. Hoping to visit, she pitched an idea for a freelance piece on Siberia, and she and Bob applied for visas. But their applications were turned down, despite the fact that Jane herself worked for the government. It was the late 1940s and hysteria over Communism was building. America's creative classes were increasingly suspected Communist sympathizers.

One day, casting about for article ideas, Jane stumbled upon an author named Saul Alinsky who interested her. Alinsky had been the first to organize the poor and exploited against the inhumane working conditions in the meatpacking plants of the Union Stockyards in Chicago (brought to shocking life in Upton Sinclair's 1906 novel *The Jungle*). A tireless advocate for the oppressed, he was considered "the father of grassroots community activism," having given the downtrodden of Chicago's South Side an effective political voice, inspiring others in turn to do the same. Alinsky's view that community action grew from the bottom up made sense to Jacobs, as did his belief that fighting for progressive ideals meant marrying action to ideal, not just writing about social change.

But Alinsky raised red flags in certain quarters. Amidst gathering Cold War suspicions, he was a risky figure to venerate. While he never advocated violent tactics—he once threatened a "piss in" at Chicago's O'Hare airport, arranging for large numbers of well-heeled African Americans to occupy the toilets until the city came to the bargaining table—he *was* dedicated to giving "power to the people," as he often said. A potent source of "anti-government rhetoric," he was unafraid to challenge organized authority, which, at that moment, sowed fears of conspiracy, and, quite possibly, political subversion.

In 1949, Jacobs received a letter from the Loyalty Security Board, the State Department agency set up to ferret out Communist Party allegiance among government workers. In the government's inquiry, they asked for information about her visa application to the Soviet Union, her subscription to the *Daily Worker*, and her *Iron Age* supervisor's recollection of her as a "troublemaker." Jane calmly answered each question. She said she had hoped to write about Siberia, subscribed to scores of magazines, and that her boss at *Iron Age*, begrudging her for getting promoted, was a "chauvinist" who paid male workers more than women.

Apparently, the response was unsatisfactory, because the board sent her another Cold War interrogatory in 1952, this one more detailed. America was in the thrall of a fresh wave of red-baiting, fed by the Berlin blockade, the Rosenbergs' conviction for espionage, and Senator Joseph McCarthy's paranoid witch hunts.

Jacobs, infuriated by the government's overreach, pulled out her typewriter and carefully composed several single-spaced pages in response. "Upon first reading the questions submitted to me, I was under the impression that possibly I was to be charged with belonging to the union and to registering in the American Labor Party. But since neither of these has been declared illegal, I concluded, upon further thought, that I am probably suspected of being a secret Communist sympathizer or a person susceptible to Communist influence." She allowed that it still shocked her that Americans could be questioned about their associations, reading matter, and political beliefs. While she understood the necessity for such questions sometimes, as someone concerned with the preservation of traditional American liberties, she also wanted to be clear about her beliefs: "I was brought up to believe there is no virtue in conforming meekly to the dominant opinion of the moment," she said. "I was taught that the American's right to be a free individual, not at the mercy of the state, was hard-won and that its price was eternal vigilance, and that I too have to be vigilant."

Several passages followed outlining the extent of her involvement in the union and the American Labor Party. While she admitted to canvassing against a Republican congressman one evening, she said that by no stretch of the imagination was she a member or an affiliate of the Communist Party. "I abhor the Soviet system," and fear "the whole concept of a government which takes as its mission the molding of people . . . that controls the work of artists, musicians, architects and scientists . . . that centralizes into the monolithic state every activity which should properly be controlled locally or by individuals."

"I believe in control from below and support from above," she added. Bravely, Jacobs further asserted that while the Soviet Union was a threat to the American tradition, another peril lay at home in the "current fear of radical ideas and of people who propound them . . . I believe I have the right to criticize my government and my Congress. I make these criticisms within the framework of our own system and traditions."

In this, though she didn't yet know it, Jacobs was not alone. Others in America were also beginning to feel that the "mania for internal security" had gone too far, that such intrusions of official power were crushing to the human spirit, that the silencing of dissent and the demonizing of enemies—the intolerance of *difference*—was bad for democracy. The placid facade of the 1950s was beginning to crack, the first voices of resistance riding the air, rumbles of an oncoming storm.

IN OCTOBER 1955, THREE YEARS AFTER JACOBS'S RESPONSE TO THE STATE Department inquiries, a flyer for a poetry reading was passed around the North Beach neighborhood of San Francisco. "Remarkable collection of angels all gathered at once in the same spot," it announced. "Wine, music, dancing girls, serious poetry, free satori. Small collection for wine and postcards. Charming event." The reading was to take place that night at a converted auto-repair shop. It was now called Six Gallery.

The author of this fanciful missive was twenty-nine-year-old Allen Ginsberg, a maverick poet, writer, and provocateur. Over the previous decade, Ginsberg's life had swerved far from the American mainstream. In 1944, while at Columbia University on scholarship, he'd been suspended for scrawling an obscenity on his dorm window. Arrested soon after for lending his apartment to a "poetic drifter" who was trafficking in stolen goods, he agreed, in lieu of jail time, to serve his sentence in a psychiatric ward. Once out, he worked a series of menial jobs, read widely in Zen Buddhism, and drifted to Mexico, where he wandered wide-eyed among Mayan temples and hiked naked in the hills. He hitchhiked to Florida next, immersed in Rimbaud; caught a flight to Cuba, hoping for erotic adventure; and eventually migrated back to Greenwich Village, where he began spinning his dreams and laments into reams of verse. By the time he appeared in San Francisco in the mid-1950s, he was part of a small band of artists and hipsters who called themselves the "Beats," a rebel brotherhood searching for freedom from a regimented culture and a new, more personal voice. The Beats were still obscure figures at this point. Except for a few articles—the *New York Times Magazine* had run a piece on the "Beat Generation" in November 1952— few Americans knew of the group's icon-smashing beliefs. But this was about to change.

As Ginsberg stepped to the stage of the Six Gallery, the North Beach crowd, much of it fortified with cheap red wine, was rapt. With wild enthusiasm they listened as the rumpled poet in horn-rimmed glasses began reciting a long, hallucinatory poem entitled "Howl." From the opening lines to the mesmerizing last, what they heard was "a declaration of independence" from the suffocating constraints of postwar American culture, the conformity and sexual repression that Ginsberg and his cohort believed were strangling the nation's spirit, the Cold War hysteria that had muzzled an entire generation—the blacklists and hearings and baseless interrogatories such as Jacobs had received.

The crowd whooped, and the breathless torrent of incantatory images continued. The way out, they could gather, was to stand in opposition to the age, to rebel against all forms of oppression, all social taboos, to *feel* in numbing times, rejecting the culture's spirit-killing materialism—what Ginsberg called "Moloch." "Moloch whose eyes are a thousand blind windows! Moloch whose skyscrapers stand in the long streets like endless Jehovahs!"

At the root of the Beat credo was a deep discontent with American ideals of progress and power, a yearning for "authentic" experience, unfettered freedom, "alternate routes" to transcendence. "To dance beneath the diamond sky with one hand waving free," as Bob Dylan would sing.

This liberating advice would help shape the sensibilities of an entire generation. It would signal the first green shoots of a new consciousness, a cultural change of heart that would bloom, profusely, into the countercultural 1960s, separating those who embraced the fruits of postwar industrial affluence and those who dreamed of a freer and more feeling world. The only desires worthy of modern man were not, perhaps, those of "Moloch": the striving, restless technocratic order signified by men like Robert Moses. The modern spirit was beginning to part ways with the modern corporate contract, the cold imperatives of postwar technology. Ginsberg, like Jacobs, was seeing another way.

AS ANNOYED AS JANE WAS BY THE SECOND GOVERNMENT INTERROGA-tory, she wasn't especially concerned about losing her job. The State Department was relocating *Amerika* to offices in Washington, D.C., and she had no interest in leaving Greenwich Village. She already had something else in mind.

One morning that spring, Jacobs pedaled her bike up to Rockefeller Center. At the foot of the soaring Time-Life Building, she locked the bike, smoothed her skirt, and then strode into the splendid lobby. She made a beeline for the elevator bank, and when the first door slid open, she stepped in, pressed the button for the penthouse, and after the elevator ascended, stepped purposefully out, proceeding now to the elegant, glass-walled reception area of *Architectural Forum*. This was the magazine she wanted to write for.

No one was expecting Jacobs that morning. She didn't have an appointment, nor did she know anyone there. No matter; it was a small obstacle, a thing to be talked through. Indeed, within minutes Jacobs was sitting in the office of its editor, a Yugoslavian-born writer named Douglas Haskell.

Haskell liked Jacobs immediately. He could see she was ambitious, that she was canny and knew her own mind, and he admired her fearlessness. She was charmingly forthright, up front that she hadn't been trained as an architect, that she hadn't even graduated from college. Haskell wasn't troubled by her lack of credentials; he hadn't been formally educated as an architect either. It was often an advantage to write without the encumbrance of academic theory, he felt. He decided he would give her an assignment, a trial story about a candy building at Herald Square, and see how she did. The piece arrived quickly, as Haskell expected. It was offbeat and original, also as expected, peppered with thoughtful detail. Within days, he had hired Jacobs as an associate editor. "You'll be our schools and hospitals expert," Haskell told her.

"I was utterly baffled at first," Jacobs later recalled, "being supposed to make sense out of great, indigestible rolls of working drawings and plans, but my husband came to my rescue. And every night for months he gave me lessons in reading drawings, learning what to watch for as unusual, and discovering what other information I needed to have." Jane *did* learn what to watch for, in the drawings and plans before her, but also, more usefully, in real life.

IN THE SPRING OF 1956, LESS THAN A YEAR AFTER HER FIRST TOUR OF East Harlem with William Kirk, Jacobs stood outside the auditorium of Harvard University's prestigious architectural school. She felt edgy in her own skin, jittery and unnerved. People were beginning to arrive: graduate students and dapper men in well-made suits, many of them friends it appeared. Or at least acquaintances, to judge from their quick, genial smiles and cordial handshakes, their nods of recognition as colleague spotted colleague from across the room. Jane watched the happy fraternity—for indeed it was mostly men here today—stalling another moment before finally stepping in. It was the last place she expected to find herself, and she wasn't altogether convinced she should be there at all. It was Haskell, her boss at *Architectural Forum*, who had accepted the speaking engagement. It was months ago. But Haskell had fallen ill, and asked her to speak in his stead. She had tried to politely decline. She had only spoken before a large audience once before, and, on that occasion, suffered terrible stage fright. She didn't think she was the right person to speak before an academic conference on urban design, she insisted. Especially this one, at Harvard's Graduate School of Design. Fiercely egalitarian, as a rule she was unimpressed by academics, who often seemed to be talking to themselves, she thought, more interested in theories than practice. In the audience today would be some of the country's most ardent proponents not only of modernism, but also of urban renewal: Edmund Bacon, whom she had met in Philadelphia; Josep Lluís Sert, the dean of the design school; Victor Gruen, who had designed the towers for Boston's bulldozed West End. Harvard was in the grip of a near messianic belief in the power of modernism to solve every ill of the modern city; Walter Gropius, a founder of the Bauhaus and one of the pioneering masters of the modern movement, was a revered professor here, his influence at the design school enormous. The conference's attendees would be expecting a speech from one of their own. But Haskell had kept pressing, and finally she had agreed, although only on one condition: that she be allowed to say what she really thought. Haskell gave her the green light and so, like it or not, here she was, standing now at the podium.

She took a deep breath and began:

"Sometimes you learn more about a phenomenon when it isn't there, like water when the well runs dry—or like neighborhood stores which are not being built in our redeveloped city areas," she said. "In New York's East Harlem, for instance, 1,110 stores have already vanished in the course of rehousing 50,000 people."

"Planners and architects are apt to think, in an orderly way, of stores as a straightforward matter of supplies and services. Commercial space. But stores in city neighborhoods are much more complicated creatures which have evolved a much more complicated function," she added. "Although they are mere holes in the wall, they help make an urban neighborhood a community instead of a mere dormitory."

The candy stores and bars, the fruit stands and butchers, even the empty storefronts occupied by the "hand-to-mouth cooperative nursery school" or political club served a vital function, she continued: they were social centers, "institutions" that people created themselves. When neighborhoods were leveled and rebuilding took place wholecloth, these hole-in-the-wall places disappeared, and with them the "vestigial" areas that housed them. There was a subtraction of commerce and culture. The cavernous supermarket built into the ground floor of a drab housing tower was no substitute for the thirty mom-and-pop shops it replaced. A neighborhood stripped of these familiar and casual places went dead, and its residents went elsewhere—sometimes blocks away—to find services akin to what they'd lost on their old blocks.

"Do you see what this means?" she asked. Some very important sides of city life were getting lost because there was literally no place for provisional and eclectic little enterprises "in the new scheme of things," no place for people to make easygoing social contact. "This is a ludicrous situation and it ought to give planners the shivers," she added.

Determined to leave her audience with a few positive ideas, Jacobs went on to suggest how things could be improved. Planners needed to take their cues from the "lively old parts of the city," she said, to heed their jumble and diversity. "Notice the stores and the converted storefronts. Notice the small buildings called 'taxpayers' and up above, the bowling alley, the union local, the place where you learn guitar."

The idea wasn't to copy these units outright, but to understand the intimate, informal qualities they added, however ugly or makeshift they appeared. It was important, she insisted, for designers to become more socially astute. Parks and outdoor spaces needed to be designed to function as social spaces "at least as vital as the slum sidewalk." Seemingly trivial conveniences like laundries and mailbox areas needed to be thought of as potential social hangouts.

Planners, she explained, needed "to respect—in the deepest sense—strips of chaos that have a weird wisdom of their own not yet encompassed in our concept of urban order."

"We are greatly misled by talk about bringing the suburb into the city," she added. "The city has its own peculiar virtues and we will do it no service by trying to beat it into some inadequate imitation of the noncity."

Jacobs looked up. She hadn't dared meet the eyes of the audience until that moment, assuming they would be hostile to everything she said. How could it be otherwise? She was telling the top designers in the country that everything they believed about urban renewal was misguided; that the monolithic towers they were erecting to revitalize the slums were strangling the very neighborhoods they purported to be saving; that for all their high-minded ideals, they were failing on a monumental scale.

But the audience broke into thunderous applause. As she stepped away from the podium, several people rose and approached to introduce themselves. One of them was the great man of American letters Lewis Mumford, who commended her on her clear-eyed critique, expressing his sympathy for all she had said. He too deplored the "tower-in-the-park" designs, he confided, and had said as much in his column for *The New Yorker*. He hoped, he added, that she might write more on the subject, and urged her to send her work to some of the more mainstream magazines like *The Saturday Evening Post,* to enlarge the audience for the topic. She should expect hostility from the urban design community, he warned; her ideas would most certainly incur their wrath. "Your worst opponents are the old fogies who imagine that Le Corbusier is the last word in urbanism," he added in a letter soon after. Nonetheless she should "keep hammering."

Word of Jacobs's Harvard speech spread quickly. Back in New York, one man whose interest was especially piqued was William H. Whyte Jr., a writer and senior editor for *Fortune Magazine*, another jewel in Henry Luce's sprawling media empire, and a sister publication to *Architectural Forum*. Whyte was a Princeton man who had served in the Marine Corps, joining *Fortune* after the war. His father, a Philadelphia blue blood, had hoped his son would pursue a corporate career. But Whyte's interest in corporate life was strictly philosophical. Fascinated by how environments shaped human behavior, his first big story for *Fortune* followed the lives of a white-collar community living in a typical middle-class suburb in Illinois, where the men were corporate strivers and the women stay-at-home wives, idle and sidelined. The piece, written after months of close observation, grew into a groundbreaking book, *The Organization Man*, published in 1956, which documented the stifling conformity required of a corporate career. Wildly popular, it sold over two million copies, vaunting Whyte to prominence and joining several other seminal books that challenged the status quo in 1950s America.

Whyte was a friend of Haskell's, and for some time had been concerned about the state of American cities. He was assembling a series on the challenges facing urban communities, and as always was on the lookout for fresh perspectives. Intrigued by what he'd heard about Jacobs, he reached out to her, asking if she might contribute an article based on the Harvard talk. Jacobs, though flattered, initially demurred, claiming she lacked the background for such a piece. And she wasn't sure she could write to the length he was proposing. It was longer than anything she'd done. Whyte didn't press. His colleagues had pronounced Jacobs "a most inappropriate choice," advising him to assign the story to another writer, which he did. But when that writer fell through, Whyte approached Jacobs once again, and this time she agreed.

"Downtown Is for People," published in *Fortune* in 1958, was a breakout success. Almost overnight, it established Jacobs as a highly astute critic of urban planning, a fresh and credible voice. Jacobs's thesis was at once bold and simple: Downtown redevelopment plans across America were threatening the life of cities. They were monstrous failures, demonstrating no understanding of how cities worked, or how people actually lived in them. "These projects will not revitalize downtown; they will deaden it," she argued. "They will be stable and symmetrical and orderly. They will be clean, impressive and monumental. They will have all the attributes of a well-kept, dignified cemetery."

Readers responded almost immediately. Jacobs had given shape and legitimacy to what many were feeling, but few had found the words to say. Letters of praise poured into *Fortune* in unprecedented numbers, many from the general public, but some, also, from unlikely corners of academia and the architectural world. Whyte sent on the best to Haskell, scrawling on the corner of one, "Look what your girl did." Whyte was feeling vindicated for having stuck to his guns. Jacobs, in his estimation, represented all the qualities he looked for in a writer: a lively and inquisitive mind, a healthy distrust of dogma, a plainspoken voice. She had a powerful sense of narrative, and a nose for pretension, an ability to touch the emotions as well as the mind. He could already see that hers would be a long and interesting career.

"DOWNTOWN IS FOR PEOPLE" TOOK THE DESIGN WORLD BY SURPRISE. IN it, Jacobs not only blasted planners for particularly egregious projects, but also for their flawed and paternalistic thinking. Almost without exception, she asserted, projects coming off the drawing boards have "one standard solution for every need." Whatever the activity—commerce, medicine, culture, government—"they take a part of the city's life, abstract it from the hustle and bustle of downtown, and set it, like a self-sufficient island, in majestic isolation."

These "fortress" settings, she went on to say, worked at cross-purposes to the city. They banished the "street," which was the connective tissue that gave urban spaces their life and spontaneity. These projects killed "the gaiety, the wonder, the cheerful hurly-burly" that made people want to come into the city in the first place. "Urban ecosystems," she wrote, were "fragile" in the same way natural ecosystems were; they could easily be knocked out of balance and destroyed.

Current plans for Lincoln Center were a case in point, she added provocatively. Lincoln Center would require demolishing eighteen lively blocks to make way for what was being called "a world-class" center for the performing arts on New York's Upper West Side.

"This cultural superblock is intended to be very grand and the focus of the whole music and dance world of New York," she observed. "But its streets will be able to give it no support whatever." Operagoers disembarking from taxis and cars will find themselves facing the *back* of the Metropolitan Opera House. The eastern side of the complex will open onto a major trucking route, where cargo trailers roar so loudly that sidewalk conversation will require shouting. Lining the other side of the street will be "the towers of one of New York's bleakest public-housing projects."

It was a bold assertion. At the time, Lincoln Center was an appealing project to many, and the man behind it, Robert Moses, a darling in New York's planning circles. When C. D. Jackson, the publisher of *Fortune*, read the draft of Jacobs's piece, he was livid. "My God, who is this crazy dame?" he fumed, storming into Whyte's office. "How could we give aid and comfort to critics of Lincoln Center?" Whyte immediately called an editorial meeting so Jacobs could defend her position.

The two argued. But Jacobs gave no ground. Vast sums of money were being spent on this project, she insisted. It was important to honestly assess if the Lincoln Center plan delivered on its promises, and she firmly believed it did not. Later, Jacobs would deliver an even more scathing critique at the New School, calling Lincoln Center a "piece of built-in rigor mortis."

"DOWNTOWN IS FOR PEOPLE" DELIVERED A GALVANIC JOLT TO THE DEsign world. But it also did something more. It changed the conversation. Like *Silent Spring*, it set the terms for a larger debate—in this case about urban policies—that would deepen as the fifties gave way to the activist sixties, raising important questions about state planning and the "hubris" of unchecked technology.

On one side stood autocrats like Robert Moses, who believed in "elitist, top down policies" based upon efficiency, dogma, and control. On the other, citizen activists like Jacobs, who abhorred "central control," operating by observation, common sense, respect for process, and a belief in the capacity of cities to self-organize and self-renew. The Moses camp relied on central planning and order superimposed from on high; the Jacobs side saw "regenerative potential" in the complexity of the urban fabric as it existed, viewing the city as a web of relations and interdependencies that drew its vitality from this messy, ever-changing dynamic. Where Moses saw "static form," Jacobs saw "process." Where he sought to level old neighborhoods wholecloth, she advocated for an urban evolution that grew organically, from the ground up, building upon what already existed. The humane world, writes Roberta Gratz, was challenging the "top-down command economy" and the primacy of the machine—the mechanistic mind-set of men like Moses, who saw in "the industrial model" an answer to the problems of cities.

It was a cultural clash that was just beginning, and that would play out in city after city, as well as in the highest precincts of government and industry, an urban version of the same struggle between those who sought to control and reengineer nature, and those—like Carson—who sought to understand and protect it. Carson, who was just beginning *Silent Spring* at that moment, drew attention to the "interconnectedness and fragility of the natural world" in the same terms that Jacobs was using to describe the human ecosystem of the city. The idea of interdependency ran counter to the notion that elements in nature or the built world could be examined—or understood—apart from the larger system.

THE ENVELOPE CAUGHT HER EYE IMMEDIATELY. FINGERING THE LETTER, she pulled it from the mail slot. "Save Washington Square Park," it said in big blocky letters below the return address. Curious, Jacobs tore open the envelope and began reading. She was aware that Robert Moses had designs on the park. She had seen mention of it in the *New York Times*. But then the story had dropped from sight, and she assumed the matter was dead. Though apparently not. Moses was trying to ram a road right through the middle of Washington Square Park, this letter said, to cleave it in two with a highway.

She read on. As it stood, Fifth Avenue ran from Harlem to the triumphal Washington Square Arch, where it abruptly stopped. A small carriageway allowed city buses to curl around and then head back up Fifth Avenue, which at the time was a two-way boulevard. Now Moses wanted to extend Fifth Avenue by cutting a four-lane road straight through the park, joining Fifth Avenue to West Broadway, which he would rename Fifth Avenue South. The Fifth Avenue extension, he was promising Villagers, would relieve traffic.

But there was a second, more veiled reason for why Moses wanted the road, as the letter explained: it was essential to his larger plans for the Village. As chairman of the mayor's Committee on Slum Clearance, Moses controlled all of the federal Title I funds flowing into New York, which was being billed as a sort of "Marshall Plan" for cities. There were a number of areas he saw as ideal targets for urban renewal, and the portion of Greenwich Village directly below the park was one. With the flourish of his pen, Moses had recently designated ten city blocks south of Washington Square Park a "blighted slum" to make way for new construction. Demolition was just beginning in preparation for a sprawling new complex of high-rise towers to be called Washington Square Village. The plan was calling for four thousand identical "egg-crate" apartments.

The blight designation, of course, was a convenience. Once the heart of New York's hat industry, the area south of the park was perhaps a bit tatty and shopworn, but it was hardly a slum. The 130 or so buildings slated for the wrecking ball were home to a mix of low-income families, immigrants, and light manufacturing. Some were shabby and poorly maintained, in need of paint and facade work, but the buildings themselves were solid and well built, in some cases even stately.

Moses clearly thought otherwise. By his measure, the key to the city's revitalization was "modern construction and wider streets." The newly erected towers would be surrounded by open space, sweeping clear the dense urban street grid that had latticed the city since the days of the Dutch. To his mind, the Fifth Avenue extension through the park was progress; it would provide easy access by car to the sleek new superblocks of Washington Square Village, whizzing traffic south, where, if all went as planned, it would meet yet another roadway he had in mind, a vast crosstown expressway that would provide high-speed, east–west travel between the Hudson and the East River. "Five minutes from Wall Street and Times Square, in the heart of Manhattan, is rising a dream of the future that genius, skill and inspiration have made today's reality," read a pamphlet promoting the urban renewal project. "It is Washington Square Village—destined to be a spacious new concept of city living in New York and the world." It was all of a piece: enormous redevelopment and a modern network of roads—better, faster access to all points of the city by automobile. Now that the car was king, the city needed to modernize. The four-lane road through Washington Square Park was only the first link.

Jacobs frowned. She loved the park. Like so many Greenwich Village mothers, she took her children there almost every day, even if only for an hour, when she returned from work. It wasn't just the sandbox and the swings in the designated playground, the grassy spots where her kids dove and swerved, chasing each other in some imaginary game; it was the entire scene: the maze of bench-lined walkways and the Good Humor wagons making the rounds; the chess and checkers tables of poured concrete; the central fountain area—the feeling of community there. The people in the park were Villagers and their friends. Most of the faces were familiar. The fountain was a kind of "theater in the round," she'd often remarked, with some question as to "who [were] the spectators and who . . . the show." Everyone was both. On hot days the fountain was full of children, squealing at the sting of the water jets. Dogs barked and tricycles raced about the perimeter. Sunbathers sat on its rim, baking. Toddlers chased pigeons. There were singers and beatniks with drums, guitar players and artists with easels, photographers and impromptu dancers, children and tourists and goateed young men flinging Frisbees, and, mixed in with them all, "a bewildering sprinkling of absorbed readers." The park had a casual, disheveled feel that she liked. It was open to everyone.

From her research for *Amerika*, Jacobs knew some of its storied history. Henry James and Edith Wharton had strolled there; Edgar Allen Poe, Stephen Crane, and Walt Whitman lived in flats nearby, as had she, ten years before. She and Bob had spent their courtship on its benches and paths. Once a place of peat bogs and pine barrens, the park had been a burial ground after a yellow fever epidemic, a dueling area, and a public gallows for a time. In the early nineteenth century, the scions of New York's oldest families had settled there, their fine brick town houses circling the square. Later it was the grandees of the Gilded Age, the Vanderbilts and Astors, who took up residence, their swirling costume balls the stuff of legend. But even then, the park was never the exclusive province of the privileged. In the nineteenth century, hobos and streetwalkers were as common a sight as parading blue bloods.

A popular site for protest and rebellions, the park had also been a locus of free speech and civil disobedience since its first beginnings, a place where New Yorkers of all stripes came to air their grievances, or to flaunt their lack of inhibitions. Suffragettes and war veterans had marched its grounds; sober-faced bankers had picketed, militating for new currency standards. In the 1910s and '20s, bohemians and radicals had reveled there, settling into cold-water flats at its margins. Stonecutters and Irish laborers had rioted, red-faced and hurling stones.

These days, according to the *Village Voice*, the park was an outdoor stage for "the beats, the hips" and the folkies. Pete Seeger, Dave Van Ronk, and Woody Guthrie regularly performed there; Jack Kerouac and Ginsberg lounged on its benches when they weren't haunting the dive bars at its fringes, or ducking into some jazz joint to catch the riffs of Charlie Parker and Thelonious Monk. The fountain by night was used for conga lines and parties.

Washington Square had all the trappings of what was respectable and iconic in New York. Yet it also stood for rebellion against authority and the status quo—and always had. It was both street theater and respite from the drumming pressures of the city, an on-going carnival and a much-loved communal backyard. The park wasn't formal; it was comfortable and unpretentious. One didn't have to pay for a seat or the view. The green space belonged to everyone. It was emphatically democratic, which she liked.

The park, Jacobs knew, had been contested space before. Private interests had plotted periodically to make a piece of it their own. But residents had repeatedly fought these incursions with success. In 1898, when a law finally passed stating that Washington Square Park was to be used "in perpetuity for the public as a public park," the matter seemed settled: the park was never to become a developer's plaything. Given all this, Jacobs reflected, it certainly shouldn't be overrun by a highway now.

THE "COMMITTEE TO SAVE THE PARK" HAD ENCLOSED A LETTER TO SIGN and send back, but Jane decided to draft her own as well. Dated June 1, 1955, the same summer of Ginsberg's inaugural reading of "Howl," she penned a note to Mayor Robert Wagner and the Manhattan borough president, Hulan Jack. Then she signed her name to the committee's form letter, making note of the person spearheading the campaign.

Shirley Hayes, the woman behind the fight, was an energetic mother of four who lived on East Eleventh Street. An attractive blonde with an open smile and a starlet's wavy, chin-length hair, she was a whirlwind organizer: magnetic, outgoing, well-liked. A native of Chicago, Hayes had trained as an actress and a painter before buying a one-way ticket to New York with hopes of a Broadway career. Married now to an ad man, she and her husband had decided to raise their four sons in the Village, resisting the siren's song of the suburbs, just as the Jacobses had done.

For the last few years, Hayes had been watching the changes taking place near the park with growing dismay. The apartment towers going up at the foot of Fifth Avenue were ugly and oversized, and Moses's urban renewal project south of Washington Square portended more of the same. She worried that the quirky charm of the Village was being lost, its human scale dwarfed by faceless towers. The road through the park seemed the last straw: it would kill the special intimacy of the neighborhood for good.

The plan, in fact, had years of scheming behind it. In 1935, Moses had proposed lopping off the park's corners, ripping out the fountain, and rebuilding the space as a giant traffic circle. In a nod to the former park, a miserly swatch of garden would be planted at the center. Furious, the residents of the neighborhood rose up in revolt and Moses, never one to be crossed, responded with a quick counterpunch. If the Villagers refused to cooperate, they would get nothing, he allowed; he would hold back all funds for park mainte-nance and future improvements, which as city parks commissioner he had the power to do. The matter quietly simmered. Biding his time, in 1939 Moses returned again with almost the same plan, with the small addition of a lily pond at the center. One resident likened the look of Moses's new proposal to a "bathmat." The "bathmat" plan, as it came to be called, was again beaten back by Village activ-ists. Then, before Moses was able to make his next move, WWII broke out. The road plan was temporarily shelved, along with sev-eral other public works projects he was pushing.

But Moses was tenacious. After the war, shifting his attention to the area south of Washington Square Park, he met clandestinely with NYU officials to forge a new and even more sweeping redevel-opment plan, using the powers accorded him by Title I to smooth the way. And so, in 1952, the road scheme rose once again, this time as four lanes of blacktop that would bisect the park to serve the new towers rising to the south. Shirley Hayes and another concerned mother, Edith Lyons, decided it was time to organize.

Hayes was a dynamo, and her activism soon earned her a place on the Greenwich Village community planning board, where she began pushing for alternative plans for the park. A tireless coalition builder, she churned out letters of recruitment by the hundreds, pressing priests and rabbis to announce park meetings in their ser-mons. She reached out to reporters and circulated petitions, garnering thousands of signatures. She stationed volunteers on street corners to make traffic counts, determined to refute data compiled by Moses's "traffic engineers." She even sent out personal pleas to Robert Wagner himself, then New York's soon-to-be mayor.

Jacobs was still only peripherally involved at this point. But she was watching Hayes with interest, impressed with her political acumen, her efforts at outreach, her flair as a strategist. Hayes urged Villagers to stand firm in their resistance. She insisted that the park should be closed to all car traffic permanently. There could be no deals or compromises, she argued. A smaller, less harmful road would spell defeat; once a road was there, it could never be dialed back.

The Villagers had seemed to be making headway in 1952, when the road plan was withdrawn for further study. But by 1955, when Jacobs entered the fray, they were losing ground again. Moses was feeling testy and impatient. His redevelopment scheme south of the park was finally a go, but he needed to get past this last bit of resistance. More determined than ever to see his dream-road realized, he offered Villagers what he thought was a major compromise: the four-lane roadway would be submerged, he now promised, with a pedestrian overpass built over it. The idea had worked near Grand Central Station, he argued; it would work in Washington Square Park. The Villagers were vehemently against it. Moses, they claimed, had made no concession whatsoever. The submerged-roadway plan was unacceptable, little better than the original surface roadway proposal. And so the war of ideas drummed on.

By 1957, Shirley Hayes's letter-writing campaign had become a well-oiled machine. Every morning, officials were met with a barrage of new anti-roadway missives, all of them vocal in their opposition to any and all traffic through the park. Manhattan's borough president, anxious to find some middle ground, decided to withdraw his support for the Moses plan, proposing a smaller, two-lane road through the park instead. Moses was livid, rejecting the idea outright. The defection by one of his own was infuriating. The roadway would be four lanes, with a strip of trees in the middle, partially submerged "if necessary," but there would be no further compromises, he announced.

The Villagers were feeling weary and pushed to the wall. Six long years had passed, and the battle for Washington Square Park still raged. Though the neighborhood had managed to stall the road, Moses was still pressing hard, unwilling to give up. The Villagers sensed that they needed a new strategy, fresh blood. And then, as if in answer to their prayers, a new player entered the fray in 1958, and the picture began to shift. Raymond Rubinow was an eccentric economist and political gadfly whose address was not, in fact, Greenwich Village, but Gramercy Park, some fifteen blocks north of the contested parkland. He came by his populist leanings both by instinct and family tradition. (The concept of Social Security was said to be his father's brainchild.) Passionate about protecting New York's historic buildings and neighborhoods, in 1955 he had joined the J. M. Kaplan Fund, a philanthropic concern dedicated to preserving the city's heritage, including the endangered Carnegie Hall. A friend of Jacobs's, he was unafraid of a messy fight. With Hayes's blessings, he took over the coalition she'd built. And one of the first things he did was to persuade Jacobs to take a more active role, both as a strategist and a bridge to the press, knowing that she had an insider's knowledge of how it worked. Jacobs agreed to sign on despite still holding a full-time job.

Their first move was to tighten the battle's focus. To this end, they decided to rename the group "The Joint Emergency Committee to Close Washington Square Park to Traffic."

"We weren't trying to embrace all kinds of points of views about the Village, all kinds of political groups, all kinds of anything. We were trying to collect and concentrate on this issue, the people who felt as we did on that issue," Jacobs later explained. "In order to dramatize . . . and clarify this, a name like that was necessary." The point wasn't to get into ideology, which inevitably divided people; the point was to concentrate the focus. "People knew what they were getting into . . . [We joined] people who believed in a particular thing, and might disagree enormously on other things."

Though Hayes had assembled a large and eclectic coalition of Villagers, Rubinow and Jacobs now upped the ante, adding national figures to their ranks. They asked Eleanor Roosevelt and the anthropologist Margaret Mead, both of whom lived in the Village, to join the committee, and the publisher of the new downtown paper the *Village Voice*. Jacobs pressed her new friend, William Whyte, author of the recent bestseller *The Organization Man*, into service, and they enlisted Norman Vincent Peale, a prominent pastor.

Jacobs believed their best hope was also to limit the scope of the discussion, to concentrate their energies on saving the park, and not debate Moses's plans for Washington Square Village, which was where he would want to take the battle, knowing he had powerful developers behind him. The Villagers must insist that Washington Square was a park, and that a park was an inappropriate place for a highway. Whatever contiguous development was going on, the park should remain a park, and no vehicles of any kind should be permitted. Following Hayes's lead, "there should be no negotiation" and no trade-offs, she argued, no caving to a smaller, less harmful roadway. Once a road was built, it could and would be widened. Only blocking the Washington Square roadway outright would achieve their goal.

Their next move was to seek high-profile press advocacy, and for this, Jacobs approached Lewis Mumford, the architectural critic for *The New Yorker*, who had kindly offered his support after her slashing Harvard speech on urban design.

Some years earlier, Mumford had called Moses's plan for redesigning Washington Square "absurd" and "a process of mere sausage grinding." Now, at Jacobs's urging, he supplied the committee with a statement they could use as a press release. "The attack on Washington Square by the Parks Department is a piece of unqualified vandalism," Mumford declared. "The real reason for putting through this callow traffic plan has been admitted by Mr. Moses himself: it is to give the commercial benefit of the name 'Fifth Avenue' to the group of property owners who are rehabilitating the area south of Washington Square, largely at public expense. The cause itself is unworthy and the method used by Mr. Moses is extravagant. To satisfy a group of realtors and investors, he is as ready to change the character of Fifth Avenue as he is to further deface and degrade Washington Square."

Washington Square, he added, "has a claim to our historic respect . . . It was originally used as a potter's field for paupers; it might now prove to be a good place to bury Mr. Moses' poverty-stricken and moribund ideas on city planning."

Mumford's claim that the roadway was a giveaway to real estate developers resonated. Within days, other influential friends of the park committee were expressing similar misgivings. Eleanor Roosevelt used her "My Day" column in the *New York Post* to speak out against the plan: "I consider it would be far better to close the square to traffic and make people drive around it . . . than to accept the reasons given by Robert Moses . . . to ruin the atmosphere of the square." Norman Vincent Peale, calling Washington Square "an island of quietness in this hectic city," stressed the importance of the park as a sanctuary. "Little parks and squares," he added, "especially those possessing a holdover of the flavor and charm of the past, are good for the nerves, and perhaps for the soul." But the most dramatic statement came from Charles Abrams, a celebrated Columbia professor. "Rebellion is brewing in America," he declared at a packed Village meeting in July 1958. "The American city is the battleground for the preservation of [economic and cultural] diversity and Greenwich Village should be its Bunker Hill . . ."

The support from people in high places was proving helpful. Public enthusiasm for the roadway seemed to be eroding. But Jacobs knew that Moses was a ruthless foe, a cunning, bare-knuckled fighter, and a splendid propagandist. He had a well-documented history of crushing his opponents, and he possessed enormous official power, which he had never been shy about wielding. The emergency committee, she cautioned, shouldn't assume anything. They needed to step up their outreach and pull in more allies, press local officials harder, stage more anti-roadway events to attract media attention. It would be folly to believe they were gaining the upper hand.

There was an election coming up, and those politicians already in power, while nominally sympathetic to the Villagers, seemed unable to stand up to Moses. The Villagers needed someone politically connected, a prominent insider as an advocate, a figure who was both powerful and media savvy. One evening, as Jane prepared for bed, Bob Jacobs suggested that they link closing the park to traffic to the approaching election. The state assemblyman, Bill Passannante, was in a close race against a Republican challenger who had come out publicly against the roadway through the park. If Passannante could be persuaded to go even further, by insisting that barriers be installed at the park's perimeter, to prevent all but bus and emergency vehicles, the committee would throw its substantial voting bloc his way; if he declined, those votes would fall to his opponent. Passannante, doing the numbers, quickly agreed, becoming the first elected official to support banning the park to all vehicular traffic. Soon after, another young politician, John Lindsay, who at the time was running for Congress, came out against the roadway too, followed immediately by his opponent.

Strategically, they were gaining ground, winning new political allies and influential friends. But Jacobs was convinced that to keep the issue front and center in the public eye, to exert real pressure on city hall, they needed still more press attention, a nonstop blitz of media coverage that would keep officials' feet to the fire. A journalist herself, this was a realm in which her instincts were especially sure.

The newspaper business in New York had always been a blood sport. But at that moment it was especially unsparing. Reporters regularly vied for stories, terrified of being scooped—or worse still, denied access to the crisp-suited men working the levers of power. On this score there was no one more vindictive than Robert Moses, who made a pariah of any reporter he deemed disloyal, barring him—and generally it was a him—from the hallowed inner sanctum where news was made and stories strategically leaked. Aware of these fears, Jacobs used them to her advantage.

Rather than courting the *New York Times*, which followed the controversy, but always seemed to build their coverage around quotes from Moses, she reached out to the downtown press, most particularly the *Village Voice*, the scrappy new paper that had been founded in 1955 by Dan Wolf, Ed Fancher, and the novelist Norman Mailer. The *Voice* was known for its arch commentary on city politics and the downtown arts scene. It had a distinctly sardonic edge and in general devalued authority, priding itself on its opposition to "borrowed ways of thinking." (At one point every member of the *Voice* sales staff was a poet.) One of the first places to champion stories written unapologetically from a subjective point of view, what would soon be called "New Journalism," the fledgling paper looked for writers whose prose led with their personalities. Irreverent, highly opinionated, and unafraid of controversy, it regularly ran hard-hitting stories about neighborhood issues, privileging the voices of ordinary citizens rather than the power elite.

Jacobs was easy to like, and was soon close friends with several leading *Voice* reporters, as well as its editor, Dan Wolf. She quickly made a habit of passing along insider information whenever she could, and the favors were returned in kind. Articles in the *Voice* regularly gave space to the Villagers' indignation over the city's disregard for their community. And Wolf himself often weighed in with editorials: "It is our view that any serious tampering with Washington Square Park will mark the beginning of the end of Greenwich Village as a community," he wrote. "Washington Square Park is a symbol of unity in diversity. Within a block of the arch are luxury apartments, cold-water flats, nineteenth-century mansions, a university, and a nest of small businesses. It brings together Villagers of enormously varied tastes and backgrounds. At best, it helps people appreciate the wonderful complexity of New York. At worst, it reminds them of the distance they have to cover in their relations with other people." After a Moses underling complained that the "awful bunch of artists" blocking progress in the Village were "a nuisance,"

Wolf responded accordingly: he hoped, "there are thousands of nuisances like that within a stone's throw of this office."

Jacobs had been studying earlier citizen skirmishes with Moses, searching for points of vulnerability. Few groups ever seemed to get the upper hand, she noted. But in 1956, after Moses tried to snatch a piece of Central Park, hoping to expand the parking lot of Tavern on the Green at Sixty-Seventh Street, a group of mothers had prevailed. Planting their strollers in the path of oncoming bulldozers, they bravely stood their ground, and the photograph in the next day's papers of the "little soldier"—a tiny wisp of a girl holding back a team of construction workers—worked its magic: the park standoff became a cause célèbre and Moses was forced to capitulate.

Using children to make the argument made sense, Jacobs reasoned. It was the young people, after all, who stood to lose the most if the park road went through. Soon Jacobs was dispatching kids—dozens of "little elves," as she called them—as foot soldiers in the fight. She set them up at card tables to collect signatures in the park; she sent them to put up posters, and to knock on neighbors' doors. Deploying children to get out the message, she saw, all but guaranteed press attention. Kids were like catnip to newspaper photographers. Too delicious to ignore, they presented a nearly irresistible photo op. It was a tactic Jacobs would return to again and again, one of the surest arrows in her quiver—not only for this fight, but for others to come.

One afternoon, while at Macy's shopping for long underwear for her sons, the saleslady, eyeing the two boys, asked if it was for hunting or fishing. "It's for picketing," Jacobs said, not batting an eye.

"She would bring the three children to the square on weekends to collect signatures on petitions demanding that the highway plan be canceled and the park permanently closed to traffic," remembered her son Ned, who in the spring of 1958 was seven. "This was during the beatnik era, and my brother and I were outfitted with little sandwich boards that proclaimed, 'Save the Square!' That always got a good laugh because people knew that 'squares' would never be an endangered species—even in the Village. These were also the dying days of McCarthyism. People were afraid—even in the Village—to sign petitions for fear they'd get on some list that would cost them their careers. But I would go up to them and ask, 'Will you help save our park?' Their hearts would melt, and they would sign. Years later, Jane recalled that we children always collected the most signatures."

The pressure was beginning to pay dividends. On June 25, 1958, in deference to the residents' continuing opposition, the city agreed to close Washington Square Park to traffic on a temporary basis while the roadway issue was studied further. The following day, a photo of Mary Jacobs, three and a half, and Bonnie Redlich, four, appeared in the *New York Daily Mirror*. They were holding up a ribbon that had been tied to represent a "reverse ribbon cutting." The caption below the shot read: "Fit To Be Tied."

Moses immediately lashed out, predicting that the area would be choked with traffic, utterly impassible. "There is something to be said," he sneered, "for letting unreasonable opposition have its way; find out by experience that it doesn't work."

The critical turning point finally came that fall, in the heat of campaign season. The emergency committee decided to approach Carmine De Sapio, an old-style party pol who also happened to live in Greenwich Village. De Sapio wasn't an obvious ally. But if De Sapio could be persuaded to oppose the road, they knew he would be hard to override. An emissary was sent to plead the Villagers' cause, and was rewarded with success. A plan was put in place and within days, De Sapio was calling for a Board of Estimate hearing, announcing that he intended to present "some rare public testimony." On the appointed day, a throng of supporters assembled at city hall wearing green "Save the Square" buttons and waving parasols on which was printed "Parks Are for People." A Village resident presented De Sapio with a scroll signed by thirty thousand people opposed to the road plan. Press cameras flared and De Sapio, wearing a crisp suit and his signature dark glasses, stepped up to speak. "Washington Square Park," he roared, "is one of the city's most priceless possessions and as such it belongs to every one of our 8,000,000 fellow

New Yorkers . . . To change the character of this beloved central symbol of the Village would be, ultimately, to eradicate the essential character of this unique community." The crowd erupted in a storm of clapping, pierced with cries of support.

Moses, learning of the speech, knew he had been outmaneuvered; De Sapio's political clout was legendary. A month later, to Moses's chagrin, the Board of Estimate closed the park to all but emergency vehicles and the bus carriageway. Speaking before the Board of Estimate that fall, Moses was still livid. "There is nobody against this," he sputtered. "Nobody, nobody, nobody but a bunch of mothers." Jacobs stood amidst the crowd, watching the spectacle.

THE VICTORY PARTY TOOK PLACE ON THE FIRST SATURDAY IN NOVEMBER 1958, at the foot of the Washington Square Arch. The revelers, out in full force, began assembling in late morning. Many were carrying their pink "Parks Are for People" parasols again, or wearing green "Save the Square" buttons. The children, "Square Warriors," as Jacobs called them, clutched balloons. Well-wishers from other neighborhoods joined the festivities. To memorialize the moment in the press, the emergency committee staged a symbolic ribbon tying—a send-up of the usual ribbon cutting. De Sapio, Hulan Jack, Bill Passannante, and Rubinow stood before the flashing cameras, jubilant and grinning as they held the green ribbon. Reporters jotted notes. Jacobs, charmingly self-effacing, hung back, avoiding the limelight, as was her habit. Letters of congratulations from New York's governor Averell Harriman, Mayor Wagner, and Lewis Mumford were read out loud. Then at noon, to cap the festivities, a resident of West Eleventh Street drove a dented minibus draped with a banner that read "Last Car Through the Park" under the arch and out onto Fifth Avenue. The gleeful crowd cheered.

Seven months later, the Villagers celebrated the temporary park closing again, this time with a costume ball that drew more than a thousand partygoers, with politicians, newspapermen, and hordes of local artists in attendance. At midnight, a life-size cardboard car was set on fire, drawing whoops and dancing from revelers momentarily illumined by the flames. In true Village fashion, the merry-making continued until morning.

In the weeks and months that followed, the traffic snarls Moses predicted didn't appear. The network of nearby streets gave drivers other options, actually *reducing* the traffic flow, rather than increasing it. Though it seemed counterintuitive, congestion in the city, Jacobs would observe, often *deterred* car traffic. Faced with knots of traffic, people tended to opt for walking, bikes, or the subway instead. Whereas as soon as a new highway was built, it filled up with cars.

Though it would take several more years before buses were finally barred from using the turnabout near Washington Square Arch, and a few more after that until *all* motorized vehicles were permanently banned, with continued grassroots pressure, the park was finally closed to all vehicular traffic in November 1963, a few weeks before the assassination of JFK. The "bunch of mothers" had prevailed, becoming a potent example of what community activism could achieve.

For Jacobs, the triumph was doubly sweet. That same spring of 1958, just as the Washington Square Park battle was peaking, her hard-hitting article for *Fortune*, "Downtown Is for People," appeared, laying out her critique of urban renewal projects across America. The interest it generated led to its inclusion in an anthology called *The Exploding Metropolis*, published by Anchor Books and coedited by Nathan Glazer, the eminent Harvard sociologist, and Jason Epstein, who went on to found *The New York Review of Books*. For Jacobs it would be a turning point. Soon after, she was invited to a conference on urban-design criticism hosted by the Rockefeller Foundation. Like the Harvard symposium two years earlier, it was another all-male convocation of the country's leading design theorists, exclusively insiders. But this time Jacobs was more relaxed. During a break in the program, she chatted easily with Chadbourne Gilpatric of the Rockefeller Foundation, who followed up with a phone call shortly after. He wondered if she had a larger book project in mind, perhaps based on "Downtown Is for People." Jacobs did.

"There seem to be two dominant . . . mental images of the city," she wrote Gilpatric that summer. "One is the image of the city in trouble," the metropolis as an inhuman mass of chaos and random growth. The other is of the "rebuilt city," a planned vista of projects and green spaces, where chaos is replaced with perfect, regimented order. Both narratives are "disastrously superficial," she said. What I would like to do, she added, is to create "another image of the city, not drawn from mine or anyone else's imagination or wishes," but from "real life."

She went on to elaborate. What she hoped to show the reader was how to see the city in a more holistic way: as a dynamic web of relationships, a system of astonishingly intricate and complicated interconnections. "Where it works at all well," she explained, "this network of relationships" requires a "staggering diversity of activities and people, very intimately interlocked . . . and able to make constant adjustments to needs and circumstances." She would begin, she added, with the city itself, walking her readers through the physical streets, just as Kirk had done with her.

Sufficiently intrigued, Gilpatric awarded her a $2,000 grant, which was enough to get the project started. This support was shortly followed by more: Jason Epstein, who by now had moved from Anchor Books to Random House, managed to convince his colleagues that a book by Jane Jacobs about cities would be popular—this despite the fact that she was still an unknown author. "You sort of fell in love with Jane when you met her," Epstein later remembered. "She was exuberant, original, strong-minded—and a very kind woman." Epstein sent Jacobs a $1,500 advance, which arrived that November, right on the heels of the ribbon-tying ceremony at Washington Square Park, a nice bit of synchronicity.

All that fall, on leave from her job at *Architectural Forum*, Jacobs sat at her desk on the second floor of 555 Hudson Street, chipping away at the book-to-be. She was finding it lonely work, harder than she anticipated.

By now, she had toured much of the country, looking at dysfunctional redevelopment projects in cities across America: St. Louis's much-touted Pruitt-Igoe project, based on Le Corbusier's utopian vision, and already a case study in vandalism and crime; Boston's once vibrant West End, where an Italian working-class neighborhood was being razed for high-rise residential towers; Chicago's Robert Taylor housing project, another moribund planning scheme despised by its occupants. She knew all too well what didn't work: monolithic housing towers set in anonymous, windswept plazas; fortresslike civic centers shunned by everyone but bums; planned commercial districts that were pale imitations of standardized suburban shopping malls. This was not the "rebuilding of cities," she wrote; it was the "sacking of cities."

What she needed was a counterpoint to these flawed projects, an example of a thriving city neighborhood that worked on every level. She knew what she was looking for: short blocks and a blend of new construction and old; a good concentration of people on the street at different times, for different reasons; buildings of varying sizes and a diversity of uses—residences interspersed with shops and restaurants, bars and coffee shops with commercial trades. What she still hadn't found, however, was some real-life illustration of *why* this blend of elements was so important, why the casual social life of an urban sidewalk was so essential to a neighborhood's life and safety.

And then one evening a real-life illustration found her. It was the depths of winter when she heard it: the wail of a bagpipe one February night. Jacobs padded over to the window and raised the blinds. There on the sidewalk was "a little man in a plain brown overcoat," squeezing out a Highland tune on his pipes. Swiftly, quietly, a little crowd was suddenly gathered, "a crowd that evolved into a circle with a Highland fling inside it." Jacobs could see the dancers on the shadowy sidewalk, but the bagpiper himself was nearly invisible, his presence now solely the music. When he finished, there was applause from the dancers and watchers, joined by clapping from a dozen raised windows and open doorways on Hudson Street. And then the little man vanished as quickly as he'd appeared. The windows closed and the little crowd dispersed, absorbed again into the night street.

This was the sort of event she had been searching for, an illustration of what made a well-used city neighborhood like hers so different from a superblock of anonymous towers, where, in the "blind-eyed" spaces between buildings, there were no witnesses to what was going on, no casual public life to act as a support. The scene on her block had been spontaneous and unscripted, residents and strangers drawn together in a moment of serendipity in which people felt safe because no one was alone; the presence of a crowd had protected everyone. "Eyes on the street" was the term she would coin for this phenomenon. It was an idea that would play a leading role in her book. That night no one had felt in danger because there were so many different "eyes on the street."

JACOBS CONTINUED TO DRAW STORIES FROM JUST OUTSIDE HER DOOR-step as she burrowed deeper into the book. She watched the daily rituals on her block with increasing fascination: the ebb and flow of arrivals and departures, the changing choreography throughout the day. Morning and the clatter of garbage cans, the barber unfolding his sidewalk chair, the deli man stacking crates, the locksmith sweeping his shop. Next the primary school children dribbling through; then elegant men and women, briefcases in hand, emerging from doorways, striding toward the subway, hovering on curbs, poised to flag down cabs. Midday, and a different set of characters appeared: longshoremen slipping into the White Horse Tavern for beer and talk, "business lunchers" thronging the local eateries, meat market workers headed for the bakery lunch room. These were followed shortly by the after-school crowd: kids on roller skates, toddlers on tricycles, women pushing strollers; women returning from work or running errands, zigzagging from the butcher's to the fruit man. Dusk then, and another tableau altogether: beautiful girls getting out of MGs, motor-scooter riders with beards and long-haired girlfriends, teenagers primping before shop windows. And at last the night shift. As streetlights blinked on and bars came alive, a new set of arrivals again: well-dressed couples departing for the theater, musicians ducking into basement clubs, night workers stopping to pick up salami at the deli.

Though things settled down for the evening, the life of the street never stopped. Even deep into the night, Jacobs observed, there were people passing through. Sometimes, waking after midnight to tend a child, she sat in the dark, listening to the sounds from the sidewalk: pattering snatches of party conversation, singing at 3 A.M., the bellows of a passing drunk.

The life of her block was like an ongoing dance, Jacobs decided, alighting on an image that would be quoted frequently throughout the years. Every day, this little stretch of Hudson Street, she observed, "was the scene of an intricate sidewalk ballet, in which the individual dancers and ensembles all have distinctive parts which miraculously reinforce each other and pose an orderly whole." The ballet of a good city sidewalk, she added, "never repeats itself from place to place and in any one place is always replete with new improvisations." For all its seeming disorder, the block where she lived, peopled as it was by a changing guard of workers, strangers, and neighbors, had a fundamental order all its own, a marvelous capacity for maintaining the safety and freedom of the street.

One night a boy fell through a plate glass window and a stranger emerged from a bar and swiftly applied a tourniquet. A woman sitting on the steps ran over to the bus stop, snatched a dime from the hand of a waiting stranger, and raced to the phone booth to call an ambulance. Nobody remembered having seen the man or the woman before or since.

Another day, a struggle ensued between a man and a little girl of eight or nine. The man seemed to be trying to get the girl to go with him. The girl was making herself rigid, as children do when they resist. As Jacobs watched from her window, considering how to intervene should the drama escalate, she saw it wouldn't be necessary. From the butcher shop emerged a woman who ran the shop;

she was standing within earshot of the man, a look of determination on her face. At about the same moment, a man from the deli materialized and stood stolidly to the other side. Several heads poked out from the windows of a tenement, and a moment later the owner of one of the heads appeared in the doorway, behind the man and the girl. The man did not know it, but he was surrounded. "Nobody was going to allow a little girl to be dragged off, even if nobody knew who she was," Jacobs observed.

The strangers on Hudson Street were "different people from one day to the next." They used the block at different times and in different ways. But it was their eyes, along with those of her neighbors, that made her block feel safe and comfortable, Jacobs reflected. "Impersonal city streets make anonymous people," she wrote. "Lowly, unpurposeful and random as they may appear, sidewalk contacts are the small change from which a city's wealth of public life" grew. Jacobs had found the archetype of a successful urban neighborhood just outside her doorstep. Hudson Street possessed just the sort of diverse, fine-grained street life she was after.

IN THE WEEKS AND MONTHS THAT FOLLOWED, JACOBS STEADILY ADDED to these ideas. She studied her own city and others like an anthropologist fresh to the field, watching to see if there were analogous patterns of behavior from one place to the next, continuously asked questions. What made a successful park? What was a slum? What was the role of the sidewalk, and how did it work? How did time factor into a neighborhood's stability? What made one block flourish while another went dead?

True to her promise to the Rockefeller Foundation, she walked her readers through unloved, unused parks eaten with decay, and those where life whirled, down lively city sidewalks that pulsed with human interest and others edged with menace. She visited low- and middle-income housing projects that were "marvels of dullness," and vibrant immigrant neighborhoods that teemed with life; she toured bustling cultural hubs like New York's Carnegie Hall, well-woven into the urban fabric, and civic centers severed ruthlessly from the larger city by snarls of arterial highways. "Project prairies," she called these failed places. Again and again, she contrasted scenes from cities that worked with those that were moribund, poking holes in orthodox planning theory every step of the way.

A city cannot be a work of art, Jacobs wrote. To treat a city as a problem capable of being ordered and arranged was "to make the mistake of attempting to substitute art for life." Cities in real life were perpetually in flux. They were vibrant living systems, immense laboratories of "trial and error, success and failure." Planners had it wrong, she insisted. "Incurious about the reasons for unexpected success," guided by notions derived from everything "but cities themselves," they had neglected the value of simple observation; they had failed to look at life itself.

If you moved people from tenements into high-rise housing towers, crime would fall, they said. If you built highways, traffic would ebb. If you created malls of culture, the arts would flourish. If you separated work, play, residence, industry, and retail, the city would grow; if you thinned out high-density neighborhoods, they would thrive. If you discouraged children from playing on city sidewalks, urging them instead into parks, they would be safer.

No, Jacobs demonstrated, again and again. The opposite was true; crime increased in high-rise housing projects, isolated islands of culture became urban wastelands, traffic worsened with more and wider roads. The economy of a neighborhood stagnated when work, living, and entertainment were separated. Thinning out a neighborhood didn't stem decay. Children shooed from city sidewalks into underused parks were in *more*, not less, danger.

Conventional wisdom at the time was that densely settled old neighborhoods were a problem, a red flag indicating unhealthy conditions. Jacobs thought otherwise. Population density in a neighborhood was a sign of vitality and health, she wrote, a quality to cultivate, not discourage. And density with *diversity* was even better. Neighborhoods that drew together lots of people with different tastes, backgrounds, skills, needs, and opinions were incubators of new enterprises; they grew a city's economy. Similarly, it was folly to believe, as most planners did, that sorting cities into neat little zones—residential here, commerce there, industry somewhere else—improved cities. This was nonsense, Jacobs said, "dangerous nonsense." Neighborhoods thrived when work, culture, and residences were tossed together. "Intricate minglings of different uses in a city are not a form of chaos," she wrote. "On the contrary, they represent a complex and highly developed form of order." Natural ecosystems, and those created by humans, "have common fundamental principles," she observed. "Both require diversity to sustain themselves."

Planners were equally deluded about parks and playgrounds, Jacobs contended. "Ask a houser how his planned neighborhood improves on the old city and he will cite, as a self-evident virtue, More Open Space." More open space for what? she asked. "For muggings? For bleak vacuums between buildings?" City parks were not automatic repositories of virtue or uplift, any more than sidewalks were. The success of a park depended on the number and sequence of users passing through it at different times of the day. If a park was empty for large tracts of time, then it became unsafe. "In defective city neighborhoods, shooing children into empty parks and playgrounds is worse than useless. That it is done in the name of vaporous fantasies about city childcare is as bitter as an irony can get," Jacobs wrote. In underused parks and project playgrounds, kids often had to "run a gauntlet of bullies." To think this was an improvement was "pure daydreaming."

Children playing on lively urban sidewalks, even slum sidewalks, Jane argued, were actually safer. If a child was threatened, a storekeeper could intervene. The ratio of adult eyes to children was high. Besides safety, vibrant city sidewalks also presented other advantages. They gave children an occasion for "unspecialized" play, invented games and imaginary spaces, a break from organized activities: time, in other words, to drift. "I know Greenwich Village like my own hand," Jacobs's younger son told her, and then described his secret hiding places along the street, including a crack between two buildings where he hid his "treasures." Busy sidewalks were places for children to form notions about the world, to mingle with people of all ages and walks of life, people, in other words, beyond their mothers or other female caregivers. If children were to escape the "matriarchy" that planners seemed to have in mind for them, Jacobs asserted, they needed to see men in the course of their daytime life, men like those who worked on or near Hudson Street. Putting in an "occasional playground appearance" as a substitute mom wasn't the same.

The ethos at that moment, of course, was for large-scale interventions, supplanting whole districts of old buildings with new structures. Jacobs again disagreed: Well-worn buildings that varied in age and condition served a critical function. A healthy, growing neighborhood needed to have a good proportion of old structures, not because they were inherently better, but because they helped to spark new economic activity. In a neighborhood with only *new* buildings, the sole enterprises that could exist there were those that could support the high cost of new construction, as reflected in steep rents. "Well-established, high-turnover, standardized" businesses—banks, box stores, and chain restaurants—could survive in new buildings. But the many hundreds of ordinary enterprises—the neighborhood bar, the butcher, the deli—places "necessary to the safety and public life of streets" and "appreciated for their convenience" needed the lower rents of old buildings to survive. As for new ideas and initiatives of any kind—small, untried, still-tentative ventures, no matter how ultimately profitable—"there is no leeway for such chancy trial, error and experimentation" in the high-overhead economy of new construction. "Old ideas can sometimes use new buildings," Jacobs wrote. "New ideas must use old buildings."

The economics of time, Jacobs cautioned, worked by decades, even generations, not short-term corporate models. In a thriving district, decrepit older buildings were replaced each year by new ones, or rehabilitated to an equivalent degree of finish, so that over the years, there remained a constant mix of buildings of many ages and types. Once-new buildings gradually became old ones, in a perpetual cycle of change and regeneration, she said. "Time makes the high building costs of one generation the bargains of a following generation." Time made certain structures obsolete, but also allowed for "ingenious adaptations of old quarters to new uses": the warehouse that became a living loft, the "town-house parlor that became a craftsman showroom," the garage that became a theater. Jacobs's concern, like Carson's, was ultimately *sustainability*. For her, this meant "things that are inherently organic, rooted to their environments in such a way—through use and contribution—that they will naturally endure and evolve through adaptation," notes Mary Rowe, project coordinator for *Ideas That Matter*. Historic preservation and adaptive reuse were not the *only* ways a city grew and thrived, of course, but such natural processes needed to be respected, as they were fundamental to sustaining the larger entity.

JACOBS WORKED ON. SPRING GAVE WAY TO SUMMER AND THEN TO FALL. A year passed. Then, another. Slowly, mindfully, she shaped her argument, writing in layers, collecting stories, continuously measuring her conclusions against the city itself, until at last the book was done. It was January 1961, and she had titled it *The Death and Life of Great American Cities*. She sent it to Random House.

It felt good to decompress. Now, with the manuscript in her editor's hands, Jacobs could allow herself the luxury of sleeping in on occasion. It was a bitter February, snowy and raw. She was glad she still had a little time before her scheduled return to *Architectural Forum*, glad to be indoors, cozy in her own home, reading at her kitchen table. Normally, she got down to work right away. But on this particular morning, feeling lazy, she let herself linger a bit longer over the *New York Times*. Which was when she saw it, buried at the back of the paper: "Two Blighted Downtown Areas Are Chosen for Urban Renewal." Jacobs stopped and read the headline again, squinting this time at the accompanying map. One of the two "blighted" areas was on the Lower East Side near Tompkins Square Park; the other spanned fourteen blocks of the West Village, including, she now saw, her own. She frowned and read on. The area where she lived was described by the city as a run-down industrial zone with beat-up tenement buildings housing at best six hundred families. The implication seemed to be that such an insignificant number of households would be easy to relocate. Before moving forward, the article said, the city would spend $350,000 to conduct a study to determine if the two neighborhoods were eligible for urban renewal.

Jacobs stared at the article in disbelief. Her house and her friendly, easygoing neighborhood, the lively place she had just described so glowingly in her book, now stood to be destroyed by the same "inane, anti-city forces" she had spent the last two years picking apart. She studied the map again. It was an affront to everything she knew about cities, everything she loved about her neighborhood, about as "neat a case study," she said, of the "intellectual idiocies" driving urban renewal policies these days.

It was also underhanded. To characterize what was going on as a "routine study" was disingenuous, deeply cynical on the city's part. From all the reporting she had done, she knew that the outcome was already a foregone conclusion: these studies always ended with a blight designation. She could even guess at the scope of the project. The money granted for the study was always a given percentage of the cost of the whole project. The price tag for this study was "just what you would expect for wiping out . . . fourteen blocks and putting in high rent apartments."

Jacobs had seen the cycle again and again. The city would single out an area and designate it a slum. Then landlords, anticipating the wrecking ball, stopped repairing their buildings, unwilling to throw good money after bad. Homeowners who *could* sell their buildings *did*, always at sharp discounts to what values had been. And so the slum designation soon became a self-fulfilling prophecy: the targeted neighborhood swiftly went downhill. After that things moved liked clockwork. Developers swooped in like vultures, picking up the depressed properties at bargain-basement rates. Soon after, the city issued eviction notices to households still living in the redevelopment zone, giving them ninety days to leave. Three months later, the bulldozers rolled in, leveling the neighborhood.

The entire dishonest affair was especially vexing given the promises Robert Wagner had made in his recent mayoral run. Mindful of the political fallout over Robert Moses's heavy-handed slum-clearance policies, Wagner had pledged that urban renewal policies going forward would be more open and transparent. Appeals for historic preservation would be considered, and better relocation policies put in place. The public would be given a legitimate voice in the planning process. But apparently Wagner didn't mean it. Despite recent scandals over brazen financial improprieties in several of the city's high-profile renewal projects, along with widespread public outrage over how officials had handled residents' pleas to save their brownstones in the run-up to construction at Lincoln Center, nothing had changed. Nothing except the new name the city was giving to the program, which was now being called "community renewal." As if semantics fooled anyone, Jacobs thought. It was still the same routine: urban renewal couched in softer language.

Even the "new" chain of command was a sham. While Wagner had replaced Moses with two new lieutenants, James Felt and J. Clarence Davies Jr., both hand-picked to carry out the city's ambitious redevelopment plans, Moses still stood in the wings, a puppeteer poised to pull strings. The truth was, neither Felt nor Davies represented any change in philosophy. Privately, Felt was known as a Moses protégé. A slight, balding man given to quoting from the Talmud, he embraced the Moses model with a missionary's zeal: thin out the urban grid, expand roads, raze districts that didn't have a well-defined "single function." Davies, while more charismatic than Felt, was little better. Though active in charity circles and putatively a member of the Landmarks Preservation Commission, he had been a key player in developing Lincoln Center. Both hailed from prominent New York real estate families. Both had cut their teeth on big-city development projects. Neither had ever displayed any real interest in citizen participation, which of course always interfered with a developer's time line.

As far as Jacobs could see, it was hardly "the new era for urban renewal" that Wagner was claiming. It was business as usual. There had been no public hearing to debate the idea of this West Village blight study, no promised citizen participation. The city had drawn up plans behind closed doors. And now it was trying to rush them through before neighborhood opposition could be raised.

Jacobs put down the paper. She wondered if there wasn't even some payback involved.

In 1960, after Moses surrendered his post as chairman of the Committee on Slum Clearance to become chief executive of the 1964 World's Fair, he had bequeathed two files to Felt and Davies: one of urban renewal projects still in progress, the other a dossier of new neighborhoods to consider for future clearance. Moses had never been a man to cross, and losing the battle for Washington Square Park had been publicly humiliating. It wasn't far-fetched to imagine him leaving his successors with a few last thoughts about the next neighborhood to clear. How convenient to have suggested the West Village, the site of his waterloo, as the place to target.

Jacobs glanced out her window. She had hoped to get back to magazine writing as soon as the book was done. But the threat of a blight designation in her own neighborhood had to be fought. She picked up the telephone and made calls for the rest of the morning. En route to the grocer's that afternoon, she stopped in to alert Mr. Lacey, the locksmith, and Mr. Slube, at the cigar store; she warned the mothers she saw at the park, and the neighbors she passed going home. By nightfall, more than two dozen local residents had come together to form the "Committee to Save the West Village," with Jacobs and a dentist from the neighborhood as co-chairs.

Their first order of business was to draw up a petition demanding a one-month delay in the blight designation. The idea was to buy enough time to blunt the city's claim. The committee fanned out to collect signatures, beginning that very night. Bob Jacobs swung by the gritty local bar favored by the Irish-Catholic longshoremen; Jane hit the White Horse Tavern. Block by block, over the next several days, they canvassed the district.

Their next move was to appear unannounced at a Board of Estimate hearing two days later, catching the commissioners off guard, as citizens rarely if ever attended such meetings. Why had the blight-study proposal been put in place? the Villagers demanded. A Davies deputy stood up and assured the group that there was no cause for worry. The city was simply following through on a "routine request" for a survey, he said. There were no plans in place. The Villagers pushed back. If this was correct, then why had the *Times*, the paper of record, run an article announcing "a project" on the docket for the West Village?

Anticipating just such a denial, Jane now held up a telegram from Representative John V. Lindsay to Mayor Wagner, charging that the Villagers had not been given adequate notice of the city's intentions. It wasn't just a few upstart residents who were objecting to the city's backroom dealings, she said.

A FEW DAYS LATER, MORE THAN THREE HUNDRED RESIDENTS STREAMED into the auditorium of St. Luke's School for the first official meeting of the Committee to Save the West Village. Jacobs stood and read aloud parts of the *Times* article, explaining to the assembled what was *not* being said. The issue wasn't really about affordable housing, she warned. It was about making the buildings available to developers, who stood to make a lot of money. She had witnessed the city's tactics again and again while researching her book.

The study, she went on to explain, "was the opening fraud of a government racket to clear the neighborhood of buildings and tenants to be replaced by expensive versions of both, who paid higher taxes," recalled her neighbor Erik Wensberg, a friend and fellow Villager. "If we liked our busy, friendly, frowzy hood, urban renewal had to be fought."

They mustn't assume defeat, Jacobs counseled her neighbors. They had won the Washington Square Park battle, she reminded them. The park was now permanently closed to traffic. They would defeat this project in just the same way: by standing their ground, refusing to negotiate for better terms. If they let down their guard, they would invariably lose. She had seen this too. In earlier urban renewal battles, those civic groups open to compromise left the bargaining table empty-handed, having secured only minor adjustments in the plans.

"The aim of the committee is to kill this project entirely because if it goes through it can mean only the destruction of the community," Jacobs announced. Once the blight study is shelved completely, "we will look for an alternative. We want enforced conservation of the buildings, not their destruction."

The Villagers dug in for a fight. In the weeks and months that followed, Jacobs and her neighbors plotted their war plan, returning to many of the tactics they had used in the park battle. They forged alliances with young, upcoming politicians eager to expand their constituencies. They courted the media, making sure there were always reporters on hand to cover contentious city meetings, especially those in which the public was forbidden to speak. Jacobs had observed that when residents raised objections in defiance of the rules, they were likely to get quoted in the press, which helped get out their message.

With the help of friends in city hall, they collected intelligence on the city's next moves, then lay in wait for the right public moment to confront officials, hoping to expose inconsistencies, or to catch someone in an outright lie. The idea, which was Bob Jacobs's, was to ask the officials to be specific about their ultimate plans. Then, if the officials denied the existence of a plan, the Villagers had it on record. This way it was harder to press forward without being guilty of deceiving the public.

As the clash shifted into gear, Jane also added a new tactic. She turned for the first time to the legal system to confront Felt and Davies directly. The idea was to undermine their credibility, to publicly shame both men by exposing their indifference to the law. Wagner had instituted new urban renewal procedures that were legally binding, which Felt and Davies had willfully ignored. The Villagers had been given insufficient public notice of the redevelopment plan and Jacobs was now suing them for having violated the mayor's own rules. Felt and Davies, accustomed to being insulated by the city's Kafkaesque bureaucracy, were now on notice. Further violations would not go unchallenged.

One of the Villagers had a friend, Lester Eisner, who was in charge of overseeing all Title I programs in New England and also New York. He knew the ins and outs of how the urban renewal law worked, and agreed to tour the neighborhood one afternoon, to see if he could help. Eisner was struck by how lively and well-loved the neighborhood felt, as they took him around. The buildings, even the warehouses and tenements near the river, weren't terribly deteriorated. And the community felt cohesive, despite the mix of incomes and ethnicities of the residents there. It didn't seem a natural fit for urban renewal, he offered. For a district to be "officially" designated a slum, it had to meet specific physical criteria. There were set formulas on crowding, for instance, ratios of abandoned to occupied properties, standards of hygiene. What the Villagers needed to do, he explained, was to document the physical conditions in the fourteen-block area that the city proposed to condemn, refuting the city's case for slum clearance by supplying hard data that disproved it.

The residents decided to go door-to-door themselves, rather than spending money they didn't have for a formal survey. And Eisner, now their unofficial consultant, advised them on the kinds of statistics they needed to collect: the number of occupants per building, its condition, the number of kitchens and bathrooms, the rent paid. He also warned that they must never reveal to the city what they wanted, even if it was only small amenities, as they would be considered complicit in the city's plans, or, in city parlance, "participating citizens." This would open the door to official claims that there was neighborhood support for the urban renewal plans.

Emboldened by Eisner's advice, the Villagers hit the streets, clipboards in hand. "The city wants our houses and businesses torn down," read the leaflet they slipped under door after door, in both English and Spanish. "To save them we urgently need detailed infor-

mation on the area. Please cooperate when our representative calls."
Everyone, it seemed, was in on the effort, even the children, who
scurried about the neighborhood, putting up posters and collecting
signatures.

People put in long hours, often at night, the only time they had
available, Jacobs recalled. They met at 555 Hudson Street, plotting
their next moves over martinis and cigarettes. As more and more
people dropped by, the Jacobses disengaged the doorbell and began
leaving the door unlocked. Soon their numbers were such that it
was necessary to meet at the Lion's Head, a coffee bar that would
later become a favorite haunt of reporters, columnists, and city editors at the dailies. Like Jacobs, Leon Seidel, the proprietor of the
Lion's Head, was passionately committed to the fight. Leon, Jacobs
later recalled, "got information almost sooner than anyone else and
then spread it almost sooner than anyone else. You just dropped in
at Leon's to get the news, even if you weren't getting a coffee or dinner or anything." Seidel was what Jacobs called "a public character."
Reporters could always count on him for a quote. He had a printer
friend who worked miracles: leaflets could be left with Leon at 1 A.M.
to be copied, and they were ready by 8 the next morning. Everyone,
Jacobs later recalled, must have been suffering sleep deprivation.
"But we were, everybody was, in on this," she said. "Either their
second, or their third job, was saving the neighborhood."

The survey conducted by the committee was clear and thorough, the best he had ever seen, Eisner told the Villagers when they
finished. And it proved, unequivocally, that the West Village did
not, in fact, fit the legal criteria for a slum designation. The ratio of
people to rooms was perfectly fine, the per capita number of bathrooms and kitchens more than sufficient. But that March, when a
delegation delivered it to the Board of Estimate, the city refused to
give any ground.

Their findings were irrelevant, Davies told the Villagers, waving
his hand dismissively. The West Village qualified as a blighted area
because of the "mix of uses in the neighborhood," which under
the guidelines of Title I constituted "a deficiency." Rooming houses,

outmoded building types, extreme density—all were part of the Village landscape, and all were "indicators of blight." The neighborhood, he reiterated, remained a candidate for urban renewal.

The Villagers were outraged. "They brought this proposal out of a clear blue sky and left us less than a month to present an alternative plan, but they have not produced any evidence that the neighborhood needed renovation," Leon Seidel told the *New York Times*, waving reams of assessment notices indicating an increase in property values. "'Does [this] look as if the neighborhood was decaying?' he asked. 'It takes years to establish yourself [as a business owner] and pay off your initial cost . . . and it is all wiped out if you are forced out by redevelopment. You don't just move it somewhere else. You have to start again.'"

The newspapers were beginning to smell a good story. As reporters descended on the Village to cover the revolt, residents eagerly showed them around. "We couldn't go two steps without children appearing with handfuls of signed petitions," wrote Priscilla Chapman in the *New York Herald Tribune* after a tour of the Village with Jacobs, whom she described as "a down-to-earth . . . mother who's bringing up three children" and a respected urban critic. The fight to save the West Village had sticking power, Jacobs told Chapman, "precisely because of what was valuable about the neighborhood"—a printer willing to produce leaflets for free, shop owners volunteering their time, residents on the sidewalk looking out for one another. It was a vital close-knit community.

But the city, Jacobs would later recall, "had gotten all their ducks in a row." For months, officials had been plotting behind the scenes to divide the community, promising certain residents housing and favors if they came out publicly in support of the city's plan. A group called Micove, led by Charlotte Schwab, a self-described "young housewife from Maine," appeared one afternoon at a Board of Estimate hearing, professing to represent thousands of citizens who were strongly in favor of the West Village redevelopment plan. Micove had been formed some years earlier to lobby for an affordable housing co-op slated for Soho. When that project bombed, the organization had folded. Now Schwab claimed she'd become interested in the West Village proposal after reading about it in the papers. When the city suggested new housing would be part of the West Village urban renewal plan, she announced, as press cameras flashed, "we decided to support it."

Jacobs quickly fired back. Schwab had been coached by city planning officials and promised an apartment, she told reporters. Micove was "a puppet" that had been "invented and nurtured by the Housing and Redevelopment Board or its handpicked site sponsor." According to the law, the city needed to have citizen "participation," and Micove, conveniently resuscitated by the city, fit the bill.

But to Jacobs's dismay, another civic group, the powerful Citizens' Housing and Planning Council led by Roger Starr, now came forward, claiming that the Villagers had it wrong. The city did not "intend to level the area, to destroy its notably good housing, to tamper with that neighborhood character of which the residents are justifiably proud," Starr told the press. The official plan should be embraced.

Jacobs and her neighbors had sources inside city hall, "moles," she called them, who kept them apprised of the city's latest moves, including the times and locations of last-minute meetings Felt convened, with hopes he would evade notice. They confirmed that the urban renewal scheme made no mention of preservation, "only clearance and redevelopment."

Jacobs, joined now by four other civic groups, swiftly refuted Starr's claims, calling his statements wishful thinking. There was nothing on record to indicate that the city planned to save anything, she said. No mention of selective clearance.

But appearances, sadly, can trump fact, and on this count, the Villagers were at a marked disadvantage. Affordable housing in the Village was in short supply and the city's repeated claim that their mission, however disingenuous, was to provide more middle-class housing was playing well in the press.

Though Jacobs suspected there were conflicts of interest involved, that certain members of Starr's council were connected to powerful real estate interests that stood to profit from the city's re-development scheme, she couldn't prove it.

The Villagers were feeling exhausted. It was particularly discouraging that the *New York Times* repeatedly cast them as "negative." One reason, of course, was they were following Eisner's advice. Whenever officials asked them what they wanted, they refrained from revealing anything beyond their stock answer: "remove the slum designation." It was all part of their strategy not to appear, in Eisner's words, "complicit in the city's plans," but it was still enormously frustrating, Jacobs remembered.

And then they got some bracing news. In the spring of 1961, word came down that the Villagers had won the lawsuit Jacobs v. New York City. On April 27, a New York Supreme Court judge issued a ruling declaring that the city had failed to "comply with the requirement to hold a public hearing on the plans," as mandated by Wagner himself. An order was issued demanding that city officials justify the proposal for the West Village blight designation.

With Jacobs in the lead, a mob of neighborhood stalwarts descended on city hall to present the court order to Wagner himself at a meeting of the Board of Estimate. In a brilliant bit of stagecraft, the eight hundred Villagers wore cheap sunglasses with large white Xs taped across each lens, an allusion to the dreaded mark of a condemned building. Recognizing a good photo op when he saw one, an Associated Press photographer caught it all on film. By the end of the day, the image of the protesting Villagers had gone viral, picked up from the wire service by papers across the nation—and the world.

The Villagers' plight made great copy. People were reading about a plainspoken mother of three living in Greenwich Village, the author of a soon-to-be published book about cities, who was leading a David-and-Goliath struggle against urban renewal, which uprooted long-settled neighborhoods and ruthlessly displaced people against their will. Jacobs, as always, could be counted on for pithy quotes: "An irresponsible boondoggle which will gratuitously jeopardize a sound and healthy community composed of people with a great love and pride of neighborhood," she told a *Times* reporter when questioned. City officials, she charged, were practicing "vast deception in their procedures."

It was a time when people's faith in authority was beginning to erode, and Jacobs's suggestion that New York's political leaders were riding roughshod over citizens' rights was hitting a nerve. Jacobs had begun to receive letters from people across the city and the nation, sharing their own struggles with callous and disengaged officials. The Village story resonated. "Years before the body counts and secret invasions of the Vietnam War," observes Anthony Flint in *Wrestling with Moses*, "Jacobs was portraying a government that, far from being there to help, could not be trusted."

WAGNER GREETED THE VILLAGERS CORDIALLY, EXPLAINING THAT HE couldn't speak to the court order directly, as he had just received it. But the Villagers sensed that now that the legal verdict was in, the winds had shifted.

Publicly chastened, Felt and Davies immediately launched a public relations blitz, promising to hold a citywide hearing on the "new" and more sensitive urban renewal program. Citizen voices would be heard, Felt vowed. While some areas would need to be completely rebuilt, others, he claimed, could be more selectively redeveloped. "The backbone of renewal is in conserving and improving our existing structures and relating new development to the character and needs of the community," he told a *New York Times* reporter, the suggestion being that the city understood the Villagers' concerns.

At a May hearing convened shortly after, several prominent organizations publicly commended the city's "renewed commitment" to citizens' public participation. Jacobs didn't believe a word of it. The sole voice of dissent that day, she stood up and said as much.

Official rhetoric aside, the fight was far from over. The slum designation hadn't been removed, the Villagers reminded themselves. It was important not to let down their guard. In June, they went on the offensive again. As the planning commission debated the West Village renewal plan, they delivered a petition to the mayor's office calling for the resignation of both Felt and Davies, who had "discredited their offices" by ignoring the wishes of the people. The state assemblyman Louis DeSalvio explained that the Villagers had been pushed by circumstances to this extreme action. "They don't want or need bureaucratic interference in their way of life," he said.

At first Wagner balked, unwilling to swallow the notion that his hand-picked appointees were guilty of misconduct. But in August, eyeing the upcoming primary election, he began to back away from his initial support of the West Village redevelopment plan. "I want to say for the record that I shall vigorously oppose any study which would contemplate a change in the basic character of Greenwich Village," he said. "The bulldozer approach is out." Any new improvements, he told one of the downtown papers, "must conform to Village tradition."

The mayor's avowals were "pious platitudes," Jacobs swiftly retorted. "If the mayor cares about the wishes, character and well being of Greenwich Village and its citizens," she told the *Times* on August 18, 1961, "he will have the urban renewal proposal killed outright."

The pressure was building. As the election drew near, Wagner's opponent announced that he would kill the plan if elected. Carmine De Sapio, vying for leadership of the local district of the Democratic Party, insisted that the mayor take a clear position in advance of the primary. Feeling the political heat, Wagner finally buckled. Declaring that he was "deeply concerned and sympathetic with the people of the West Village neighborhood in their desire to conserve and to build constructively upon a neighborhood life," he asked the City Planning Commission to "shelve" the redevelopment project.

Felt assured the mayor he would give his request "earnest consideration." But privately he had already come to another decision. Several weeks later, on October 18, 1961, in a stunning public statement before a crowded city hall meeting, Felt announced that in a closed-door meeting two weeks before, the commission had voted to designate the West Village a blighted area ready for urban renewal. Oozing sanctimoniousness, he held up an accompanying report defending the commission's right to "independence" from the mayor's office.

"Down with Felt! Down with Felt!" Jacobs and the Villagers shouted, jumping from their seats and surging forward. Felt had made "a backroom deal with a builder, David Rose Associates," who stood to make enormous profits from the city plan, someone cried. The commission's action was illegal, another shouted. The mayor had been "double-crossed."

Felt banged the gavel and called for order, as several Villagers twisted away from police officers moving in. The Villagers were horrified. Felt and his cronies had acknowledged the public's opposition and then, with apparent relish, flagrantly disregarded it, declaring the will of the people to be of no consequence.

Felt leaned into the microphone, insisting on silence. This was not a public hearing and the public was not allowed to speak, he warned. But Louis DeSalvio was making his way up to the podium now, and Felt, relenting, stepped aside so the Village assemblyman could speak.

The crowd went silent. "By this reprehensible and strange decision," DeSalvio declared, "you have sent the urban renewal program of this city, state and federal government back to the dark ages of Robert Moses, and his arbitrary and inhuman procedures."

The audience roared. Flustered, Felt suggested to DeSalvio that he and other elected officials could meet "in private" to work things out quietly by themselves. The Villagers now erupted in rage.

"You are not an elected official!" cried Stephen Zoll, a resident and an editor at the Macmillan book publishing company. "You have made a deal with David Rose!"

Felt called for the police to remove him. "Your name will be remembered in horror!" Zoll shouted as two officers spirited him away.

"You belong with Khrushchev!" cried a woman. "How dare you assume such authority? Who the hell do you think you are making decisions in the interest of builders?"

Felt blanched and pounded the gavel, calling for a recess. Once behind closed doors, he immediately called for more police backup, unsure how to navigate this new political landscape. Later in the 1960s, of course, public outcries of this sort would be common. In 1968, when two Jesuit priests, Daniel and Philip Berrigan, invaded a suburban Maryland draft board, setting fire to files naming the young men most likely to be drafted and sent to Vietnam, their civil disobedience was immediately broadcast around the world, becoming a touchstone for an entire generation. But in 1961, such protest actions were unfamiliar, especially at a meeting of New York City officials.

"This is the most disgraceful demonstration I have ever seen," Felt announced, pounding the gavel again. "We cannot operate on the basis of disorderly conduct," he told reporters.

An hour later, the room now heavily fortified with police officers, Felt reopened the meeting. The Villagers immediately resumed their accusations that Felt had made an insider deal, that he had deceived the mayor. They insisted that there be a new hearing.

Felt turned to the police. "You will be obliged to arrest anyone who interrupts the meeting," he said. When an actor who lived on West Street refused to stop chanting, two officers carried him out "feetfirst," followed by several others, also forcibly removed. The police, it was clear, were not entirely comfortable. They made no arrests and whispered their apologies to the Villagers as they hustled them out.

Felt pounded the gavel again and adjourned the meeting. There will be "no more hearings," he boomed, as the Villagers leaped from their seats in fury. Then he slipped quickly out a side door.

IN THE IMMEDIATE WEEKS TO FOLLOW, ANGRY OFFICIALS DECRIED what the newspapers called the Villagers' "near riot." One commissioner called the Villagers "ignorant, neurotic, dishonest, slanderous, disorderly and disgusting." Another described the demonstration as "riot tactics" and "an attack on the democratic processes."

But it was the early 1960s and the cultural mood was shifting, not only in Greenwich Village but also in other liberal pockets of the country. People were beginning to openly challenge their political leaders, unafraid to confront those who neglected the public interest. In this outspoken new age, writes Flint, "it was not the protesters that were the outrage," but the public officials with their brazen disregard for the will of the people. As Jacobs and her neighbors made plain, civil disobedience had become their only choice once their polite dissent went unheard.

"We had been ladies and gentlemen and only got pushed around," Jacobs said. "We were not violent. We were terribly alarmed at what is happening in our neighborhood and our city . . . so yesterday we protested loudly."

Jacobs had long suspected that some of the resident groups supporting urban renewal were pawns of not only the city, but also the private developers who stood to profit once the slum designation was in place. She had noticed characters that she surmised were developers' agents—"creepers," she called them—lurking about the West Village asking questions about living conditions. To monitor things, she recruited the neighborhood kids as spies, telling them to keep an eye out for creepers and to follow them whenever they could. It was in this way that she was able to trace several of the "creepers" back to David Rose Associates, the developer chosen by the city to rebuild the district once the West Village was leveled.

And then came the piece of the puzzle she had been searching for. Leafing through her correspondence, she noticed that one of the Rose firm's employees had sent her a letter asking about a travel fellowship. There was something faintly familiar about the typeface, and when she pulled out press releases from the various citizens' groups supporting the city's plans, she immediately saw what it was. In both the firm's letter and the press releases, whenever the letter r appeared, it was dropped slightly. Jacobs consulted a forensic expert, who confirmed that the developer's correspondence and those of the many supportive neighborhood groups, including Micove, had been composed on the same typewriter. Within days, the typewriter was tracked to an office near Columbia, and one of Jacobs's friends, Erik Wensberg, dropped by to do some further detective work. As he stepped into the office, he saw a telegram from David Rose Associates on a desk.

The implications were stunning. The pro–urban renewal citizens' groups who'd gone door-to-door through the Village collecting signatures early in the struggle were stooges not only for the city, but also for the firm that stood to gain the most from the redevelopment project. The Villagers who had signed the pro–urban renewal petitions had been doubly deceived—promised amenities far surpassing anything proposed, and left in the dark about who was really behind the citizens' groups canvassing for redevelopment. Jacobs interviewed the residents who had signed the petitions, collecting hundreds of notarized statements attesting to the false promises and lack of transparency.

It was time to share her findings. At a brainstorming session at the Lion's Head the day after the insurrection at the City Planning Commission, Jacobs went public with her allegations. City officials had been in collusion with the developers throughout the entire process, she charged, holding up a folder of papers documenting the city's duplicity. The paper trail, she added, even included boasts from a consultant who claimed he had already drawn up plans for the West Village urban renewal project, which under the Title I rules was illegal. Reporters' pencils bore down on notebooks as indignation swept the room. The rules were common knowledge: the city was supposed to designate a slum first, and only then put out calls for proposals from the private sector. In this instance, said Jacobs, the developers had been in on the real estate grab from the start. The entire scheme was designed to line the pockets of real estate moguls at the expense of the people. "It's the same old story," Jacobs told the *New York Herald Tribune*. "First the builder picks the property, then he gets the Planning Commission to designate it, and then the people get bulldozed out of their homes."

The city and David Rose Associates denied that plans had been illegally drawn up. But the taint of wrongdoing attached to the fourteen-block slum designation stuck. Beyond the city's duplicity and subterfuge, there were now allegations of corruption, to which no politician could turn a blind eye.

Davies's stature was the first to suffer. Weakened by the negative press picturing law-abiding New Yorkers being dragged from meetings by the police, embarrassed by the revelations linking David Rose Associates to sham neighborhood groups who themselves had been deceived, he conceded that the West Village plan had become too politically volatile to pursue. On January 31, 1962, a year after Jacobs first read of the city's designs on her neighborhood, the City Planning Commission removed the slum designation by unanimous vote. Against all odds, the Villagers had prevailed.

It was a stunning moment. A group of private citizens had thwarted a previously unassailable coalition of public officials and real estate speculators in the so-called "redevelopment" of their own neighborhood. Jacobs had proved that autocratic control and top-down planning could be defeated by populist efforts. She had shown that everyday people could challenge authority and triumph. It was "a pioneering act," writes Anthony Flint, "that would resonate throughout the 1960's." It was also a portent of conflicts to come, another variation of the rift Rachel Carson had identified between business rights and the rights of the community—soon a banner issue for the gathering counterculture.

Six months later, Felt stepped down as chairman of the City Planning Commission. Davies would shortly follow.

THE BATTLE AT LAST OVER, JACOBS FELT TREMENDOUS RELIEF. FOR more than a year, the press had been covering the West Village struggle in granular detail, including Jacobs's pivotal role in the fight. *Vogue* described her as "Joan of Arc"; another magazine as "a sort of Madame Defarge leading an aroused populace to the barricades." She was the "*madonna misericordia*" to the West Village, "who has probably bludgeoned more old songs, rallied more support, fought harder, caused more trouble, and made more enemies than any American woman since Margaret Sanger," Jane Kramer wrote in the *Village Voice.*

Jane gently scoffed at these descriptions, none of which seemed terribly apt. "There seems to be a notion that I run these people," she told a reporter. "But I wouldn't have dreamed of telling them how to behave . . . I was an instrument of what the neighborhood wanted to do." It still seemed odd, she told friends, to be such a polarizing figure. She looked forward to slipping back into her old life, resuming her old routines, enjoying the crisp fall air as she biked to work, more or less invisible again. But this, it turned out, was not to be.

In late September 1961, Jacobs found herself the subject of controversy for a new reason: Random House released her bold and culture-rocking book, *The Death and Life of Great American Cities.* Capitalizing on her notoriety as a neighborhood firebrand, a wily conquering angel who had outwitted city hall, the publicity department decided to market it aggressively. *Death and Life,* they announced, was an "explosive" new work that laid waste to the powerful forces that were systematically destroying America's cities. "The City Planners Are Ravaging Our Cities!" boomed the headline of the publisher's first ad. It was an apt summary of her argument: that the men who claimed to be saving America's cities were actually savaging them.

Jacobs didn't mince words: "This book is an attack on current city planning and rebuilding," she announced in her now-famous opening line. "The economic rationale" of current city planning is "a hoax," she went on to say, the "means to planned city rebuilding are as deplorable as the ends." Planners were reducing city and countryside alike to a "monotonous, unnourishing gruel," all coming from the "same intellectual dish of mush."

Jacobs led her argument with the premise that the men responsible for current urban renewal theories—Le Corbusier, with his utopian skyscrapers in the park; Ebenezer Howard's decentrist Garden City; Mumford's low-density, quasi-suburban variations—actually hated and feared cities; all were at heart *anti-urban,* she wrote. (They "hammered away at the bad old city . . . They were interested only in failures.") Then, in answer to these downbeat visions, she went on to mount a spirited defense of what made cities great and worthy of saving—and what planners had gotten wrong.

Cities, by their very nature, were complex organisms that made themselves up as they went along, Jacobs insisted. They were "thoroughly physical places," agglomerations of incredible numbers of different people with vastly different ideas and purposes that were perpetually adapting to circumstances. Urban life couldn't be reduced to "engineering models for traffic" or highly planned, mechanistic strategies for living. Cities were too complex for such reductive approaches, too various and particular. Cities in real life didn't need artificial order imposed from above; they already possessed an "innate functioning order" of their own. "There is a quality even meaner than outright ugliness or disorder, and that meaner quality is the dishonest mask of pretended order," she said.

Jacobs wrote with crackling wit and unflinching candor. Her prose was crisp and chatty, leavened with humor and breathtakingly clear. Above all it was accessible. At the time, urban planning wasn't part of the national conversation. Whatever talk there was took place in architecture schools, or among the design elite at planning conferences. Jacobs made it essential reading. She gave ordinary readers a vocabulary to describe their feelings: reasons for the anomie they felt standing amidst a superblock of faceless towers, or crossing a barren plaza; the comfort they took in the physical variety of existing old neighborhoods. Jacobs mistrusted jargon, which too often concealed illogical thinking, she felt. "It is hard for muddled thoughts to hide in plain English," she told an interviewer. She worked impressionistically, folding in pithy anecdotes to illustrate her points, real-life scenes that got people thinking. No matter how well-made her arguments, she never talked down to her readers, never lectured. Her voice was warm and playful, suffused with homespun observations and common sense.

A populist at heart, Jacobs deplored what she called the "statistical city," the tendency of planners and policy makers to see cities as mathematical abstractions, rather than as living communities. She didn't hesitate to link urban renewal, "with its obvious elements of regimentation," to the authoritarian impulse, to the "repression of all plans but the planners'." Conventional approaches to urban problems were "paternalistic," she said, because they treated city dwellers, whom she called "great informal experts," as incapable of acting in their own self-interest. As she saw it, urban renewal was both anticity and antidemocratic. "People who get marked with the planners' hex signs are pushed about, expropriated, and uprooted much as if they were the subject of a conquering power. Thousands upon thousands of small businesses are destroyed . . . with hardly a gesture at compensation." Urban renewal not only leveled old neighborhoods, destroying the yeasty human ecosystem of those living there, it reduced communities to mathematical constructs. "In the form of statistics, these citizens could be dealt with intellectually like grains of sand, or electrons or billiard balls," she wrote. The ramifications in terms of social justice were clear. But urban renewal's greatest drawback, she insisted, was that it drained the economic life from cities by fixing conditions. What the planners failed to see was that change was the animating spark of vital cities, its essence. To freeze conditions in a city was to snuff out its life.

Indeed, perhaps the most controversial assertion Jacobs made was that cities couldn't really be planned at all. Vibrant cities were continuously adapting over time, in response to the external environment, just like other natural systems. They worked by the "same universal principles that the rest of nature used." Vital cities "have marvelous innate abilities for understanding, communicating, contriving and inventing what is required to combat their difficulties," she wrote. They are dynamic places that "contain the seeds of their own regeneration."

Like Carson, Jacobs was viewing time and human endeavor through a longer lens. She was seeing the city as an astonishingly complex and reactive human ecosystem, rather than an inert "problem to be solved." The balance of nature "is not a status quo," Carson would write in *Silent Spring* a year later; "it is fluid, ever-shifting, in a constant state of adjustment." Jacobs was saying the same of cities: a vital city was never static; it was perpetually evolving, an ongoing churn of dynamic adjustments. Jacobs was taking "a biological view of the built environment versus a mechanistic one," observes Roberta Gratz. "It was human ecology versus the machine." "The alternative isn't to develop some other way," said Jacobs. "Some other way doesn't exist."

DEATH AND LIFE NEVER MADE THE BESTSELLER LIST, BUT IT WAS AN IMmediate sensation, a subject of extravagant praise and prickly debate. Critics writing for the general press almost universally embraced it. A "brashly impressive tour de force," pronounced the *New York Times*. "Seminal" wrote another publication, "a major work," one of those "rare books that make a difference in world history." *Architectural Forum*, where Jacobs remained a senior editor, called it "a revolutionary and revelatory volume of uncommon sense." A writer for *Progressive Architecture* predicted that it "will have as much impact on urban thinking as the Armory Show had on art."

Even those uneasy with Jacobs's dismissal of the great visionaries of urban planning—figures who loomed as gods to an entire generation—recognized the book's power. "It won't matter that what this author has to say isn't fair or scientific. Few significant works ever are," wrote Lloyd Rodwin, an MIT professor reviewing the book for the *New York Times*. "Jane Jacobs' book should help to swing reformist zeal in favor of urbanity and the big city. If so it might well become the most influential work on cities since Lewis Mumford's classic *The Culture of Cities*."

But the planning establishment took a dimmer view. Attacks from government officials, academics, and urban planners followed hard on the heels of the glowing press. *Death and Life*, they said, was not only "flawed" but also "dangerous."

Moses was one of the first to weigh in, though he did so privately. Throughout *Death and Life*, Jacobs had called Moses to account for his autocratic methods, his flawed thinking about highways and public housing, and for his egregious misuse of public funds: "Robert Moses . . . has made an art of using control of public money to get his way with those whom the voters elect and depend on to represent their frequently opposing interests," she wrote. This is, of course, "an old sad story of democratic government," she added. "The art of negating the power of votes with the power of money," is "easiest when the electorate is fragmented."

"Libelous, intemperate, and inaccurate," Moses shot back in an angry note tossed off to Random House's publisher Bennett Cerf, who had sent him a copy of the book. "Sell this junk to someone else," he added.

Other critics attacked more publicly—and more personally. Jacobs was called a "demagogue" and a "crank," a hack and "a gadfly," a "wild bohemian" and an "anarchist." She was "misinformed" and out of touch. Her writing was filled with "transparent gaps" and errors, with "blind spots" and "oversimplifications." She was guilty of "blasé misunderstandings of theory," of gathering only evidence "congenial" to her thesis. "Mrs. Jacobs has presented the world with a document that will be grabbed by screwballs and reactionaries and used to fight civic improvement and urban renewal projects for years to come," charged Dennis O'Harrow, director of the American Society of Planning Officials. "The Jane Jacobs book is going to do a lot of harm. So batten down the hatches boys, we are in for a big blow!" Jane Jacobs, bellowed another detractor, would do best to consider the social costs that accrue to society in disease, crime, and poverty before venturing any further opinions about cities. "Unfortunately, this cannot be achieved by star-gazing from the second floor window of a Greenwich Village flat—while anxiously awaiting the 3 A.M. closing of the neighborhood pubs as an omen that all is well in the land," he smirked.

And then, predictably, there was the matter of Jacobs's sex. She was dismissed as a "housewife" who lacked credentials; "an angry young woman" with "no formal training"; an inept amateur with "seemingly limited experience in her chosen field." She was a "mother from Scranton," a "sentimental Hausfrau," a misty-eyed incompetent.

"Mistakes in City Planning Get a Housewife's Panning," boomed a headline in Baltimore's *Sunday American*, despite the fact that Jacobs was by now a respected architectural critic and author. That a self-taught woman with a keen eye and a rapier tongue, an outsider operating by instinct and observation, had dared mount a polemic against the effete, all-male planning profession was rankling.

Perhaps just as unsettling was the realization that hers was only the first of several powerful challenges to the status quo being mounted by uncredentialed women, all reflecting a similar humanistic trend. At a time when even college-educated women were being told that getting their husbands' shirt collars white was "a peak experience," that a woman's place was to make dinner and not decisions, Jacobs, Betty Friedan, and Rachel Carson—all effective amateurs—were producing books that were rocking the culture. It was a period, it should be added, that also saw significant social critiques by male writers asking parallel questions about the skewed priorities driving America's postwar culture. Michael Harrington exposed the nation's yawning poverty in *The Other America*; Ralph Nader spotlighted safety issues in the auto industry in his 1965 book *Unsafe at Any Speed*; Paul Goodman wrote about youth and delinquency in *Growing Up Absurd*. All of these writers, in calling into question deeply entrenched ideas tied to powerful corporate interests, were articulating views—and drawing upon energy—already in the air in the early 1960s. But none were as culturally transformative, nor resonated as deeply in the collective imagination, as those by Jacobs, Carson, and Friedan (perhaps because each put a personal face on a problem). And none were subject to the same degree of censure and fury as that visited upon the women.

"How have the planners reacted to the toppling of their dogma?" wrote planner Edward Chase in *Architectural Forum*'s April 1962 issue. "Rather badly, I would say. There is resort to ridicule; there is patronizing dismissal of Mrs. Jacobs as a crackpot anti-intellectual, anti-planner; and, there is malicious misrepresentation of her book."

Lewis Mumford, seething over what he felt were unforgivable misrepresentations of his ideas, waited as long as he could before venting his rage. "I held my fire for a whole year," he wrote to a friend, "but when I got down to write I discovered that the paper burned, in spite of the long cooling period."

In a stinging, ten-page review for *The New Yorker* (which was cut from twenty pages), Mumford called Jacobs a "sloppy novice" whose analysis was filled with "schoolgirl howlers," and accused her of oversimplifying urban design theory and grossly distorting planning history.

"She describes her folksy urban place on Hudson Street with such spirit and womanly verve that she has made a considerable number of readers believe it really exists," wrote Roger Starr, whom Jacobs had exposed as a "stooge" for the real estate industry in the West Village fight. It was as if "Mrs. Jacobs had visited Pompeii and concluded that nothing makes a city so beautiful as covering it with ashes," he sneered.

Jacobs's urban utopia, Starr continued, was a place "where factories nestle beside homes, and never give off smells or smoke . . . No mere socio-analyst, but a reformer too, Mrs. Jacobs tells us we must cast off the wicked spell" of high-minded thinkers and forswear "sunlight, clean air," and open space. "The rest is easy, chum. You takes $20,000 out of your savings . . . and you buys a house in a part of the city known as a slum, but has potentialities. Then, says Mrs. Jacobs, you and your like-minded (and like financed) neighbors 'unslum' it. This means, Mrs. Jacobs admits, reducing the neighborhood populations; i.e., throwing out the people who were living there.

"Mrs. Jacobs is an advanced thinker and cannot keep herself from acknowledging, though briefly, that there are race and economic problems in the city," he mocked. Jane Jacobs has described paradise "so well your mouth waters," he added, "She's just a teeny bit hazy about how we're all a-gwine to get there."

Starr, who would later advise three successive mayors and become an editorial writer for the *New York Times*, eventually grew disenchanted with the results of large-scale planning. In his obituary in the *Times* on September 11, 2001, his son was quoted as saying,

"Sometimes we would drive past a Mitchell-Lama project, and he [Starr] would say, Mea culpa, mea culpa."

But at the time *Death and Life* appeared, he was still an ardent champion of slum clearance and public housing. And while his charges were unjust, he had zeroed in on a weakness in Jacobs's thesis. For all her fresh insights, such critics said, Jacobs was short on shovel-ready prescriptions for pressing urban problems, most especially for housing the poor, where solutions were desperately needed. It was a banner issue for those on the left, who claimed Jacobs's vision of organic neighborhood growth would lead to gentrification. "Her book had no recipe for changing the city," Starr charged. "Her goal was to keep the present neighborhoods alive, on the theory that they were charming, attractive and desirable."

Which to Jacobs was precisely the point: "I was interested in process, not in the notion that you can wave a wand and control things and get some end result," she said. "I just want to know how to keep the life going (which in my mind is the purpose of life)." Ideologies, she would later tell a reporter, are "one of the great afflictions . . . they blind us to seeing what is going on" and "to what is being done."

As for Mumford, whose review *did* allow that Jacobs's mind was "big with fresh insights and pertinent ideas," she took his barbs in stride, chalking them up to a mix of "arrested development" and a bruised ego. "I thought his reaction to the book was not quite rational," she explained to an overflow audience at City College years later. "I believe now that he felt hurt and betrayed that I didn't position myself as a disciple of his. I didn't take this personally. I think it had to do with his time . . . Maybe if he'd lived at a different time he would have understood that women didn't necessarily aspire to be patronized. He believed that women were sort of a ladies' auxiliary of the human race."

DESPITE THE POTSHOTS, OR PERHAPS *BECAUSE* OF THEM, JACOBS WAS becoming a sought-after figure. Editors from across the political spectrum now approached her for articles. Lady Bird Johnson invited her to the White House for a luncheon; Jacobs agreed, providing she could speak about pressing urban issues, rather than gladiolas and gardens. Soon after her White House outing, Jacobs got a note from John Lindsay, who had received a glowing account of her talk from a fellow congressman: "Dear Jane, What have you done to [Congressman] Widnall? I think you've bewitched him. Both of you have good taste."

Widnall wasn't the only one bewitched by Jacobs's words. With the book selling and on everyone's mind, Jacobs was being hailed as a streetwise urban sage, mixing nuanced observation with boots-on-the-ground activism. *Vogue* dubbed her "Queen Jane." *Newsweek* and *The Saturday Evening Post* ran lengthy profiles. Diane Arbus photographed her for *Esquire*. Jane Jacobs has become the "prophet and leader of a great neighborhood revival," wrote the *Village Voice*. "In the past year and a half she led the people of Lower Manhattan into innumerable battles . . . She has turned her causes into hot-potato issues and is lately the terror of every politico in town."

Civic groups from across the country now invited her to their cities, anxious for her input about revitalization projects in neighborhoods where they lived. "Forget the big parking garages," she told an audience in West Palm Beach. "Just keep the stores and cafés open at night." Asked what she thought after a tour of several public housing projects in Pittsburgh, Jacobs described them as "bleak, miserable, and mean." Pittsburgh, she told an interviewer, "is being rebuilt by city haters." She was equally forthright in Philadelphia. "Planners," she told an interviewer, "always want to make a big deal of everything they do . . . In urban renewal you need new buildings—I have no quarrel with that—but there are plenty of good buildings here. Why tear them all down?"

And it wasn't just in white neighborhoods where she was having an impact. African Americans across America were using *Death and Life* to fight demolition plans for their neighborhoods too. Black leaders, who had initially supported urban renewal, now despised the "projects," as they called them. "They are hated almost as much as policemen, and this is saying a great deal," wrote James Baldwin in *Nobody Knows My Name.* "Their administration," he added, echoing a Jacobs theme, "is insanely humiliating."

"This culture," Jacobs told a reporter, "with its huge housing developments and its planned cities—it's fantastically materialistic, you know. You're poor? You go to a low-income project. Make a little more money? You move to a middle-income one. I can't think of anything more demoralizing to the life of a great city than this kind of economic and social stratification, than this no-roots pattern of life."

The deck was especially stacked against the black community, said Jacobs. As Nikole Hannah-Jones of the *New York Times* has written: "White residents used Federal Housing Administration–insured loans to buy their way out of the projects" and to move to shiny new middle-income enclaves. "This subsidized home-buying boom led to one of the broadest expansions of the American middle class ever." But the FHA's "explicitly racist underwriting standards, which rated black and integrated neighborhoods as uninsurable, made federally insured home loans largely unavailable to black home seekers." This effectively meant blacks couldn't get loans to buy into, move out of, or fix up their old neighborhoods. In 1962, Jacobs would testify to this effect before a Senate subcommittee. Banks routinely withheld loans in neighborhoods targeted for urban renewal or highway construction, she asserted. These blacklisted areas, she added, are selected "often because Negroes have moved in," and the neighborhoods "deteriorate because it is impossible to get money for improvements." It was a practice she deplored.

TIME PASSED, AND *DEATH AND LIFE* ONLY CONTINUED TO GROW IN REPutation. Like *Silent Spring*, it entered the public imagination to the extent that people who had never read it drew from its ideas. Not only did popular notions about *rebuilding* cities shift under the pressure of Jacobs's pen, ideas about *living* in urban spaces—attitudes about community, democracy, grassroots activism, sustainability—changed too. Like *Silent Spring*'s, the book's impact went beyond simple engagement; it was potent and culture changing.

In Thomas Kuhn's *The Structure of Scientific Revolutions*, Kuhn brilliantly argued that science often moves forward not by "deductive" argument, but through "imaginative leaps," ideas that are "pursued because a scientist intuits unsystematically that they are right and only later proves this to be so." These creative hunches of "rightness" eventually turn into new paradigms, but only with time. *Death and Life* produced just such a paradigm shift. It changed perceptions about cities by calling attention to the overarching idea of "connection," the web of interactions and dependencies that knit together the urban fabric. The built world, like the natural world, was profoundly interconnected, Jacobs showed.

Kuhn's book, like *Silent Spring*, came out in 1962, a year after *Death and Life*, and their concurrence was no accident; all three spoke powerfully to the cultural moment, which was teetering on the brink of epochal change. The sixties had opened with millennial hopes, an intense and palpable sense of promise and expectation, but also, more darkly, an airing of grievances long tamped down. It was a time, increasingly, of convulsive upheaval and sharp divisions: face-offs between black and white, hip and straight, doves and hawks, the counterculture and the old guard.

In the South, race relations were exploding. Blacks marching for voting rights in Selma were being tear-gassed and bludgeoned by state troopers, buses carrying black and white Freedom Riders doused with gasoline and torched. Viewers watching national TV were horrified to see young, neatly dressed blacks "pinned" to the ground by high-pressure hoses "strong enough to strip bark from trees." In Birmingham, white vigilantes were planting bombs in black churches, conducting a terror campaign against nonviolent blacks who dared defy the color line. "We will wear you down by our capacity to suffer, and in the process we will win your hearts," pronounced Martin Luther King Jr. It would take more time than anyone knew.

Kennedy was assassinated and the women's movement began.

Demands for equal pay and an end to sexual harassment were followed by legal access to birth control and eventually abortion. American bombs began dropping on Southeast Asia, and, despite denials, "American advisers" were being sent to Vietnam to fight a proxy war against Communism that few wanted. In the spring of 1962, student radicals met in Port Huron, Michigan, to draft what became known as the Port Huron Statement, a declaration of generational identity that soon was circulating across college campuses. Among its tenets was a call to end "militarism, racial injustice, and poverty." A year later, students at Berkeley rose up in protest against the university's close ties to federal defense contractors, giving birth to the Free Speech Movement.

An insurgent youth movement was progressively building. In college towns and other bohemian-friendly bastions, from Madison, Wisconsin, to Washington Square Park, Harvard Square in Cambridge to South Campus in Berkeley, folk icons—Joan Baez, Bob Dylan, Pete Seeger—were becoming staples at every civil rights rally, conjoining popular music with the civil rights struggle and the growing antiwar movement. When a song by Dylan called "Blowin' in the Wind" hit the charts in 1963, its wistful refrain, "how many years can some people exist before they're allowed to be free," spoke for a counterculture just beginning to find its voice.

That voice increasingly saw the "machine" of industrial progress pitted against those who embraced the "less tangible, sensory, aesthetic and environmental qualities of life," as Leo Marx observes in *The Machine in the Garden*. When the Free Speech Movement leader Mario Savio implored student demonstrators at a Berkeley rally to "put your bodies against the Machine, and make it stop," he intended his image of the machine to "conjure the military industrial complex." In effect, he was asking his fellow students "to mobilize a force of nature—their own frail bodies"—in a last-ditch effort aimed at blunting "the misuse of official power"—surely a basic Jacobs theme if ever there was one. "I saw a newborn baby with wild wolves all around it," Dylan sang in a song invoking a world devastated by nuclear war—perhaps the darkest manifestation of a world in thrall to the machine. "I saw a highway of diamonds with nobody on it."

Though it wasn't necessarily her intention, *Death and Life* would ally Jacobs with this growing cohort of activists and the young, a loose coalition which, if it didn't always speak as a unified voice, shared a common antipathy to the increasingly technocratic direction of American life: its deference to specialists, its mass industrialization, its counterfeiting of nature. Ours is a civilization with an "unshakable commitment to genocide, gambling madly with universal extermination of our own species," warned Theodore Roszak in *The Making of a Counter Culture*. Indeed, in its widening resistance to the country's militarism, its grassroots activism, its newfound respect for nature and natural systems—a world less harnessed to the machine—this gathering insurgency was echoing ideas whose origins were embedded in Carson's and Jacobs's brave and influential books. Whether by direct influence, or because these women's thinking so quickly entered the river of discourse, this emergent movement found common cause with their ideas—and inspiration in their actions.

JACOBS, LIKE CARSON, WASN'T ALTOGETHER COMFORTABLE WITH HER celebrity. Strolling along Hudson Street with her children, or lingering over a martini at the White Horse Tavern, she was often interrupted by admirers seeking an autograph or a bit of strategic advice for some community initiative. Jacobs was always polite, but she disliked the intrusion. "I like attention paid to my books and not me. I don't know who this celebrity called Jane Jacobs is—it's not me," she protested. "My idea is to be a hermit . . . and do my work. I just detest when I'm around somewhere and strangers say, 'Are you Jane Jacobs' and engage me in conversation in this obsequious way. That has nothing to do with me. You either do your work or you're a celebrity; I'd rather do my work."

IN THE SUMMER OF 1962, JACOBS WAS KNEELING IN HER BACK GARDEN, yanking out weeds, when she heard the doorbell. Her first impulse was to ignore it. It was a relief to sink her fingers in the soil, to be doing something physical again. After a year of living in the public eye, of talks and tours and travel in the wake of her book, she was hungry for simple pleasures, for privacy and solitude and time to work. She had just decided to leave her job at *Architectural Forum* so she could devote herself full-time to writing. She had a new book in mind that she was eager to explore. But then the bell sounded again, this time more insistently, and she overrode her own resistance. She hauled herself up and went to answer it.

The man on the other side of the front door was dressed in a monk's cassock. Fresh-faced, with a steady gaze and a mop of thick brown hair, he introduced himself as Father Gerard La Mountain, pastor of the Church of the Most Holy Crucifix, on Broome Street. He apologized for calling on Jacobs cold, but he needed some advice. The city was planning on building a superhighway across Lower Manhattan along Broome Street, he said. It would level his church and uproot the lives of his parishioners. For the past two years, he had been leading the fight against the expressway on behalf of his flock. But he was beginning to lose heart. He wondered if Jacobs had a moment to talk. Perhaps she had some insights that would be helpful. Jacobs nodded kindly, inviting him in.

Jacobs knew of the Lower Manhattan Expressway, or "Lomex," as it was being called. It had been on the back burner since before WWII. Much of the thrust behind Moses's push to extend Fifth Avenue through Washington Square Park, in fact, had been to tie in with Lomex, despite official denials. For years Moses had been talking about building a series of elevated expressways that would crisscross Manhattan like "laces" on a boot. Lomex, she assumed, would be one more link in this vast "spaghetti dish" of arteries looping over and around the city—all so drivers could speed through New York's boroughs without ever having to stop their cars. Moses's aerial expressway over Thirty-Fourth Street, joining the Lincoln Tunnel to the portal of the Queens Midtown Tunnel, had been thwarted. So now, it seemed, he had his eye on Lower Manhattan again.

Lomex, La Mountain explained, would be a hulking, ten-lane monstrosity stretching two and a half miles along Broome Street. It would loom fifty feet overhead, joining the Holland Tunnel on the west to the two-lane Manhattan and Williamsburg bridges on the east. On its sweep eastward, it would slash through the heart of Soho, as the neighborhood would soon be called, and then rip through Little Italy, Chinatown, the Bowery, and the Lower East Side, leveling everything in its path: apartment buildings, tenements, warehouses, churches. It would force 2,200 families from their homes, including La Mountain's flock, and obliterate 800 small businesses, leaving some 10,000 blue-collar workers without jobs. The price tag for this boondoggle was estimated at $150 million, but as part of the federal highway system, the expressway would be 90 percent federally funded. From Moses's perspective, this was part of the draw. Although he argued there were also other advantages: in addition to relieving nettlesome traffic at the tunnel and East Side bridges, Lomex would spur economic development, he said, raising property values, leading to higher tax revenues, new commercial activity. It would infuse fresh life into a neighborhood in need of a face-lift. Or so the official line went.

The residents who lived and worked in the area, of course, saw it differently, as La Mountain explained. Local merchants and business owners had been speaking out against the plan since 1959. Early on, they had conscripted several local assemblymen to their cause, including Louis DeSalvio, the state legislator who had been so supportive in the West Village battle. The heavyset, balding DeSalvio, a good Catholic and a neighborhood stalwart, had filed a bill in Albany to remove state authorization for the expressway, arguing that two thousand evictions would create "great hardship and suffering to the families involved." The city, he added, was already swamped with relocation cases it couldn't solve.

Moses had responded with menacing threats that postponement would jeopardize the federal money. If too much time elapsed, the feds would walk away from their commitment; the city stood to lose tens of millions in funding. DeSalvio grumbled that Moses was "hitting us below the belt" by frightening the city with the prospect of lost funds. City officials, shifting gears, began assuring residents that there was "ready availability" of loft space nearby for housing evicted families and businesses. It was a dubious assertion.

La Mountain and DeSalvio dispatched several busloads of citizens to picket city hall at a subsequent hearing. But a flood of fervent pro-Lomex testimony unleashed by the New York automobile club had taken the day, swamping their voices. It was discouraging. Moses persisted in portraying opponents of the expressway as "naysayers taking pot shots at progress."

The hearings continued. "Don't approve this road if you want to remain in office," Leon Seidel of the West Village Committee, now also a soldier in the Lomex fight, warned at a June meeting of the Board of Estimate. DeSalvio implored the board to "kill the mad visionary's dream" to "cut the city's throat with this stupid idea." A protester read aloud a letter from Eleanor Roosevelt, urging the city not to build Lomex. The board members sat stone-faced and listened, apparently unmoved.

Through all this, approvals for property condemnations were proceeding apace, La Mountain said. Building values along the expressway route, once relatively stable, had plummeted since the neighborhood death sentence, creating a sense of futility and despair. No one was sure how to move forward. People felt doomed, knowing that their homes and businesses would soon be lost, that eviction notices would be arriving, subjecting them to a relocation process no one believed would be fair. Some in his congregation had stopped showing up for mass, too distraught to sit in the lovely old church they had attended all their lives. The church they were now losing to a highway.

Father La Mountain broke off now. He was worried and discouraged.

He had done everything he could since his arrival in 1960, fresh out of seminary: organized meetings, joined neighborhood commit-

tees, brainstormed with local assemblyman. He had even managed
to arrange a face-to-face meeting with Mayor Wagner, to plead the
neighborhood's case. But the fix seemed to be in, and he was out of
ideas. Having heard of Jacobs's success in saving the West Village,
he hoped she might help.

Jacobs bit her lip. She identified deeply with La Mountain and
his flock. From everything he was describing, it was a neighborhood
much like her own: a little rough at the edges, a mostly working-
class district that was home to a mix of immigrant families—Italian,
Eastern European, Puerto Rican, and lately, a smattering of professional people like herself. The rents were still reasonable; the small
warehouses and simple, four- and five-story walk-ups perfectly serviceable, if a bit run down. These people were salt of the earth,
blue collar and hardworking. They were raising families as best they
could. For most, their homes and small businesses were all they
had. They had nowhere to go, no one to turn to for help. She knew
the gnarl of emotions these threats raised: fury, fear, paralysis. Having just come off a similar fight, the memories still seared. But she
also knew how much raw energy and time these battles required,
how much perseverance and mettle. She wasn't sure she could make
a commitment, she explained to La Mountain, as she walked him to
the door. But she promised to give it some thought. And she agreed
to come and sit in on their next meeting, if in some way it would
be helpful.

Standing at the open door, watching La Mountain's robed figure
recede down the block, she resolved to at least walk the route where
Lomex would run. She owed La Mountain that. She would get a feel
for the blocks, see what she thought.

IT WAS EARLY MORNING, AND THE AIR WAS STILL COOL WHEN JACOBS
left the house. The light was brilliant, the sky a cornflower blue. She
walked south, along Hudson Street. Sunlight flecked the sidewalk
beneath the trees; her shadow bounced along. There was still little
traffic. After about fifteen blocks, the peace of Hudson Street began

to give way to a more industrial feel. She crossed Houston Street and the streetscape changed; then Prince Street, then Spring. When at last she reached Broome Street, she turned left and started east.

The air reeked of diesel and hot tar now, spices and axel grease. The cobblestone street was teeming with men, hectic with shouts and the whine of machinery. Delivery trucks pulled in and out, workers heaved boxes and barrels onto handcarts, doors to freight elevators clattered open and closed. She stopped and took it all in. The loading docks thrummed with activity, every building seemingly in use. Though the five- and six-story warehouses were utilitarian, home to textile manufacturers and other light industry, they were not without architectural interest. Many, she noted, had cast-iron facades, lavishly detailed columns and cornices. The buildings had an air of elegance that was unusual for an industrial district. She'd read that in the mid-nineteenth century, when these warehouses were built, they were known as "the Palaces of Trade." She could see why.

She walked on, stopping in front of the six-story Gunther Building at 469 Broome to admire its curved corner and rolled glass facade; then again at the corner of Broome and Broadway, to study the E. V. Haughwout Building. It had once boasted a department store that was considered the "Tiffany of its day." These were not insignificant buildings, she reflected. If necessary, they could be refurbished and reused. And they marked a critical moment in the city's history. The lovely old firehouse between Mott and Elizabeth Streets was built in the Renaissance Revival style. The police headquarters at the corner of Broome and Centre was a classic Beaux Arts building. She and Bob had stood at this very spot and admired it years ago. Odd as it seemed, there was something almost reminiscent of market areas of Paris here. It was the blend of elegance and utilitarian purpose, she supposed. The scale was human, a reprieve from the skyscrapers farther downtown, the facades varied and interesting, some quite ornate. She shook her head, walked on. All this would be lost if the expressway was built: the hole-in-the-wall pastry shops and the spaghetti joints of Little Italy; the live poultry market and noodle houses of Chinatown; the street vendors and open-air food stalls, even the shady park beside Chrystie Street, with its benches and ball fields. She stopped again. Demolishing this swath of Lower Manhattan made no sense. The pulsing, multiethnic character of the area felt earthy and inviting, busy and commercially vital. Standing on the sidewalk, watching the swirl of humanity, the eddies and momentary bottlenecks as people moved along, she tried for a moment to imagine this same route were the highway to go in: the march of heavy steel risers as far as the eye could see, the girders overhead, the garbage and the stench below. No one would want to walk anywhere near this awful highway, she reflected. It would kill the adjacent blocks in every direction, eviscerate the community. And the area south would be amputated from the rest of the city. At night, this once-lively neighborhood would be a wasteland.

Jacobs stiffened, her sense of outrage rising. Here was another example of an area, not so different from her own neighborhood, that had all the bustle and stir she had described as essential to a thriving city. And once again, arrogant city officials were out to destroy it, in this case sacrificing a flourishing neighborhood for the false idol of the car. For all the official promises, she didn't see revitalization here; she saw ruin, wanton and pointless destruction. By the time she met La Mountain at the meeting a week later, Jacobs had decided to give the Lomex battle her all.

THE TINY CHURCH AUDITORIUM THAT EVENING WAS PACKED, THE COLlective energy palpable. Glancing around, Jacobs was impressed by the diversity of the crowd. It was reminiscent of the West Village fight: a gathering that cut across social and political lines. There were Republicans and reform Democrats, artists and elderly Italian grandmothers, professionals and blue-collar laborers, shopkeepers, restaurant proprietors, housewives, Village bohemians, immigrant families, blue blood New Yorkers, and even, it was rumored, a few representatives of the Mafia, whose turf had traditionally been Little Italy. She recognized some of the faces.

La Mountain took the floor and welcomed the crowd, offering a brief review of where they stood. Then, without further ceremony, he presented Jacobs, who by now needed no introduction. There was a burst of applause. Jacobs and DeSalvio were quickly appointed co-chairs of the newly anointed "Joint Committee to Stop the Lower Manhattan Expressway" and they set to work.

Jacobs was an inspiring leader: open, inclusive, down-to-earth. She was flexible and emphatically democratic. Organizations, like cities, worked best when power bubbled up from below, she said.

A masterful strategist, within weeks she had organized a series of guerilla actions, each designed to grab press attention. For one, she furnished residents with gas masks, which they wore to a city hall hearing, to underscore the soot and fumes that would foul the air were the expressway built. Press cameras flared, and the front pages of the next days' papers were splashed with theatrical images of grandmothers holding children's hands and wearing gas masks. For another, she staged a mock New Orleans funeral march along Broome Street, accompanied by a live band playing Dixieland music. Hundreds of protesters marched the route. Dressed as skeletons and other ghoulish characters, they carried tombstone-shaped pickets, some painted with skulls and crossbones, others with the slogan "Little Italy—Killed by Progress," to symbolize the death of the neighborhood. Local politicians from both parties, including then-congressman John V. Lindsay, joined the spectacle, as did hippies and longshoremen, old ladies, children, businessmen, and artists. By now, two hundred community groups had committed themselves to the fight. It was hot and humid that evening, but spirits were high. Halfway along, a light drizzle began to fall on the marchers, who stopped in Little Italy for speeches. Jacobs, DeSalvio, and several other officials stepped onto a makeshift dais, denouncing the highway amidst a chorus of rowdy cheers. And then Jacobs and her merry band of protesters pressed on, marching the length of Manhattan until they reached Gracie Mansion, on the Upper East Side, where they delivered a letter to Mayor Wagner, imploring him to "cancel land acquisitions" for Lomex. The press picked up the story in all its lavish and satiric detail. And, as before, it played well. Jacobs, as always, was ready with a pithy quote that evening. Approached by a TV reporter, she leaned into the microphone. "The expressway would Los Angelize New York," she declared. The sound bite made the evening news, just as she had hoped.

The anti-Lomex cause was gaining traction. Eager to aid the effort, a young folkie by the name of Bobby Zimmerman, not yet the superstar Bob Dylan he would become, collaborated with Jacobs on a protest song that they called "Listen, Robert Moses," repeating the street names in the area—Delancey, Broome, Mulberry—in a soulful refrain.

But Moses and his allies were intransigent. In no mood to be bullied by a few fanatical activists, they cranked up the propaganda mill. The opposition was "anti-progress," they asserted. "Every delay gives added hope to those . . . who make municipal progress a dirty word," pronounced the president of the Automobile Club of New York. If officials buckled every time a sputter of protest was raised, no redevelopment project would ever move forward.

"It was no more possible to rebuild a city without dislocation than it is to make an omelet without breaking eggs," Moses declared.

Wagner had put a young Hispanic politician, Herman Badillo, in charge of compiling a report with an explicit timetable for relocations. There would be no expressway, the mayor promised, "if no feasible relocation plan . . . was produced." It was a pledge Jacobs was quick to remind the press of when Badillo, as hoped, missed the deadline. City officials now began to worry. As a hundred protesters with boldly painted signs picketed city hall in opposition to the highway, the mayor and the Board of Estimate postponed a vote on condemnation hearings. It seemed wiser to wait until after the November 1962 elections, they decided.

Anticipating as much, Jacobs was immediately ready with a response. The residents are "outraged at this cat-and-mouse game of postponement," she told the press.

Finally, in December 1962, Badillo presented his report, promising that housing would be found for displaced families, without actually providing specifics. The city was hedging, Jacobs shot back. The report that the city should be releasing for public review, she insisted, was "engineering and economic data that prove the highway is feasible and necessary . . . especially given the other east-west routes that were now available."

Nervous city officials were now looking for ways to discredit Jacobs, and as it happened, Lewis Mumford's damning *New Yorker* review of *Death and Life* appeared that same month. It seemed just the blunt instrument they needed. Voices dripping with derision, city representatives now quoted from the review whenever—and wherever—they could. Jacobs knew, however, that for all Mumford's fury over her critique of him in *Death and Life*, he shared her anti-expressway sentiments. It seemed important to swallow her pride and approach Mumford, whatever the outcome. "This isn't about *The New Yorker*," she told him when she called. "Of course, I don't agree with you, but that doesn't matter." She asked if he'd write a letter of opposition to Lomex that could be read aloud at a critical upcoming hearing. Mumford obliged, writing a "wonderfully effective letter," Jacobs would later recall. "Nobody could have exerted the influence that he did . . ."

Mumford's letter was read aloud that same December. The expressway, he wrote, repeating an argument he had made in the *New York Times*, "would be the first serious step in turning New York into L.A. . . . Since L.A. has already discovered the futility of sacrificing its living space to expressways and parking lots, why should New York follow that backward example?"

Forty-four other people rose to speak out against Lomex that day, including seven elected officials, representing both the Democratic and Republican sides. "Except for one old man, I have been unable to find anyone of technical competence who truly is for this so-called expressway," Louis DeSalvio roared into the microphone. "This cantankerous, stubborn old man," he added, referring to Robert Moses, should "realize that too many of his technicians' dreams turn out to be a nightmare for the City." Jacobs, biding her time, was among the last to weigh in. The proposed expressway was a "monstrous and useless folly," she declared. The rationale for it amounted to "piffle."

Chatting with a reporter later, she was even more explicit. Lomex, she told the *Village Voice*, was but one link in a sprawling interstate system that was getting approved "a piece at a time, so people won't be able to grasp the whole picture." If Lomex wasn't stopped, she said, "we'll be fighting the tentacles of this stupid octopus forever."

The hearing had gone well, Jacobs reflected the next morning as she perused the papers. Better than expected, in fact. The press was now soundly on their side. The only disappointment was that La Mountain hadn't been there to see it. Unbeknownst to the committee, La Mountain had been summoned one afternoon for a tête-à-tête with a church official. Chastened for his neighborhood activism and told their little talk was to remain "confidential," he was advised that it would be best if henceforth he removed himself from the Lomex battle. While no one could corroborate where the order had originated, Cardinal Spellman, the archbishop, was a friend of Robert Moses; it was suspected it had come from him.

But it was already too late. Five days later, the Board of Estimate shelved the expressway plan by unanimous vote. The press coverage of the highway had grown increasingly negative, and the mayor indicated that he had no appetite for a fight.

"The most spectacular demonstration of the new-found powers of local citizens' groups," hailed the *New York Times* the following morning, noting that the expressway battle had lasted only a few months. "We won! Isn't it marvelous!" Jacobs scribbled beside an article in the *New York Daily Report* she promptly sent on to her mother. "You can well imagine what went on behind the scene!" she penned in a second note to her brother, this one beside a story in the *Village Voice*. A "traffic man told me this was the first time in the whole United States a federal interstate highway had been killed. Love, Jane."

Elated over their victory, Jacobs and La Mountain planted a tree in front of the church on Broome Street as jubilant parishioners looked on. But their victory lap, as Jacobs would learn, was premature. There remained one item of business still left undone: Lomex was still on the official city map, despite its defeat. The possibility existed that the hydra-headed monster could rise again, at some later date. "The rule of thumb is that you have to kill expressways three times before they die," Jacobs would later say.

LOMEX DID, IN FACT, RISE AGAIN, AS JACOBS PREDICTED. IT WAS FOUR years later, and community opposition again ran high. Although this time what followed was more troubling.

"Well, here I have been arrested again!" Jacobs wrote to her eighty-nine-year-old mother in mid-April, 1968. "I hope you won't think too badly of me," she added. But it was "better than just lying down and taking all this crookedness."

The arrest had seemed unwarranted, even at the time, said on-lookers; the provocation unequal to the charges. Two days earlier, a public hearing on Lomex had been held at a high school on the Lower East Side. A few state officials and a stenographer were seated on the stage, two hundred or so citizens on fold-down seats in the auditorium. One by one, residents had risen to testify against the expressway from a floor mike that faced not the officials, but the crowd.

"We want Jane, we want Jane," the crowd began to chant half-way through. People were being cut off by an arbitrary time limit, and the officials seemed disengaged. Though required by law to collect public opinion, it was clear they had no intention of acting upon what they heard. And so Jane had made her way up to the mike, setting in motion the events that followed.

"It's interesting," Jacobs observed, "the way the mike is set up. At a public hearing you are supposed to address the officials, not the audience." This was a sham hearing, she now charged, as "even the speaking arrangements indicated."

Promptly one of the officials leaped from his seat and turned around the mike. "Thank you, sir," Jacobs said, turning the mike back toward the audience. "But I'd rather speak to my friends. We've been talking to ourselves all evening as it is." Chortles rose from the crowd.

If the expressway went through, Jane continued, the community would be devastated. It was time, she added, to send the "errand boys" holding this bogus hearing a message to take back to Albany. Those residents who wanted to should join her in a march across the stage in a peaceful show of their opposition, she suggested, since no one seemed to be paying any attention to their words.

And so down the aisle they had come—elderly couples, small shopkeepers, Catholics, Jews, Italians, businessmen, artists. As the marchers reached the stage, the stenographer worriedly picked up her machine, spilling steno tape to the floor. Yards and yards of testimony. Jacobs, who had been leading the march, exited the stage. Others, following the pre-agreed-upon plan, picked up the tape and began tearing it into confetti-like pieces. Without the steno-typed notes, the officials couldn't prove they had gathered public comment.

"Listen to this!" announced an apparently surprised Jacobs. "There is no record! There is no hearing! We are through with this phony, fink hearing." A moment later, the official chair called for Jacobs's arrest. She was the only one among the fifty or so marchers detained.

The charge was disorderly conduct, she was told at the precinct house. Jacobs was fingerprinted and photographed and then led to the courtroom. On the way, the elevator operator asked her to identify herself. "I'm the prisoner," she said. The policeman tapped her on the shoulder and said, "Don't say that, dear, say, 'I'm the accused.'"

Jacobs had been arrested once before, in December 1967, in a nonviolent protest against the Vietnam War. On that occasion, she and 259 others locked arms on the sidewalk in front of the White-hall Induction Center in Manhattan. Among those charged that day were Benjamin Spock, Susan Sontag, the writer and activist Grace Paley, and Allen Ginsberg. Sontag was Jacobs's cellmate as they

awaited arraignment. In January 1968, Jacobs had pleaded guilty
to disorderly conduct and was given an unconditional discharge (as
were her cohorts). This time, however, the legal proceedings were
less transparent.

On April 17, when she appeared in court, Jacobs learned that
the charges had been raised from disorderly conduct to four misde-
meanors: rioting, inciting to riot, criminal mischief, and obstruction
of public administration. Collectively, they could carry from fifteen
days to a year in prison.

The charges, she told reporters, "bear no relation to what happened." She was being singled out as a troublemaker to spook the
community into silence. "The inference seems to be," she added,
"that anybody who criticizes a state program is going to get it in the
neck."

The community immediately rallied to Jane's cause, and by mid-
summer, public ire, a spate of pro-Jacobs editorials, a reasonable
judge, and some plea-bargaining took the day. After months of legal wrangling, Jane's charge was reduced to disorderly conduct and
she was allowed to go free. But the ordeal had taken its toll: Jacobs
was tired of fighting battles; she and Bob had decided to move to
Canada.

"I resent, to tell you the truth, the time I've had to spend on
these civic battles," she told Susan Brownmiller in an interview for
Vogue. The new book, she added, "was begun two years later than it
should have been because of that expressway and the urban renewal
fight in New York's West Village. It's a terrible imposition when the
city threatens its citizens in such a way that they can't finish their
work."

Jacobs was weary. But she also had another reason for leaving
the city: her sons had come of age and she was unwilling to see
them drafted to fight in Vietnam, a war that everyone in the family opposed. "They would have preferred to go to jail than to go to
war," said Jane. "And my husband said, 'You know, we didn't raise
these boys to go to jail.'" And so, in late summer 1968, to the dismay
of many in the Village community, the Jacobses packed up a VW

bus and drove north to Toronto, where they settled into a two-story house on a quiet, tree-lined street. Bob had a commission to design a hospital; Jane was eager to get back to writing.

LITTLE CHANGED, DESPITE LIVING IN A DIFFERENT CITY. JANE ADAPTED to her new life quickly, moving through the world as she always had: looking hard, fixating on what was interesting, clipping articles, reading voraciously, stashing away notes in her files, all with the hope that some of it, someday, might find a place in one of her books. "I don't make up my mind about things and then look for examples," she explained to a reporter. "I look for examples of behavior first and then eventually, when I start to see patterns in them, I begin to generalize."

Jacobs would go on to write five more books, none as popular as *Death and Life*, but each in its own right breathtakingly original. In *The Economy of Cities* (1969), she took on urban commerce, probing everything from Neolithic trade to changes in European weaving between the eleventh and fifteenth centuries to the development of the brassiere industry in New York. In *Cities and the Wealth of Nations* (1984), she questioned the wisdom of both Adam Smith and Karl Marx ("Ideology is narrowing and limiting," she said. "It prevents us from seeing the order.") In *Systems of Survival* (1992), she looked at the moral foundations of business and politics, moving seamlessly from Plato, Lao Tzu, and the Bible, to Italian mobsters and software pirates at modern universities.

True to form, she would continue to startle and challenge, rocking the status quo when it least expected it. During a visit to New York, where she was accepting an award at the UN building, Jacobs happened to receive from PEN, the international writers' association that works for freedom of expression, a new list of writers imprisoned in various countries. "I thought how ironic it was that I was getting an award for exercising our right of free expression in front of so many representatives of these countries where writers are being persecuted," she remembered. "So I started naming the countries and of course, a number of people began walking out." In an interview with *Vogue*, she flummoxed her interlocutor by refuting the popular notion that underdeveloped countries need population control to overcome poverty. "Take a look at the world," she countered. "If densely populated Japan and Western Europe were poor, and thinly populated Congo and Brazil were prosperous, a nice case might be made that people reinforce their poverty by their own numbers. But this is not so."

Jacobs's fascination with systems, both man-made and natural, would also continue, not only in the context of urban problems, but as she encountered them in the world. On a trip to Grand Canary Island, she described the self-sustaining ecology of a farm she visited with palpable excitement. While touring Scotland, she spent an hour in a bagpipe shop, learning the intricacies of how the pipes were engineered. On a visit to Goethe's house in Germany, you could "see the evolution of the piano in the instruments there," she wrote, and then proceeded to explain how it happened. "Everyone talks about the canals of Venice," and they are wonderful, she reflected, "but the streets are more so. Of course the streets and canals are together a web." Venice, she added, was easier to navigate than New York. "This is because every single place is its own shape and no other, and the shapes are easily comprehended. Even if you lose your place on the map, you look at the shape of the space you are in and find it, like finding the right piece of a jigsaw puzzle."

In the 1980s, Jacobs was approached by a local entrepreneur from Eugene, Oregon. The town was experiencing a severe economic downturn, the woman explained, and she was eager for Jacobs's thoughts on a grassroots development strategy for the area, which, like so many others, was facing the "jobs versus environmental dilemma." Corresponding by letter, the two women brainstormed over time about the need for "community development banking," which would make credit available to environmentally responsible farmers, fishermen, foresters, and real estate entrepreneurs. The idea eventually led to the establishment of an "environmental bank," which proved to be a great success, reinvigorating the once-foundering economy. So much so that in 1997, Jacobs, by then eighty-one, would join the group's board, curious, as always, to think more deeply about the nature of development, the mechanics of self-regulating systems, and the often disorderly looking processes that make such innovations tick. More than thirty years after the publication of *Death and Life*, she was still inspiring new generations of urban activists, planners, and entrepreneurs to think in fresh ways. It was one of countless examples.

Brilliant and quixotic, an intellectual adventurer to the last, Jacobs never stopped asking questions, or militating for more livable cities. She did so principally through her writing, unless, as her friend and editor Jason Epstein once observed, some "vital nearby neighborhood [was] at risk," at which point she got involved. (Soon after arriving in Toronto, in fact, she would jump into the fight against a proposed highway there, one of many grassroots battles she would join, despite the impositions on her time.) Preternaturally observant, she continued to mine the general implications she found in the particular. An architect tells the story of attending a seminar on cities to which Jacobs had been invited. All morning she listened patiently as the attendees spoke about street life, city planning, and urban policy. At the end of these wide-ranging and seemingly weighty discussions, Jacobs was asked to respond. To people's astonishment, he remembered, "she began by talking about the changes that Metro Public Works should be making to the design of the city's street drains. Their design, she explained, was a threat to bicycles because the bars and openings are placed in the same direction as the cycle wheels, frequently resulting in accidents as the front wheel gets stuck in the openings. She went on to discuss the need for design changes that would make [these] drains safe . . . , the need for cycle routes through city streets, the importance of alternate modes of transportation within the city, [and] the implications of urban form. She showed us the essential linkages between the small things that directly affect people's lives, and how the larger issues of the city depend on them," he remembered with admiration.

Jacobs never lost her buoyant spirit, even in old age. "Internally, I'm not any different from when I was younger," she told a reporter at age seventy-seven. "It's always a surprise to me that I don't get out of bed so easily, and I can't run up and down those stairs." Of growing older she said, "It's a fact of life and also it's delightful to grow older and see how things have turned out . . . You see you're a link in the continuum of life. There's that phrase—'the unexamined life is not worthwhile'—it's true, I think. The older you get the more you can examine it—not only your own life, but the life that's come under your view . . . My only regret is that the human span is so limited. I'm just so curious to see what happens."

Jacobs *did* live to see a lot of what happened—at least to cities. Speaking before a rapt audience at Boston College Law School in 2000, her tone was hopeful. "Cities on the whole," she allowed, are "doing much, much better. Cities are beginning to heal themselves . . . [to] get back their old pizzazz."

Unsaid, but widely understood among the multigenerational crowd that day, was that much of this was due to Jacobs herself, though her humility was such that she regularly scoffed at this idea, no matter how often it was suggested. But Jacobs did indeed change our thinking—about housing developments, strip malls, car culture, the social value of street life, the virtues of preserving historic buildings, and, perhaps most significantly, the implications that natural systems have for city design. For many reasons, as Sally Helgesen has written, she was decades ahead of her time. Not least among them because computers, the fundamental technology of the twenty-first century, so closely mirror her most essential idea, which is that interactive networks, not "top-down" hierarchies, are what move the world forward.

In 1997, Bob Jacobs died of lung cancer. He was seventy-nine. Though devastated, Jane soldiered on, living as she always had: gardening, cooking, writing, visiting her children. Mary had moved to British Columbia, Ned to Vancouver; Jim and his family were settled in Toronto. Nine years after Bob's death, on April 25, 2006, Jane was hospitalized in Toronto for an infection. It was two weeks before her ninetieth birthday. She died that afternoon.

In New York the next morning, the owner of the Art of Cooking, the Greenwich Village cookware shop housed at 555 Hudson Street, arrived early to open her business. Laid before the door, she found bouquets of lilies and daisies and an unsigned note: "From this house, in 1961, a housewife changed the world."

chapter THREE

....

JANE GOODALL

In the winter of 1961, in a far-flung corner of what is now Tanzania, a young Englishwoman sat alone in a tropical rain forest. She was just twenty-seven. But like Jacobs, who was eighteen years older and whose book *Death and Life* would appear that same year, her stature in the world was about to change. She too was on the verge of rattling the foundations of what had seemed a settled field, defying all expectations for her sex in her pioneering observations of animal behavior. Though at that moment, far removed from the creature comforts of ordinary people, living an elemental life alone in the wilds, a simple tent her only shelter, Jane Goodall didn't yet know the extent of it.

It was Goodall's second visit to Gombe, the remote game preserve on the shores of Lake Tanganyika, where a year earlier, the legendary anthropologist Louis Leakey had sent her to study chimps in the wild, caring little that she possessed neither credentials nor academic training, or that in the eyes of the scientific world, she was decidedly of the wrong gender for such work.

It was February, the depths of the rainy season; the grass at its highest was more than twelve feet in places. That morning, as always, she had risen well before dawn, setting out through the dim, canopied forest as the sun was rising. She followed the animal trails cutting through the underbrush. Sometimes she moved on all fours, crawling on bare knees through the dense scrub. Other times she slithered along on her stomach, her eyes scanning the ground for snakes, relieved when the path opened up again. She paused periodically to listen for chimps, faint, faraway cries somewhere deeper in the forest. She watched for dark shapes, a commotion in the leafy canopy, a hairy arm flung out in the course of feeding. She readjusted her direction according to what she heard and sensed. Quietly, painstakingly, she ascended the steeply forested slope, her body like a finely tuned antenna, until she reached the top, her perch these days. There, she settled in for a time, a pencil, her binoculars, and a warped cloth notebook slung across her lap.

The day had started well. Earlier, she'd managed to get reasonably close to a group of chimps, close enough, she jotted in her notebook, to see "sex, face color, etc." Dressed as usual in khaki shorts, a bland cotton shirt, and sneakers, she had learned to wear inconspicuous clothes; to approach slowly and in a nonthreatening way; to keep a respectable distance, according to what the apes seemed to tolerate on any given day. She had learned, essentially, to become "one" with the forest, weaving herself into its mesh as if she were no different from a bird's nest or a bushpig, alert to the rustlings in the grass, the faintest perturbation in the leaves.

It was almost a year to the day since her first arrival here. "It does seem a long time ago, in many ways," she wrote to her family. "Yet sometimes I can look at the Peaks & Valleys & see them with my early eyes." To Goodall's early eyes, the peaks and valleys had seemed alien and unforgiving, a nearly insurmountable challenge, made all the more so by the extreme weather. Sometimes the forest was like a "high powered tropical green house," she wrote home, the heat and humidity infernal, the winds fierce and wildly mercurial. There were slashing storms: biblical rains with thunder that shook the forest floor, lightning fissures that hissed and fractured the dome of the sky, howling morning gales that froze her bones.

The beginning had been the hardest. During her first confusing weeks, seeing the chimps at all, beyond a few fleeting glimpses, had been a feat unto itself. The terrain was perilously steep, with deep, plunging ravines and towering cliffs, presenting obstacles that for a human were impossible to overcome. A chimp encountering a cliff could grab a vine here, a tree limb there, scaling its face in a matter of seconds, while she, its less acrobatic cousin, was left to watch in frustration as it was swallowed into the dense underbrush.

Sometimes she hiked for hours without a single sighting. Chimps are acutely sensitive creatures, easily aroused and upset, running off as soon as they sense a human presence. And with her "minders" in tow, the two local men who the government insisted should accompany her on her daily searches, the problem was compounded. No matter how delicately they approached, the apes seemed to know they were near, erupting in panicked cries, only to vanish in an instant, as if by the wave of a wand.

She had known, of course, there would be challenges. Louis Leakey had said as much. But what those challenges were, she couldn't have imagined. Three years before, newly arrived in Africa, she had met the charismatic Leakey, her "life's mentor" quite by chance. Chatting at a cocktail party one night, she had mentioned she was looking for a job; she wanted something that would bring her into contact with animals, she said. "Then you should meet Louis Leakey," she was told. And so she had, calling up the famous anthropologist cold and making an appointment despite her lack of an introduction.

Leakey was fifty-three years old at the time, "a true giant" of a man, Goodall recalled, whose four decades of fossil hunting would help unravel the puzzle of humankind's prehistoric beginnings. Charming, wildly eccentric, a brilliant and iconoclastic scientist, he had already written eight important books by then and was head curator of Nairobi's natural history museum, the Coryndon. He knew Africa's Stone Age past like none other, having spent his life piecing together the ossified remains of ancient animal species long extinct. An enchanting host, he had given her a four-hour tour of the Coryndon that day and afterward offered her a job as his personal secretary. She had accepted on the spot.

And so, for the next year, she had learned all she could about the animals of East Africa at Leakey's knee. Not long after their first meeting, Leakey invited her and another young Englishwoman from the museum, Gillian Trace, to accompany his wife, Mary, and him on their annual dig at Olduvai Gorge in Tanganyika. Riding across the vast plains of the Serengeti, perched with Gillian on the roof of the Leakeys' overloaded Land Rover, she had been ecstatic.

They worked like demons that summer, chipping away at the hard, parched soil with pickax and shovel or hunting knives, the sun a fire on their backs. Toward midday, when the heat was most intense, they cleaned and sorted and labeled specimens, standing in the relative shade of a tarp. In the afternoon, they resumed digging and sifting, rarely knocking off until the light began to fade. She had loved it all, unfazed by the high heat and swirling dust, the dearth of water for ordinary washing and bathing, the mile-long walk to and from the site multiple times each day. When work was over, still energized, she and Gillian would hike up the sides of the gorge, roaming the plains to watch for animals before heading back to camp. "Admittedly, we don't meet lion & rhino around every bush," she wrote home. "But we do see Gazelle and . . . Jackals, mongooses & an occasional snake. I am disappointed there are not more of those." At night, pulling out their camp beds, she and Gillian slept under the stars, surrounded by the whir of insects and the occasional animal cry. It was a charmed interlude she would never forget.

It was sometime near the end of their stay on the Serengeti that Leakey first mentioned his hopes for a study of chimps in the wild. There were no guidelines or precedents for such a study, he told her. The physical challenges were legion. It would be difficult and dangerous, requiring enormous endurance and nerve, the ability to live alone for great tracts of time, to face uncertainty on a daily basis, for nothing was known about apes in the wild. Many believed such a study would be impossible, in fact. "I remember wondering," Jane later wrote of that night, "what kind of scientist he would find for such a herculean task."

But back at the Coryndon that fall, Leakey had proposed it to her all the same, convinced she could do it. Watching her work that summer, he later told her, he was struck by her unwavering focus, her energy and grit and perseverance—even in the face of numbing tasks. She had seemed to possess an unusual capacity for solitude, to thrive on minimal food and little sleep—and to be afraid of nothing. These were unteachable qualities, distinctly different from the abstract concepts that came with an academic degree.

Watching her work that summer, he'd also noticed something else: a certain inner stillness, a calm that had got him thinking. Perhaps, he'd begun to muse, a woman would be better suited than a man to a study of wild animals, quieter and less threatening, more patient as an observer, "less likely to arouse male aggression."

And on this, as on so many of his hunches, the great scientist had been correct. Despite the dangers and the isolation, the difficulties of sighting wild apes on the move, nothing had stopped his young protégé from pushing on with the chimp study. For four long months, Jane Goodall had scrambled up slippery mountain paths and slogged through waist-high grasses. She had pushed through thorns and thicket, dodged snakes and stinging insects, endured battering rains and tropical fevers, clammy clothes and moldy bedding, swarms of malarial mosquitoes and peevish minders—only to triumph. Louis Leakey's instincts about the hardworking Brit obsessed with the animal world had been spot on. Near the end of her stay, twenty-seven-year-old Jane Goodall had witnessed chimps using tools, an observation that a year later, in the summer of 1963, would turn the scientific world on its head, forever altering our sense of ourselves as a species, and our place in the natural order.

Nothing had been easy or comfortable about her time at Gombe. But then, Jane Goodall had never aspired to the easy or the predictable.

BORN IN NORTH LONDON ON APRIL 3, 1934, VALERIE JANE MORRIS-Goodall was the first of two children in a comfortable, if increasingly threadbare, household. Her mother, Vanne, was sweet natured and energetic, a gentle beauty with deep reserves of patience and sociability. Progressive for her day, she believed in reasoned discipline rather than physical punishment, the latter the norm for most British parents at the time, choosing to encourage her two daughters, rather than to shame or suppress them. Jane's father, Mortimer, was a charming if enigmatic man, vital but restless, a racecar driver who was unreliable with money and frequently away. His family, at one time wealthy, had for several generations lived off the fruits of a vast playing-card fortune without adding to the principal. By the time Jane came along, what remained of their riches was mostly gone. The greater part of Mortimer's portion had been staked as collateral so he could buy the Aston Martin that launched his racing career.

Mortimer was living in London when he met Vanne. It was the early 1930s, and Vanne had moved to the city from her family home in Bournemouth to work as a secretary for a showbiz impresario. Secretly, she longed to be a writer. After hours, she would sit in her boardinghouse room, banging out short biographies of theater people, imagining it would someday be her life's work. All this was interrupted one evening when a handsome young man who lived two floors up introduced himself. Soon Mortimer was driving Vanne to work in his Aston Martin. He was lighthearted and gentle, and he "laughed a lot." Later Vanne would say that Mortimer introduced her to a "whirly kind of life."

Mortimer was a big spender who went through every penny he earned. Though he already had one foot in the racing world, he was still working for a company that laid down telephone cable across England, drawing down a decent salary. Even so, he was perennially in debt; Vanne, giddily consumed with her whirly new life, chose not to notice. They decided to marry, tying the knot in September 1932, in a church in Sloane Square. Afterward, they drove to Monte Carlo in Mortimer's new green racing car. When they returned, they found a small town house in Chelsea to rent. Vanne quit her job so she could follow her new husband from race to race, where, flying by in an open cockpit at fabulous speeds, unprotected beyond goggles and a fitted leather helmet, he was beginning to make a name for himself. (Drivers in those days raced without benefit of safety harnesses or seat belts.) Her job was now chief enthusiast for Mortimer's burgeoning career. And when, in June 1933, he drew notice for his driving skill in the grueling Le Mans Grand Prix d'Endurance, marking his arrival as one of Britain's top racecar drivers, Vanne was as elated as her husband. Almost nine months to the day, on April 3, 1934, their first daughter, Valerie Jane Morris-Goodall, was born.

He was a disengaged father, cool and emotionally remote, almost from the beginning. Whatever the ties of home and hearth represented, they were no match for the high-stakes world of men and revved-up machines, the easy fellowship of those who shared his appetites for glamour and vertiginous speed.

Vanne treated her husband's defection with patient forbearance. Mortimer's own father had died when Mortimer was nine; he had no model, she felt, for what a father should be. It was certainly more complicated than this, but it wasn't in Vanne's nature to assume the worst. Mortimer disappeared into his work, and Vanne gently papered over his absence, seeing no advantage to concentrating on what was missing. Jane later recalled that her father "touched me only once" as a young child.

In the spring of 1935, Mortimer moved the family from London to suburban Weybridge, home to England's most important racetrack. They settled into a rambling, three-story brick house with an overgrown yard in the rear. Wild sorrel had overtaken the once-trim lawns, adding to the garden's aura of romantic neglect. Jane, who could now run out into this riotous green world, was entranced. For hours, she crouched in odd corners of the garden, enthralled by the parade of beetles and spiders and ants that trooped by. A vital, and deeply feeling, child, she expressed a powerful affinity for living creatures even at this early age. And it was her good fortune to have a mother who recognized this passion. Vanne not only tolerated her daughter's growing earth love, she encouraged it. Kneeling beside the spellbound child, she taught her the names of all they saw, how to know the birds by their plumage, how to identify their eggs and their nests.

From watching came the still more powerful desire for possession. At eighteen months of age, Jane gathered up a fistful of earthworms and took them to bed. When Vanne arrived to say good night, the little girl proudly lifted her pillow, eager to show her mother the worms. Vanne gently explained that the wriggling creatures would die if left all night in her bed. "They need the earth," she told Jane, whose sober little face was now clouded with worry. Minutes later, the worms cupped in Jane's small-child hands, mother and daughter padded out to the garden, where Jane bravely returned them to the wet soil.

A dreamy child with a vivid fantasy life, Jane had an imaginary friend named Dimmy, who she claimed could fly, a trait she envied enormously. Dimmy could be a cutup. Sometimes he said things that made her laugh out loud.

Jane loved the visits the family made to see her grandmothers, especially Mortimer's mother, whom she called "Danny" Nutt, "Granny" being too difficult to pronounce. Danny and Major Nutt, Mortimer's stepfather, lived in Kent, in a great eighteenth-century stone edifice called the Manor House, built, somewhat improbably, inside the ruins of a crumbling castle. The Manor House and castle were on the grounds of a nearby racecourse, which Major Nutt managed. The castle was a marvelous and "melancholic" place, Jane remembered, "all gray, crumbling stone and spider webs." What Jane liked best, however, were the animals on the grounds: the pack of foxhounds that her grandfather kept; the squawking geese that strutted the lawn; the hens roosting in the five henhouses, who she was allowed to feed, and where she went to collect eggs; the occasional mare and colt from the neighboring racetrack. Life for the cosseted little girl was interesting and full. And then, to her great displeasure, another creature arrived, this one noisy and decidedly unwelcome: a chubby baby sister named Judith Daphne, born four years to the day after Jane. The unanticipated visitor fussed and squalled all night, stealing Nanny's and Vanne's attention, which was vexing to the child, who already had to share her mother with a father she barely saw, except from the stands of the racetrack. Vanne went out with

Mortimer on most evenings, still immersed in their fast, whirly life. Sometimes they were away on weekends, traveling for races.

There are certain "irreducibles" in the character of each of us, Judith Thurman has written. Some children have an intensity of nature from birth—a pressing inquisitiveness—while others are wary, passive, or withdrawn. While those "original qualities" can easily be dampened, they also "define a mysterious ground of one's being that defies analysis." Jane was always electrified by her passions. She was an active and intensely curious child, fiercely independent and full of ideas, a dreamer with a capacity for sustained attention that set her apart, even as a small child.

A story often repeated from her early childhood seems to have anticipated the person she would become. It was the fall of 1939, the same cheerless September that the Swiss chemist Paul Müller made his breakthrough discovery of DDT and Rachel Carson, thirty-two at the time, was living in Baltimore, working on her first book about the sea. Jane Goodall was just five.

Earlier that year, Mortimer had moved the family to a seaside resort in France, to be closer to the European racing scene. Though war was still a dark fantasy, Vanne had begun noticing English and French soldiers in town. Then one day, she got word that the family needed to leave France immediately. That same night, Nanny and the children left by boat from Boulogne. Vanne waited for Mortimer, who was away on racing business. A few days later, he and Vanne crossed the Channel and drove to the Manor House, where all of Mortimer's extended family waited with apprehension. Hitler invaded Poland on September 1, and two days later, England declared war on Germany. Soon after, the army commandeered the Manor House and its outbuildings, and Danny Nutt turned one of the racecourse bars into a canteen, where she doled out hot tea, cigarettes, and chocolate to the massing soldiers. Vanne went there to help in the afternoons.

It was on one such autumn afternoon that Jane disappeared. An independent child, she often went off alone, usually to visit the stables, to pat the horses and feed them withered carrots. But she wasn't at the stables that day, nor had Nanny seen her playing near the house. Vanne had arrived home late that afternoon. The child, she realized, had been missing for three hours. The police were called and the neighbors alerted. Even the soldiers billeted on the grounds joined the search. But no one could find the girl. As darkness descended, everyone looked tense. And then Vanne spotted "a small, disheveled figure coming a little wearily" across the field. There were flecks of straw in her hair and clothes, but her eyes, "though darkly ringed by fatigue, were shining."

"She's found!" someone called.

"Wherever have you been?" Vanne asked gently, bending down to speak to the child.

"With a hen," she answered brightly.

She had wanted to find out how hens lay eggs, she explained, and couldn't see an opening big enough for an egg to drop through. So she had crept into the henhouse and crouched in the straw, waiting quietly for five hours. Finally a hen came in, raised herself up from her makeshift nest, and gave a little wiggle, whereupon "a round white object protruding from the feathers between her legs" appeared. Then, with a plop, it landed on the straw. Wisely, Vanne chose not to reproach her unusually curious child. Looking back on the incident years later, she would recognize it for what it was: though just five, Jane had displayed all the qualities of a born naturalist.

Mortimer went missing on almost that same day. When he returned he was wearing a starched soldier's uniform. He had joined the battalion bivouacked in the stables beside the family home. That December he shipped out for France. Vanne decided to move to Bournemouth, to live with her mother, the second "Danny," whose house they called the Birches, a spacious red brick Victorian with gingerbread trim and a lovely tree-shaded garden. Jane would later say it was her first real home.

Bournemouth was a seaside town, and thus a potential target for foreign invasion. Men were laying land mines along the beach, stringing barbed wire across the sands, planting guns in the niches of the cliffs. The town was closed to tourists and the military took over the forty largest hotels. Jane and her sister kept suitcases packed with a few necessities in the front hall, in the event the family needed to flee. They were outfitted with gas masks.

Mortimer was "somewhere in France." Occasionally he managed a brief visit home, but he was undergoing officers' training now. His letter writing dropped off, and soon after, his visits. Even on those occasions when he *did* appear, his stays were disappointing: "We had tea and still Daddy dident come," nine-year-old Jane wrote in her diary in 1943. "Then I went up stairs had a bath and dressed up as a fairy (in my nighty) . . . Then Daddy rang us and said he wasent coming till to-morrow." In 1944, Mortimer shipped out to India; the following year, he was transferred to Burma. The bomb was dropped and Japan surrendered. But Mortimer stayed in Kuala Lumpur, returning to England only briefly in 1949, so he could race at Le Mans. By 1950, he had been posted to Hong Kong. Then one day a letter arrived for Vanne, asking her for a divorce. The loss, now official, was by then a quiet one; even for Jane, his absence had already been absorbed.

Life at the Birches, meanwhile, went on, cheerful and largely unchanged. It was a lively and embracing household, ideal for a growing child, a household of women now, the men having gone off to fight the war. Vanne's two sisters, Olwen and Audrey, were also in residence. Olwen had a "bawdy" sense of humor and a mischievous streak. There were two female lodgers as well, usually single women who had lost their homes to the war's upheavals. The sole male was Vanne's older brother—kind but stodgy "Uncle Eric," a surgeon in London who visited every third weekend. It was the only time the children were told to tiptoe around.

Living in a matriarchy had distinct "advantages," Jane would later say. "I was never, ever told I couldn't do something because I was a girl." Which didn't mean that either she or Judy got an automatic pass. Quite the opposite. Danny was the "strong, self-disciplined, iron-willed Victorian who ruled over us with supreme authority" even though she "had a heart big enough to embrace all the starving children of the world," Jane wrote years later. If you coughed but you weren't really sick, you could expect no sympathy; she believed in the "stiff upper lip" approach to life.

If Danny was the titular head of the household, it was Vanne's word that held sway in matters involving the children. There were rules, Jane remembered, but there were also indulgences. Rationing began in 1940, and butter, sugar, and meat became scarce. There was very little money. But Danny was a wizard in the kitchen. She could turn one egg into a confection that satisfied six hungry stomachs.

The war left its indelible mark. Its privations taught Jane frugality and welded her family into a tight little corps. It killed "a beloved uncle" and took away her father, who vanished into the military, never really to return, except for brief appearances. The war maimed lives and blew apart nations, killing millions, incinerating cities, unleashing atrocities that were once unthinkable, from the death camps to Dresden to Hiroshima. Jane was five when the conflagration began, eleven by the time it was over. By the final months, as the Allies swept across Europe, she knew some of the worst, having seen the photographs from Dachau and Auschwitz. And yet, though the war touched her in countless ways, it didn't tamp down her curiosity or her joy. Jane found her "separate peace" in the lush green garden behind the Birches. She found it with her sister, and their two best friends, Sally and Susie Cary, who visited during school holidays. And, perhaps most significantly, she found it amidst the menagerie she kept in the garden.

There were too many creatures to count. Jane had a dog, two pet tortoises, and a succession of cats; a terrapin, two guinea pigs, and a canary; a cache of snails, a stable of caterpillars, and a "lovely big slow-worm," thanks to a gift from her aunt. The canary slept in a cage, but was free to flit about her room. The guinea pigs were given string leashes, to keep them from disappearing into the hedge. They were joined by a pet hamster, who sometimes "nested" in the upholstery of a chair. The snails, which she and her sister shared, were "racers," with numbers painted on their shells. They were corralled inside a bottomless wooden box, topped with a piece of glass, so they could dine on dandelion leaves, as the box was repositioned around the lawn.

Most children form an early identification with small animals and benign nature, Dale Peterson, Jane's definitive biographer, has written, a sympathy reinforced by their nursery storybooks. But as they grow older, this "gently atavistic fascination" gradually fades. For Jane, her identification with animals was formative, inseparable from who she was, and would become. At five she began riding a pony named Cherry. At eleven she sold her dollhouse to buy a dog. As soon as she could hold a pencil, she was posting letters to her friend Sally Cary, reporting on the foods her various crawly friends ate.

An early and ardent reader, sometimes she spent whole days with her nose in a book. Like many impressionable children, she was enchanted by the usual English classics: *The Wind in the Willows*, the Peter Rabbit stories, *The Secret Garden*. But the book that ignited her fantasies most was *The Story of Doctor Dolittle*. "I read it all the way through, and then I read it through again," she later wrote. "I had never before loved a book so much. I read it a third time before it had to go back." The child who identified so "ecstatically" with animals couldn't forget the story of the good physician with a houseful of pets who loved animals so deeply that he learned how to speak their language, and later went to Africa to save the monkeys dying of a mysterious illness. Danny gave Jane the book that Christmas. And so was born the luxuriant dream of going to Africa to live with wild animals. By the time Jane had moved on to the Edgar Rice Burroughs series, featuring Tarzan of the Apes, the orphaned son of English aristocrats raised in the African wilds by an ape mother, the girlhood fantasy had morphed into a given. Everyone in the family accepted the idea that somehow, someday, Jane would leave them to live with the animals of Africa.

From reading, it was only a small step to writing. At seven, Jane was composing sentences and stories; by ten, writing long and grandly in her journals. She sent out stacks of letters. And when the

Cary girls arrived for holidays, she wrote plays, which were staged for the household.

The two Cary girls were regular visitors, staying at the Birches for long, languorous stretches of the summer. They were a busy and mischievous little band, on their own for large tracts of the day. Jane, as the oldest, was the ringleader. Fourteen months older than Sally, she was the one who generated the ideas for what the troupe would do. When it was hot, they stripped off their shirts and hosed each other down. Sometimes, hiding in the trees, they hosed down unsuspecting pedestrians. Jane was always enormously inventive. She was bossy, but she made life gayer.

Academics, unfortunately, didn't elicit the same excitement. Jane found school achingly dull, irrelevant to her interests, and numbingly routine. She bridled at the regimentation, the arbitrary rules, the teachers she found despotic and obtuse. At eleven, she had enrolled at the all-girls Uplands school, a boarding school that also took in day students, among them Jane and her sister. The school was run by a "tall, very strict-looking woman with very thick glasses and boils everywhere," a friend remembered. The girls were required to wear uniforms, and attendance at morning chapel was expected. School was a trial to be suffered, a renunciation of everything that excited Jane's imagination: animals, wild nature, the freedom she found walking the cliffs near her house, her dog at her side. "Woke gloomily up to the dreary prospect of school, school, school," she despaired in her diary of 1949, "stretching into the future like some monster ready to swallow me up." Two years later, the dread persisted. Though she was a decent enough student, it had morphed now into something more serious: "I suppose that everyone goes through a period of utter despair in their lives," she wrote. "Well, I'm sure going through it."

The long slog into adulthood had begun, and Jane's unhappiness was beginning to express itself physically. For the first time, she was subject to migraines and mysterious fevers, insomnia and inexplicable fatigue. Her "sick days" grew more frequent. Whatever the cause—illness or depression—it scarcely mattered: the symptoms were debilitating, the sense of loss at having to let go of her childhood excitements real. She was also beginning to worry about the future. She was seventeen now and the question pressed. What would she do to support herself?

The career counselor, on her annual visit to Uplands, was baffled by the notion that a well-brought-up girl would want to study wild animals in Africa. When she learned that Jane also loved dogs, she looked mildly relieved: the answer, she suggested brightly, might be photography school. If Jane could learn to shoot pictures of people's dogs, she might make a living. The real message, of course, was that women didn't become zoologists or naturalists or scientists of any sort. These were men's professions, and even then, a male zoologist who elected to study primates might expect to do so in a zoo, or a laboratory cage in Europe or America—never in the wilds. The counselor's advice went unheeded.

And then, inexplicably, Jane's aversion to academics began to lift. At the end of the term, she was awarded a prize for writing. The honor required that she dress for the occasion, including stockings. "Foul things—stockings," she remarked in her diary. That August she had "my first Champagne cocktail." In November, she bought "her first girdle, a red one." Her aunt gave her nylon stockings for Christmas. ("I felt awful. I don't know why. Embarrassed somehow.") A new English teacher had arrived; they were reading Shakespeare now, lots of epic and lyric poetry; in biology they were delving into heredity and evolution, which seemed "rather interesting." There were other changes afoot too, these equally unanticipated. She was becoming an object of considerable male attention.

School edged to a close. During her final term, Jane won the writing prize a second time, placing well enough on her exams to qualify for university. But her family couldn't afford to send her, and it was unclear if she would have gone anyway. She was already preoccupied with what came next.

Journalism seemed a possibility, but Vanne, ever practical, suggested secretarial school, at least for the interim. With secretarial skills, she would be able to find work wherever she went. Mortimer agreed to foot the bill, and Jane, who had just turned nineteen, moved to London, where she rented a room in Chelsea and began classes at Queens Secretarial College in South Kensington. The work was "monotonous," Jane groused, but she was enjoying London's social whirl; by March 1954, she had completed the course, doing reasonably well, although the program head, writing in the school's confidential report, expressed reservations. Miss Morris-Goodall was "a clever girl, but rather smug" and "sometimes inclined to behave as if she has nothing to learn," she wrote. She was still "quite immature and not really ready for responsibility," and seems "very anxious to write." With time, she could be expected to abandon such fantasies and "settle" into a more realistic path: "Will eventually make a good secretary."

Jane turned twenty that spring, and applied for a secretarial position at Oxford. "I haven't given up the journalism idea by the way," she insisted in a letter to Sally ("but I have decided that to write anything worth anyone reading I must have . . . acquired a little experience of life, as they say"). She didn't get the job, but that August, when another clerical position opened up—this one at Oxford's Clarendon House—she had better luck. Unfortunately, the exalted setting did little to relieve the tedium of the job, which was "just very boring filing and a bit of typing. Absolutely at the bottom," Jane remembered. As her twenty-first birthday approached, Jane resolved to leave Oxford and the "boredom of this foul job." "Do you not think it is time I got something else?" she wrote Vanne.

A family friend secured her an interview at a commercial film studio in London, using his BBC connections from the war. To her delight, it went well and she got the job. By July she had moved back to London, where she found "a dingy basement room" with a bed, a gas ring to cook on, and a lone window looking onto a wall. But she adored the job, and London was thrilling. "Oh, Sally, I am having such a wonderful time," she wrote her friend. "No more typing, no more writing other people's stupid letters."

Schofield Productions was housed on Old Bond Street, above a photographer's studio and a nightclub, a stone's throw away from

Rachel Carson with her mother, Maria, her older sister, Marion, and brother, Robert, ca. 1909.
Yale Beinecke.

Rachel Carson's senior yearbook picture from Pennsylvania College for Women in 1928.
Chatham University Archives.

Mary Scott Skinker, college mentor, friend, and acting head of the biology department at Pennsylvania College for Women.
Chatham University Archives.

Rachel Carson, author of *The Sea Around Us,* in 1951.
Yale Beinecke.

Rachel Carson writing at dockside while working at Woods Hole Biological Laboratory, ca. 1951. Photograph by Edwin Gray. *Yale Beinecke.*

Rachel Carson at work in Maryland. *Yale Beinecke.*

Rachel Carson, Marianne Moore, and James Jones at the National Book Awards ceremony in January 1952. Carson won for her nonfiction book *The Sea Around Us.* *Queens Library/Archives/New York Herald Tribute Photo Morgue.*

Virginia and Marjorie Williams, Rachel's nieces, in 1942. Marjorie died at age thirty-one, leaving Rachel to raise her five-year-old son, Roger Christie. *Yale Beinecke.*

Eric Severaid interviewing Rachel Carson for *CBS Reports*, which aired on April 3, 1963. *Getty Images.*

Rachel Carson testifying in 1963 before the Senate subcommittee on the hazards of pesticide use. *Courtesy Library of Congress.*

LEFT: Jane Jacobs in about 1946, when she took a job writing for the Office of Wartime Information and then for the magazine *Amerika*, which was distributed in the Soviet Union. *John J. Burns Library, Boston College.*

BOTTOM: In 1947, the Jacobses bought a vacant, run-down building at 555 Hudson Street in Greenwich Village. They lived on the upper floors while doing most of the renovation themselves. *Jacobs family photograph.*

Jacobs read broadly and voraciously, from literary magazines and technical journals, to history and biography, to short stories, plays, novels, and murder mysteries.
Jacobs family photograph.

Jane by the fireplace with Ned and Jim in fall 1950. The house had no furnace so Jane would get up early and light the fire to warm the house.
Jacobs family photograph.

Jane Jacobs in 1960 at the White Horse Tavern, a legendary watering hole for writers, poets, artists, and musicians in Greenwich Village. © *Cervin Robinson. 2017.*

LEFT: August 1962. Jane Jacobs being interviewed by a journalist at a demonstration against LOMEX, the proposed Lower Manhattan Expressway. *Fred McDarrah/Getty Images.*

RIGHT: April 1967. Jane Jacobs (center) and her husband, Bob, with their daughter Mary (Burgin) at an anti–Vietnam War demonstration at United Nations Plaza in New York City. *Fred McDarrah/Getty Images.*

LEFT: December 1967. Jane Jacobs (fourth from left) with Susan Sontag and others in jail after their arrest following a Vietnam War protest in front of the Whitehall Induction Center in Manhattan. Benjamin Spock and Allen Ginsberg were also arrested that day. *John J. Burns Library, Boston College.*

Baby Jane Goodall with her stuffed toy chimpanzee, Jubilee. © *the Jane Goodall Institute. Courtesy of the Goodall Family.*

Jane Goodall writing up her field notes in her tent at Gombe Stream Chimpanzee Reserve. © *the Jane Goodall Institute/Hugo van Lawick*

Jane Goodall and her
mother Vanne sort
specimens in their
tent at Gombe Stream
Chimpanzee Reserve.
© *the Jane Goodall
Institute/Hugo van Lawick*

Jane Goodall at Gombe.
© *the Jane Goodall Institute/Judy Goodall*

Jane Goodall with her mentor, Louis Leakey.
© *the Jane Goodall Institute.*

Goodall during her early days at Gombe, sitting at a high spot in the mountains and searching the forest below for chimpanzees.
© *the Jane Goodall Institute/Jane Goodall.*

ABOVE: Jane Goodwall with baby chimpanzee Flint at Gombe Stream Research Center in Tanzania.
© *the Jane Goodall Institute/Hugo van Lawick*

LEFT: Goodall spent countless hours hidden in the vegetation observing the chimps through binoculars.
© *the Jane Goodall Institute/Hugo*

Jane Goodall handing a banana to David Greybeard, the first chimpanzee to lose his fear of her when she began her studies at Gombe.
© *the Jane Goodall Institute.*

Dr. Jane Goodall with Roots & Shoots members in Salzburg, Austria.
Robert Ratzer.

Alice Waters in the early 1970s, with Jerry Budrick, an early partner at Chez Panisse.
Courtesy of Alice Waters.

Driving through the back roads of Marin and Sonoma, Alice found hippies who were raising goats and beginning to make chevre as delicious as any she had tasted in France.
Courtesy of Alice Waters.

LEFT: Alice brought back mesclun lettuce seeds from Provence in the '70s and planted them in her own and in friends' gardens. At the time the delicate lettuces were unknown in America. *Courtesy of Alice Waters.*

ABOVE: Alice in the Chez Panisse kitchen in the late 1970s. *Courtesy of Alice Waters.*

LEFT: Alice with Carlo Petrini, the founder of Italy's Slow Food International Movement. *Courtesy of Alice Waters.*

RIGHT: Sometimes, before deciding on a menu, Alice would wander through the garden, sampling the fresh herbs and concentrating in such a way that she could begin to make associations.
Courtesy of Alice Waters.

LEFT: Groundbreaking for the Edible Schoolyard at the King School in Berkeley in 1995.
Courtesy of the Edible Schoolyard.

BOTTOM: Aerial view of the Edible Schoolyard twenty years later.
Courtesy of the Edible Schoolyard.

ABOVE: Hand-painted signs
made by students to
identify the crops they
are growing at the Edible
Schoolyard.
*Courtesy of the Edible
Schoolyard.*

LEFT: Chez Panisse, behind
wisteria vines and the
bunya-bunya tree that has
always bowered its modest
entrance.
Courtesy of Alice Waters.

where Vanne had worked twenty-five years before. The studio produced cinematic shorts for advertisements: films about women's stockings, men's shavers, motorcycle gear. Jane made herself available for everything. Soon she was greeting clients, projecting films, editing, splicing, helping choose music. The salary was a pittance. Sometimes, to make ends meet, she had to skip meals. But she was finding the cultural offerings of London diverting, second only to the drama—ongoing, a bit confusing—of managing what was becoming a steady queue of male suitors. Wild-hearted and vivacious, a natural flirt, Jane had become, through no action of her own, a beautiful young woman, a creature to whom men were hopelessly drawn. There were more prospects than she could manage, although Brian was the most problematic.

Brian Hovington had followed Jane from Oxford to London. He was good company, smart and certainly honorable, but the romance was hopelessly lopsided. Brian wanted a commitment, and she did not. "I have decided, even more firmly than when I spoke to you on the phone, that he is no husband for me," she wrote to Vanne. Brian, she added, was "too settled, too fond of creature comforts, doesn't like books . . . I could go on for ever, but it boils down to the fact that I just don't love him one bit."

For all the diversions of London, Jane was feeling vaguely dissatisfied. She loved her job, but something fundamental was missing, some larger purpose she couldn't quite articulate. And then word reached her that Rusty, her beloved dog, had been killed, run over by a car. She was heartsick. The dog had been her constant companion during her difficult years of school, a source of solace, even during her time in Oxford and London. Losing him made her stop and reflect more generally on what she wanted. So far, perhaps she too had been "too settled, too fond of creature comforts"—as she so dismissively described Brian. Lulled by the glamour of London, the small perks of her job, perhaps she too had opted for the "predictable" and "conventional." She needed, she realized, to extricate herself from London, Brian, her job. "I have decided that I must go abroad before it is too late," she wrote to Sally. She was "trying for some sort of job in Sweden," she added. But at the back of her mind, there lived a second, still brighter hope that she secretly nursed: a friend from Uplands, Clo Mange, had written her in the summer of 1955, just after she began working at the film studio. Clo's father had bought a farm just outside Nairobi, in Kenya Colony, and she had invited Jane to come for a six-month stay. Now she wrote to Clo again: Was the invitation to visit her in Kenya still open?

Clo's return letter arrived late that spring, confirming the invitation. Wild with excitement, Jane quit her job and moved back to Bournemouth, where she immediately found waitressing work at one of the posh hotels, living rent-free in the garden shed while she saved for the trip. It was an insane schedule: working three meals a day and two afternoon teas, with only one day off every two weeks. But by October, she had saved 240 pounds, enough for a round-trip ticket to Africa.

That New Year's Eve, Jane and Keith, her latest beau, went to a black-tie affair, where they drank Champagne and danced until almost dawn. Two and a half months later, in spite of Keith's marriage proposal that evening, Jane traveled by train to the Southampton docks with Vanne and Uncle Eric. Hugs and tearful good-byes were exchanged, and then the rapturous twenty-two-year-old with big dreams boarded the seventeen-thousand-ton passenger steamship that would carry her to Africa.

THE SEA JOURNEY TOOK THREE WEEKS. WAR HAD RECENTLY FLARED UP on the Sinai Peninsula, closing the Suez Canal, so the ship had taken the slow route by necessity, tracing the contours of the long African continent before finally rounding the Cape of Good Hope and sliding into Mombasa harbor at dawn on April 2, 1957. Jane and her fellow passengers disembarked, wandering the Arab markets in the wilting heat before returning to the boat for last farewells. Then she boarded the train for Nairobi, arriving the next morning just as Clo was emerging from the throng at Nairobi station, accompanied by her boyfriend, Tony, and her father, Roland Mange. Introductions were made and Jane's luggage collected. Then they drove due north out of the city along a rough dirt road into the highlands, arriving at the Mange farm, Greystones, in time for tea. That evening, after supper, Clo appeared carrying a large, lavishly frosted pink cake with twenty-three lit candles. It was Jane's birthday. And so began her African adventure.

Arriving in Africa, Jane would later say, felt oddly familiar, like a return to a place she'd always known. "Right from the moment I got here," she wrote Sally Cary, "I felt at home. Out here I am no longer mad—because everyone is mad." In a letter written to her family sometime during her second week, she elaborated: "I really do simply adore Kenya. It is so wild, uncultivated, primitive, mad, exciting, unpredictable . . . I am living in the Africa I have always longed for, always felt stirring in my blood."

Jane was indeed quick to adapt—to the climate, but also to the social whirl of colonial Nairobi. She intended to find work as soon as she could get her footing, not wishing to overstay her welcome at Greystones. But in the meantime, she and Clo were out nearly every night. Often they drove into town for lunch and a swim at the private Muthaiga Club before slipping into cocktail dresses and pressing on to one of the parties that lit up Nairobi's nightlife. She was mesmerized by the wildlife they passed on these drives in and out. She had already sighted springboks and a hyena, a group of colobus monkeys, and finally her first giraffe: "They are even taller and more impressive than I had imagined," she wrote home excitedly. One strode right into the road, then "walked away in a most condescending & stately fashion."

She was surprised by Nairobi's precipitous temperature swings: the way the sting of the day's heat was followed by the sharp chill of the nights, especially in the highlands, where, by necessity, a fire was lit every evening. The Kenyan climate can be "slightly degrading in its effect on some rather weak characters," she wrote home. Jane, however, loved the extremes. In this, as her biographer notes, she closely resembled her father. Jane had "surprising endurance, [an] iron stomach, minimal appetite," keen eyesight, and a capacity for sustained attention, even under extreme physical duress. But if she was her father's daughter by "constitution," she was her mother's in "sensibility" and character. Like Vanne, Jane possessed deep stores of poise and sociability, and in addition, a certain wildness of heart that men found irresistible. Indeed, within weeks of her arrival, she was already the object of several men's competing attentions.

Jane had begun seeing a man named Bob, who was part of the horsey set living in the White Highlands, a bastion of colonial privilege where riding, breeding, and racing horses was the main preoccupation. Bob was urging her to consider an equestrian future. She could train horses and play polo, he said, perhaps become a jockey. Unsaid but implied was the idea that she might also become someone's wife.

And then, quite fortuitously, Jane met a different sort of man, this one older, wilder, and considerably less conventional, a man more than twice her age who would change the trajectory of her life: the brilliant and iconoclastic Louis Leakey. Almost overnight, her attention began to pivot from the affluent White Highlands set, with their private clubs and pristine polo fields, to an Africa more in tune with her own sensibilities: the modest and more gritty milieu of scientists and researchers associated with the Coryndon, soon the beating heart of her new world.

Louis Leakey was an unusual character by any measure. Born in Kenya in 1903, the third of four children in a supremely unconventional family, he was the son of trailblazing missionaries who had raised him amidst the native Kikuyu tribe. He spoke their language, learned their games, and was schooled in their hunting skills. As a newborn, he was left outside the house in a basket, whereupon, in keeping with Kikuyu custom, all the tribal elders had trooped past to give their blessing: each stopping to "spit" on him. In early adolescence, also in accordance with tribal customs, he endured the Kikuyu's painful circumcision rites, signifying his manhood and his entrance into the tribe. At thirteen, like the other Kikuyu initiates, he moved from his parents' home into his own mud-and-wattle house, though he continued to be tutored in Latin and mathematics, and spoke French with his parents and siblings over dinner each night. At sixteen, his family returned to England for several years, where he was enrolled in a public school. He found the other boys "appallingly childish," and the school restrictions "absurd." In Kenya, after all, he had built his own hut, and for some years had been living unsupervised and alone, as an adult. And while he displayed no talent for cricket—a clear strike against him—he proved to be a passionate and brilliant student. His peers thought him an oddball, certainly. But his good looks and formidable intellect somehow saved him from becoming the butt of schoolboy taunts. He went on to study at Cambridge University, earning top honors in anthropology and archeology, despite his difficulty adjusting to the damp English climate and a culture that he continued to find infantilizing. In the summer of 1926, at age twenty-three, he returned to Africa to lead an archeological expedition, and, realizing he had found his life's work, never left. Over the next four decades, excavating tens of thousands of Stone Age tools, fossilized bones, and skull fragments, he would change the face of paleoanthropology, increasingly convinced that mankind's origins were to be found on the continent of Africa, and not elsewhere. It was a notion that contravened scientific theories of the time, but one that ultimately was borne out, largely as a consequence of his research.

Leakey's singularity went beyond his unorthodox upbringing, however. A bold and original thinker—intuitive, wildly independent, at times tough-minded—he was a maverick by nature as well as nurture. He often described himself as "a white African," which was an apt characterization. The peculiarities of perspective and circumstance that as a child had set him apart, as an adult freed him from any concerns about conforming to the expectations of polite society. As a grown man, he had simply lost track of what those expectations might be.

By the spring of 1957, when Jane first strode into the Coryndon, Leakey's eccentricity was on full display. No longer the handsome, vigorous, brown-haired Adonis of his Cambridge days, his hair had gone white, his teeth yellow, and he had developed a distinctive paunch. His standard uniform—a soiled, one-piece khaki coverall—was "missing buttons;" his pockets were "overloaded;" the knees of his coveralls ballooned. An indifferent bather and an unapologetic smoker (cheroots were his particular weakness), he tended "to stink." Yet his eyes blazed with spirit, and his charm and volubility, his raw exuberance, were infectious. As he walked Jane through the museum that day, she had been transfixed by the deep stores of knowledge he possessed. Dr. Leakey had given her "the whole morning," she gushed in a letter home. He had described experiments on lungfish that illustrated their amazing tolerance for drought. He had shown her the museum's extensive collection of snakes, about which she was "naturally very interested." Though Louis was not yet the soaring international colossus he would become, he and Mary, his second wife and professional partner of some years, had been engaged in their groundbreaking fossil excavations at Olduvai Gorge for some time. He was already an anthropologist of considerable stature, as Jane well knew. She felt enormously privileged to be in his presence, excited and lucky to be offered the job.

Louis's previous secretary, Rosalie Osborn, had left to study gorillas, he told Jane. What he was less forthright about sharing, for reasons that would later become apparent, were the circumstances of her departure. Rosalie was young and pretty and extremely smart. Louis had met her in 1954 while in England, working on a pig fossil study. The two had begun an affair, and in the summer of 1955, Rosalie moved to Kenya to become Louis's secretary. Disarmingly magnetic, even in old age, a lively and extravagant talker, Louis could hold "anyone in rapture," his son Jonathan remembered. "Women came to him like moths to a flame," a friend observed. "And he *enjoyed* it; he was a real human that way." Louis had on occasion had flings with his protégés before, though they were always casual. A shameless flirt, his weakness for women, especially those who were youthful and attractive, was no secret. Mary had learned to tolerate these occasional dalliances. (She herself, in fact, had once been the object of his errant eye.) But this one, she realized, was more serious. As his son Richard later recalled, Rosalie Osborn nearly became "the third Mrs. Leakey."

For years, Leakey had been trying to secure funding for a long-term study of the great apes. Such a study, he believed, would shed light on humankind's earliest beginnings. The forests of Africa were home to three of the four species of great apes that remained—gorillas, bonobos, and chimpanzees. For some time, in fact, Leakey had known of a particular tract of protected forest on the edge of Lake Tanganyika where a large population of chimpanzees still lived in isolation. But he had yet to succeed in launching a field study of any consequence. He had tried to send a young male researcher out as early as 1946, but little had come of it. A full decade had passed before he was able to renew his efforts once again, this time sending his pretty secretary Rosalie Osborn to Uganda for a four-month study of mountain gorillas. It was an elegant solution to his domestic troubles: a consolation for Rosalie, with whom he had broken off the affair, and a way to advance his own research into the lives of ancestral humans. But four months had proved insufficient time to collect enough data to draw any significant conclusions. And Osborn's mother, who until then had assumed her daughter was safely in Nairobi taking dictation for Leakey, somehow got wind of Rosalie's true whereabouts, whereupon she demanded that her daughter quit the gorilla study immediately. Osborn, who was just Jane's age, had returned to England in January 1957, only a few months before Jane appeared at the Coryndon, where she eagerly accepted the job.

Jane had known nothing of all this at the time. Had she understood the edgy marital dance into which she was stepping, or been aware of her predecessor's role in the drama, she would have been considerably less sanguine about the job, perhaps even turning it down. It wasn't until she'd settled into work at the Coryndon that September that she began to piece together Mary's disquiet at Olduvai that summer with the gossip in the air. The signs, however, had been there.

That May, when Louis proposed that she join Mary and him on their annual dig at Olduvai Gorge, he had predicated the invitation on Mary's approval. Jane had been on guard, unsure what to expect. Louis had told her beforehand that he wouldn't be calling her by her first name, as Mary would view this as "dreadfully familiar." "She takes violent dislikes or likes to people for no reason at all," he warned. It was more complicated than this, of course. Beyond Mary's approval, which seemed hurdle enough, they would have to see if they could find room to add food and water for another person, Leakey had added. They were already taking along another young Englishwoman, Gillian Trace. But Jane had gotten on well with Mary, whom she described in a letter home as "a small, lean woman, with blackened teeth, a perpetual cigarette, & short wavy hair." Mary had seemed "a little distant" on their first meeting, she told her family, but she quickly came around. After lunch that day at the Leakeys', Mary had turned to Jane and casually asked, "I hear you might like to come with us to Olduvai."

Louis Leakey's interest in Olduvai went back decades. In 1931, at dawn on his first day of digging, he had astonished his colleagues by uncovering a perfectly intact Stone Age obsidian hand ax, the first such human artifact ever found at the site. The significance of this discovery was tremendous, as Leakey well knew. Tools indicated toolmakers, which meant there was a high probability that in addition to further artifacts, they might *also* find the fossilized remains of ancient humans—or even prehumans, their hominid ancestors. This was the elusive quarry that Jane and Gillian were hoping to uncover that August of 1957, more than twenty-five years later. "The great aim," Jane scrawled excitedly to her family, "is to find the man who made all the tools."

They set off in the Leakeys' overloaded Land Rover six weeks after Jane got the official okay, she and Gillian squeezed uncomfortably amidst bedding, supplies, dog baskets, and the Leakeys' two dalmatians. On their second day, they ascended the high rim of Ngorongoro Crater and then dropped down to the plains of the Serengeti, which "rise & fall and are covered in golden sunbaked grass and clumps of shrub & thorn trees," Jane wrote. On their third, they awakened to "grazing herds of eland and gazelles

all around." Later, no longer on a road, but driving bumpily overland, they slowed to watch "the most beautiful cheetah who walked
along beside the car, quite unafraid, before turning rather disgust-
edly from our noisy machine." Jane was overwhelmed by the raw
majesty before her. She had found the Africa of her earliest dreams.

The living conditions at Olduvai, as Louis had warned, were
spartan, though Jane hardly noticed. Water was hauled in once a
week, arriving on a trailer hitched to a decrepit truck. To make it
last, each person was allotted a small bowl for daily washing, plus
a shallow weekly bath in a canvas tub. Jane and Gillian grew accustomed to living with "dirty, greasy hair," as she put it. They rose each
morning at dawn and walked the mile between camp and the excavation site several times a day, undaunted by the heat or the ticks,
the dust in their hair and noses, the demon wind that kicked up on
occasion—even the stinging scorpions, which they always kept an
eye out for. Day after day they worked, digging and sweating and
swinging their heavy pickaxes, surrendering themselves to the blis-
tering sun, captivated by the "mystery of evolution all around them,"
the visible trail of humankind's connectedness across the long sweep
of time.

The days that summer, Jane remembered, were long and physically exhausting, although punctuated by moments of "intense excitement," the nights soft and starlit, broken occasionally by the cry
of a hyena or a jackal somewhere in the dark: haunting "cat-like
yowls." The African staff slept in their own separate camp; Louis,
Mary, and the dalmatians in the rear of the old truck; Jane and Gillian in the tent they shared, although they often moved outside, surrounded, as Jane wrote, by "the whole immense vastness of Africa
and the Serengeti with the mysterious universe all around and very
real." A little of that first summer would always flicker in her heart.

Louis, Jane wrote home, continues to be his "utterly adorable"
self. He is "sweet" and thoughtful and "nothing is too much trouble." She thought the atmosphere of the camp very "friendly & joking & oh so pleasant—until dinner time."

Dinnertime was when Mary started to drink, pouring herself

glass after glass of brandy. Mary would wobble up to the table, quite "blotto," Jane added. It was her job to dish out the vegetables, but her serving was often comically unsteady. "One is liable to get $1/2$ a bean and 6 potatoes—whilst the cauliflower goes on the table." Had Jane known the source of Mary's brooding that summer she undoubtedly would have been more sympathetic. It was only later that she gained some insight into what fed these messy nights.

NEAR THE CLOSE OF THEIR STAY AT OLDUVAI, THE THREE LEAKEY BOYS and a couple of their friends joined the party. It was a "full camp" and "such fun," Jane wrote. All the young people slept outside, making it feel "rather like a dormitory." For weeks Jane and Gillian had sought permission to ferry their beds up the side of the gorge— for the view, and so they might see more animals—though it would put them at some distance from camp. Louis said it was too far away to be without a gun, but had finally given in, as long as Hamish, who carried a gun, was with them. One evening, while Mary and the dogs lay sleeping in the van, Louis had crept up the side of the gorge and joined them for a thermos of tea. Later that same evening, he appeared again, this time at 3:30 A.M. Jane was the only one awake. Louis had pointed out constellations as they quietly talked, trying not to wake Gillian or Hamish.

It was that evening that Louis broached the subject of a wild ape study, describing with great excitement his hopes for what such an undertaking might reveal. Before metal and fire and the emergence of language, before large-brained upright Homo sapiens roamed the Serengeti plains, what had the lives of our deepest ancestors been like? he wondered. How had they behaved with each other? How had they lived? Had they expressed love or loss, formed friendships, felt empathy? Had they possessed the capacity to forgive? Nurture their young? Play? Leakey could guess, based on fossilized reconstructions, what ancestral humans had looked like, how they moved, what they might have eaten. But behavior didn't fossilize; it couldn't be read in bones and teeth. Wild apes and humankind shared a common ancestor; years of excavations had convinced him of this. It seemed reasonable to assume, then, that if it were possible to identify any behaviors that were common to both modern apes and modern humans, then one could safely conclude that their mute common ancestor would *also* have shared these behaviors. Modern apes, in other words, held the key, he believed, to solving the mysteries of mankind's prehistoric past. This, at least, was the hope.

Back in Nairobi that September, Jane settled into life at the Coryndon. So far she was thrilled by the job. The letters she typed for Leakey were fascinating, the work hours flexible. The atmosphere was relaxed and the staff, she told her family, "charming & great fun . . . super." She liked that there was variety from one day to the next. One morning she would be out with Louis "watching ostrich courtship rituals" in Nairobi National Park, the next escorting a "charming Portuguese professor" about the museum.

But Leakey was becoming a problem. He'd begun showing up unannounced at her apartment at odd hours, one morning proffering a single red rose, a most unwelcome token of his affection. "I begin to see why Mary has taken to the brandy," Jane wrote home. At first Jane didn't take his romantic overtures terribly seriously, but when Louis continued to press, she began to worry. "Old Louis really is infantile in his infatuation and is suggesting the most impossible things," she wrote home. One impossible thing he suggested was a midnight rendezvous at Tsavo National Park, where they would camp for the night, he said, before setting out the next day to explore. Jane was horrified, as much for how his proposal appeared as for fears that he might try to pull any "monkey business." They were standing in a room full of museum trustees when he handed her the train ticket and hurriedly outlined his plan. She had been too embarrassed to gracefully decline on the spot. And when later he phoned her at the museum, wondering why she hadn't appeared at the Tsavo station, she had been sitting with another colleague, making it uncomfortable to go into her reasons for failing to show. At work that next Monday, Louis was furious and accused Jane of lying to him. She had let him down, he said, and stomped out of the room. Jane was visibly shaken, but when he appeared later to apologize, she forgave him. The incident, she assumed, was over. But that night, when she returned to her room, there was a note from Louis propped on her pillow: "I had to come back to tell you how much I love you." She was mortified. The whole thing made her feel "quite ill," she wrote her family.

Jane's romantic interests, in fact, were very much elsewhere. On the trip to Olduvai that summer, she had met a handsome young man named Brian Herne, who at the time was encased in a full-body cast. Brian had nearly lost his legs, barely surviving a horrible car accident. The driver had been less lucky, plunging to his death when the truck they were in lost its brakes and flew off a cliff. That September, finally shorn of his cast, Brian appeared one afternoon at the Coryndon in a red MG. He was stopping by to see if Jane knew the whereabouts of their mutual friend Hamish, he explained. Jane was initially unsure about Brian. He was tall and lithe, with chestnut hair and steady blue eyes—unquestionably attractive, but he seemed "very young," she wrote home, and she was put off by his profession. Brian was a big-game hunter, the youngest licensed professional hunter in Kenya. He made his living guiding well-to-do clients through the backcountry with the express purpose of killing rhinos, lions, buffalo, and other big game, a thought she could barely stomach. Though he was just nineteen, he seemed to have "an external layer of the typical hard bitten and tough white hunter about him," she told Vanne. But she had soon amended her first impression. After seeing Brian a few times, she decided that "the character underneath" was "one of the nicest I've come across out here. Loyal, honest, faithful, etc., etc." She was pleased when he kept calling.

They were kindred spirits, they found, both drawn to untrammeled nature, to the sensuous and the athletic, to animals and the thrill of adventure. Both loved the violent beauty of Africa, the emptiness and unfettered space. And both possessed a wild streak. Brian took Jane to the Serengeti whenever they could get away. He showed her untouched, "utterly remote" corners he had known as a boy. Brian sat with her under the stars. Brian was "the first person I've met out here I have liked sitting alone with," she wrote Vanne. She was falling in love.

Brian's family, Jane soon realized, was heavenly too. As were his friends. They were a crazy and uproarious lot, mostly young Kenyans, all of them "quite utterly and completely mad," she reported. But also "incredibly decent to each other," kind and loyal and enormous fun. It was "so nice to see." Soon Jane was running with the pack, hopping from party to nightclub to riotous evenings of all-night dancing. Jane was "daring and game to try anything," Brian remembered. She was unafraid of fast driving, of doing crazy things. One night they joined a gang of friends who were motorcycle fanatics. The guys kept talking about "doing the ton," which was code for pushing their bikes up to one hundred miles an hour. Jane expressed interest in doing the ton too. Soon she was straddling a roaring, deep-throated motorcycle, clinging to Brian, as they ripped along a ragged asphalt road outside Nairobi, watching the speedometer needle climb to a thrilling 105. It was hard to say who was giddier, Jane or Brian. Another evening, desperate to escape a wedding reception they deemed boring, Jane and Brian slipped out with a few friends to go dancing at a posh hotel. Once there, Jane did match tricks and walked about with a glass of beer balanced atop her head. Later, deciding it was time for a swim, Jane and a clutch of others stumbled through the dark until they found the hotel pool. Untroubled by their lack of bathing suits, Jane peeled down to her underwear and dove in, followed by Brian and another friend, pronouncing it "quite heavenly" in a letter home later.

By late fall, however, the romance had hit a rough patch. Brian was out of work and still in pain, too lame to resume his job guiding hunting parties. He was almost broke and beginning to worry. To help make up for his lost income, he and Jane had hatched a plan to partner on a series of animal-oriented articles with photographs, which they hoped to sell to American magazines. But like many such literary collaborations, this one was doomed to failure almost from the start. They squabbled about style and content, nearly coming to blows each time one of them made changes to the draft of the other. Eventually they had to abandon the project.

Still, Brian continued to delight Jane in other ways. One afternoon, he presented her with a baby bat-eared fox, an orphan from the Serengeti. The animal was immediately Jane's most beloved possession, accompanying her each day to her office at the Coryndon, where he slept at her feet. He was the first of a menagerie that soon included a vervet monkey, a mongoose, a cocker spaniel puppy, a Siamese cat, a bush baby, a hedgehog, another monkey, a rat, and a vast array of snakes. "But can you imagine," Jane rhapsodized in a letter home, "how paradisical (if that is the correct word) it is for me having all these animals around?"

By January, Jane was writing to Vanne about the "complexities" of her feelings for Brian. "Brian, in a lot of ways, I *could* marry," she confessed. "The point is that I *do* love [him]," she went on, but he would "have to change a *lot*" before she "could marry him." Even so, she admitted, "I have yet to find a more honest & essentially good hearted man than my Brian. If he didn't love hunting & if he *did* love music & literature, he would be *ideal*."

Considerably less ideal was Louis's behavior, which was beginning to be a serious problem. Even with Brian in her life, he refused to give up, sending cut roses to her room, trying to hold her hand, telling her he thought one day she might love him back. "Oh, yes. He *is* in love with me," Jane wrote Vanne. "He has been sweet & kind & helpful—& for that I am grateful. But what does he expect & hope—& with no right? Simply that one day—to quote— 'you will love me as much as I love you.' What right has a man of his age, already on his second wife, to expect & even *hope* for such a thing?" It was too grim to contemplate. "At Olduvai," she continued, "I was really & truly sorry for him . . . but I can never describe my utter & complete physical *revulsion* when I discovered that he expected me to be in love with him—*never*."

Jane continued to do battle: "We've had a little talk & he's promised to stop doing this," she told Vanne. But Louis couldn't help himself; despite his assurances, he kept going back on his word. At one point Jane threatened to leave the museum. The situation was becoming too awkward and uncomfortable. Finally she and Louis really "thrashed it out," she reported to Vanne, arriving, she added, at what she hoped was a more realistic state of affairs: Louis agreed to be "merely a father to me," she explained, and she in turn promised she would "trust him with everything as he valued my friendship more than anything else in the world." It didn't go quite as smoothly or quickly as Jane envisioned, but Louis did eventually make the reluctant transition from "suitor to mentor," though it took time.

Louis, meanwhile, had made one proposition that *did* interest her, of course, putting Jane in a position that was more complex, no doubt, than she'd initially understood. Sometime during their stay at Olduvai, he told her, he had decided that she would make "an extraordinarily promising" candidate to send on his ape study. He wasn't worried that she possessed neither scientific training nor academic degree. He preferred, he said, that his chosen researcher should go into the field "with a mind unbiased by scientific theory." What he was looking for was someone with grit and personal fortitude: "someone with an open mind, with a passion for knowledge, with a love of animals, and with monumental patience." Someone also who was able to endure months, perhaps even years, away from civilization, as he believed the project might take that long. He hoped she would accept his proposal, and she of course had.

And so it was settled. All that remained was to raise the necessary funds, which Leakey promised to begin working on at once. That fall he wrote to an anthropologist friend at the University of Chicago, soliciting his financial support. The study's duration would be four months, he explained, although he hoped it might be extended. He mentioned that "Miss Jane Morris-Goodall—from the point of view of personality and interests, and, by the time I send her, from the point of view of training also"—was unquestionably "a highly suitable candidate." Louis never heard back, nor did he receive a reply from his colleagues at the London Zoological Society, to whom he also wrote. Apparently the scientific establishment found the idea of sending an unschooled, uncredentialed amateur on such a study pure lunacy. But that the candidate was *also* of the female persuasion made it unthinkable, it seemed. Louis, of course, was dubious of such cultural biases. He had spent his life challenging establishment thinking and was impatient with stock assumptions for which he saw no grounds. The project was temporarily put on hold until he could find the funds, but he was no less determined to see it go forward.

Jane had saved up her salary and as a gift sent her mother a plane ticket to Nairobi. In early September, a bright-eyed Vanne emerged from the Nairobi airport, sinking happily into Jane's embrace. After so many months, she ached to see her daughter. But she was also concerned about what she called "the Brian problem," which, as she was soon writing home, seemed even more fraught than what she had imagined. Everyone she met described Brian as "a boy of great charm." He has "film star" looks, she reported, but he seems to breathe melodrama, "always a SITUATION!!" she added. Vanne was also looking forward to seeing Leakey again.

During a junket to England that spring, Louis had taken Vanne out to lunch, eager to meet the mother of his young protégé. The two had gotten on famously. Vanne found Leakey captivating—genial, brilliant, courtly. He had "a wonderful flow of conversation and a very strong personality, which I liked," she pronounced. She had been entirely "loath to tear" herself away. When Vanne arrived in Nairobi that September, Louis again proved himself a gracious host. He drove her into the Kikuyu reserve and "through miles and miles of real African country"; he arranged an expedition to the "uninhabited monkey-filled island" of Lolui, where, Vanne wrote, they "sailed on and on and on, under a tropical sky and landed on an uninhabited island, and slept to the sound of a million croaking frogs"; he treated mother and daughter to "a weeklong float" in his research vessel on Lake Victoria. And finally, he took Vanne to Olduvai, setting off across the Serengeti, which "far exceeded in strangeness and beauty" anything she had expected, and where they saw "cheetahs slumbering under a thorn tree, and passed of course Ostrich, wildebeest [and] giraffe." Vanne was enchanted, fascinated by Leakey's life and moved by his warmth and generosity. And so began an enduring friendship, fed in part by their mutual affection for Vanne's young and wild-hearted daughter.

Vanne had planned to stay in Kenya for three months, and then she and Jane would return to England, arriving home in time for Christmas at the Birches. Louis still hadn't found funds for the chimp study, but he was certain now about where it should take place: in Tanganyika Territory, in a piece of forest known as the Gombe Stream Chimpanzee Reserve. Tanganyika Territory was at the time a British protectorate, and the district commissioner there, a stalwart Brit named Geoffrey Browning, was proving testy. He had made it clear that "no European woman would be allowed into the Gombe forests alone." To satisfy Browning, Jane would have to have a European escort, it seemed, a second someone to accompany her on the expedition. Louis was worried that another person might compromise the research. It had to be someone relaxed, someone compatible, someone who wouldn't compete. Over lunch one day, he shared the complication with Vanne, as it was weighing on his thoughts. Before fully realizing it, Vanne had volunteered to be that second someone. The revised proposal—that two women go to Gombe instead of one—was now given the green light.

Beyond this, however, everything remained in the air. Jane would stay in England to prepare herself, in the event the money came through. In the interim, Leakey had promised to find her a job, perhaps at the British Museum, or the zoo. And she would begin reading everything she could about primatology and chimpanzees. Two weeks later, mother and daughter boarded the ship that would carry them home to England. Brian and his family were there to see them off. Hugs were shared, farewells and promises pledged, and then the ship pulled up anchor. Jane waved madly, feeling her eyes tear up as she watched the coastline of Africa recede into a faint bluish blur. She hoped she would be returning in a few months, but nothing was certain.

JANE ARRIVED HOME TO A "SMILING LONDON," AS SHE WROTE, "SUNNY & with a lovely nip in the air," and that January, after an equally lovely Christmas, moved in with her sister, Judy, to their father's London flat. Mortimer, as always, was away. It felt in some ways as if she'd "never left."

To her delight, Louis had delivered on his promise. There was a position awaiting her at the London Zoo—actually not in the zoo proper, but in the film library of Granada Television, which was housed on the grounds, and at the time producing an extended series about animal behavior. Jane was charged with cataloguing cans of films with animal shots—exceedingly boring work, she reported. But she was pleased to have a job. At lunchtime, she liked to meander around the zoo, although, as she wrote her friend Bernard Verdcourt in Nairobi, "there are some animals which I can't bear to see caged, especially the African ones."

Jane had bought a "cheap" car. And to amuse themselves, she and Judy began to hang out at a hip little coffeehouse called the Troubadour, on Old Brompton Road, Earl's Court. The Troubadour had a tatty bohemian feel, with exposed beams, antiques on the walls, and a low-lit downstairs room with pillows on the floor. It was a popular venue for poets and underground musicians, and had a small stage for poetry readings and performances. It was there, one autumn evening, that she met the charismatic Robert Young— handsome, clever, and soon a serious suitor. Robert was twenty-six and an aspiring actor. He thought Jane "an extremely interesting woman," whose ambitions, he remembered, were "way ahead of her time." Jane was equally taken with Robert, who would soon become her second real love.

The romance moved quickly. By February 1960, Jane and Robert were engaged. Robert, who still hadn't met the family, traveled to Bournemouth, where Danny pronounced him "charming." Vanne, however, was noticeably reserved, perhaps because of the financial uncertainty of Robert's chosen profession, perhaps because it was hard to imagine how Jane's plans to return to Africa meshed with Robert's aspirations to a life onstage.

Louis, meanwhile, had finally recognized he was getting nowhere in his efforts to interest the usual institutional sponsors in his proposed chimp study. It was time to pursue less orthodox channels. He would try his luck with his American friend Leighton Wilkie, an inventive, slightly wacky midwesterner who had made a fortune manufacturing cutting tools. Leighton had met Leakey in 1955, at a Pan-African Congress on prehistory, where Louis, in his inimitable fashion, had thrilled the crowd with a bit of stagecraft: performing a live demonstration of how to butcher a freshly killed antelope, using Stone Age tools. Leighton, as a tool man himself, was duly impressed. He was soon underwriting a portion of Leakey's work at Olduvai each year, dispensing small but regular grants. That February, after Leakey got word of Leighton's annual gift, he wrote to thank his benefactor, and to lobby for a second small grant, this one to fund his chimp project, which he hoped, he said, would begin in September 1959. A brief proposal was appended to the letter, in which he described the researcher he'd selected, a Miss Jane Morris-Goodall, "who has worked in Kenya with Dr. Leakey," delicately sidestepping any mention of the fact that Jane had served in a secretarial capacity rather than a scientific one. Happily, Leighton asked no questions, writing back an enthusiastic letter pledging $3,000 in seed money for the "very interesting" chimpanzee study. At long last, the project was to become a reality.

This welcome news was soon followed by an even more momentous event. At Olduvai that July, Louis and Mary had dug up a hominid skull that was soon determined to be the world's earliest man. Dubbed the "Nutcracker Man" or, as the Leakeys privately called it, "Dear Boy," the skull swiftly became international news, aided in part by a documentary aired by the BBC that captured the Leakeys' excavation in all its glory. Almost overnight the couple became global superstars, and Louis embarked on a whirlwind lecture tour that took him to seventeen different scientific institutions and universities, many of them in America, where audiences found him riveting.

But the greatest turn to come of Leakey's newfound celebrity was that it earned him entrée to the National Geographic Society, where one November afternoon, an eloquent Leakey was able to interest the society's president, Melville Bell Grosvenor, in the saga of the Nutcracker Man. By the end of their meeting, Grosvenor had agreed to grant the Leakeys $20,200 toward further research in return for exclusive American publication rights to the story. It was the beginning of an extremely auspicious partnership between the Leakeys (who up until then had been operating on a shoestring) and the National Geographic Society. Louis's article, "Finding the World's Oldest Man," would appear in the September 1960 issue of *National Geographic*, generating a groundswell of excitement. The offbeat and colorful paleontologist, as Grosvenor had sensed, possessed enormous popular appeal.

Leakey flew back to Kenya from the U.S. in December 1959, stopping briefly in London, where he gave Jane the news that the grant from Leighton Wilkie had, as promised, come through. "I saw Leakey during the past 2 or 3 days—he took Ma & I out to dinner & the theatre," Jane wrote her friend Bernard. He reiterated that the chimp study was now "all fixed . . . honestly Bernard, if I stop & think about it, I get simply terrified—not of the actual job, but of the responsibility." Leakey's reputation would now be resting on her work too, Jane realized. It was a quiet worry she tried her best to contain.

The study now real, Jane doubled down on her informal reading, quitting her job at Granada Television at the end of March so she could give her full attention to the work before her. Two of Leakey's colleagues had helped her compile a reading list, and one of them, John Napier, agreed to give her a two- to three-month private tutorial in primatology.

She was perplexed to discover how little was actually known about primate behavior. In the early twentieth century, several venturesome scientists had collected dead gorillas in Africa and then dissected them, eager for clues as to diet, tooth size, reproductive organs. There were a few behavioral studies of apes in captivity on record too, most notably by two distinguished psychologists, Wolfgang Köhler and Robert Yerkes. Other than that, not much.

Yerkes, a Yale man, had set up a research center in Florida in 1930, convinced that by observing apes, he might better understand the biological components of human psychology. Apes, he believed, could become helpful "servants of science" and "contribute importantly to human welfare." It was the same rationale that Rachel Carson's first boss had offered for their work at the Bureau of Fisheries. Animals, the thinking went, could be harnessed to serve human ends. They were auxiliary to their human overseers, a useful resource to exploit for advancing mankind's needs.

Yerkes was in many ways an outlier, however. It was his surmise that apes were "guided" by emotions that were not unlike those of humans, and that they were "probably capable of rational and symbolic thought." He saw no "obvious reason," he wrote, "why the chimpanzees and other great apes should not talk." To test out this theory, a baby chimp was placed in the household of a psychologist named Winthrop Kellogg and his wife, Luella, who for several months raised the chimp "alongside their own infant son, Donald." Soon enough, Donald was speaking while the chimp, to the Kelloggs' disappointment, "remained mute," although they were convinced that he understood at least one hundred words. But then, rumor had it, the experiment was abruptly halted. Apparently the Kelloggs' baby had stopped speaking and begun making chimp sounds. Studying primates in captivity clearly had its limits. Yerkes, like Leakey, longed for a study of chimps in their natural habitat.

In 1929, Yerkes managed to send a young Yale psychologist, Harold C. Bingham, and his wife, Lucille, into the eastern Belgian Congo. The couple hired a small army of African porters that numbered forty at one point, spending the next two months in the field, creeping through the bush, looking for gorilla prints, examining nests. Occasionally they did get close to an actual ape, but when this happened their fears often got the best of them, especially on the day when an edgy Bingham, startled by the sight of an excited gorilla, grabbed his gun and shot him dead.

The following year, Yerkes dispatched a second Yale psychologist, Henry W. Nissen, to the African bush, this time to French Guinea, for a nine-week field study of wild chimpanzees. No matter that this "one-man expedition," as Nissen described it, actually included three African assistants and a half dozen porters and guides. Nissen's chief concern was how to do the research, since there were still no earlier studies from which to draw. He built a series of "blinds" and hid expectantly behind them. But he quickly found that the "sharp-sighted chimps" were unconvinced by his ruse; they knew an interloping bit of architecture when they saw one, vanishing instantly into the bush. He pondered using "lures"— that is, leaving "chimp delicacies at the same spot day after day," but scotched this plan for lack of time. More fruitful, he decided, would be to "surround a group of chimps with a circle of hired African helpers" and impede the apes' wanderings for a day or so—this to give him time to notate basic information about "group size and compositions, age, sex, and so on." But nothing went as envisioned; the exercise was an unqualified disaster. Guns went off, grass fires were lit by forty hired African helpers, another gang of club-wielding helpers dragged a baby chimp from some trees. Understandably, the traumatized chimps—screaming, panicked, terrified—failed to yield any useful clues about their everyday behavior. Out of ideas and with little time left, the bewildered American psychologist and his entourage of porters took to tracking the apes on foot, "listening for the hoots and cries of chimps" and then moving furtively toward the sounds, usually with little success. Given all this, it was hardly surprising that the conclusions Nissen reached were at best "simplistic"—"the chimp is nomadic, having no permanent home"—and at worst wrong.

Jane was "shocked" by the image of Nissen and his club-wielding porters. As far as she could tell, he had learned little more than the skittish, trigger-happy Harold Bingham, which was almost nothing at all. With some disgust, she perused two other published field studies—one on gibbons, the other on red-tailed monkeys. In both instances, the researchers had gathered what behavioral data they could and then slaughtered their subjects to determine their "age, sex, reproductive condition"—even their stomach contents.

"More slaughter of the innocents," she thought.

Distressed by the thinness of the literature, she ventured out to the London Zoo on occasion, eager to conduct her own informal studies of chimps in captivity. "But there were only two bored psychotic individuals in a tiny cement cage with iron bars. I could learn little there," she later wrote. She was shocked by the conditions they were in.

One night Jane showed Nissen's monograph, "A Field Study of Chimpanzees," to Robert Young, a decision she almost immediately regretted. "The more he reads the more worried about me he gets," she wrote Vanne.

THE ANNOUNCEMENT OF MISS V. J. MORRIS-GOODALL'S ENGAGEMENT TO Mr. R. B. Young appeared in the *Daily Telegram and Morning Post* on May 13, 1960—a stroke of bad luck, it turned out, as the timing couldn't have been more terrible. Just days later Jane got word from Leakey that preparations for the ape study were finally in place. He had procured the necessary supplies and had two plane tickets and the final permits from the Tanganyika Game Department in hand. Marriage to the dashing Robert Young was now hastily postponed, and two weeks later, after tearful, departure-gate promises that the couple would wed as soon as Jane returned, Jane and Vanne flew by prop plane from London to Nairobi, arriving the next morning.

They encountered their "first setback" almost immediately. Apparently, a feud was brewing between two rival groups of fishermen camped along the shore bordering the chimpanzee preserve; it was making conditions at Gombe extremely dangerous, Louis said. Never without a backup plan, Louis decided to divert them instead to Lolui Island, where Jane could conduct "a short trial study" of the vervet monkeys there to get some research practice. The two women were loaded onto a train to Lake Victoria, where they were met by Hassan Salimu, the captain of Louis's cabin cruiser, who ferried them across the vast sealike lake to Lolui Island the next day.

Like Bingham and Nissen before her, Jane Goodall was effectively beginning her research without "serious precedent, established preconception, or standard method," as her biographer has written. She had nothing to build on. No one had researched wild vervet monkeys before; few had studied wild apes. It was all uncharted territory. Nissen had adopted a "hunter's technique," moving quietly and stealthily toward his research subject, until he could see what he had been tracking by sound. But stealth can go only so far, since it usually signals a predator preparing to assail its prey. Stealth telegraphs danger, imminent threat, the need to flee. Instinctively, Jane adopted a different style, doing more or less as she had done as a young girl watching the small wild animals near her home: she moved openly, rather than in secret, paying careful attention to how close she could get, and what movements were perceived as threatening, to how and where to direct her gaze. The Lolui Island study, she would later recall, "taught me a great deal about such things as note-taking in the field, the sort of clothes to wear, the movements a wild monkey will tolerate." Jane's approach was improvisational at this point, but she was learning with each encounter, noting not only what methods worked, but also those that didn't. "At 2:15 an adolescent entered the two tall bushes nearest to my tree, and I was too close for him. He started to give a warning call, and soon an adult female, a male & another J[uvenile] had joined it," she wrote. "I curled up & pretended to go to sleep. Soon they all stopped," she added, with some satisfaction.

Jane's methodology parted ways with Nissen's in another respect. Following her intuition, she chose not to approach the Lolui Island monkey population in a *general* way, as a "species," but instead to try to get an understanding of the species by observing "individual members." This meant noting and then remembering distinctive features—initially so she could differentiate one monkey from the next, making it easier to describe their individual behaviors, but soon also so she could note how and when they interacted, and, by extension, if and how they were *connected*, if indeed they were. To this end, descriptions such as "mature female with very pink, large conspicuous nipples" or "adult male—very handsome creature in his prime" soon gave way to names—the first cardinal sin of conventional scientific practice, where researchers were expected to assign numbers to their subjects. (Jane, however, didn't yet know this.) Within days, the small infant who continued to be cradled in the arms of "the huge female" had become Sammy. By the second week, Sammy's mother had been named Bessie. The "very obvious pregnant" young female with nipples "very small in comparison to the other adult females" was called Lotus. Months later, at Gombe, as Jane got to know the chimps she was watching as individuals, she would do the same, giving them names that defined them as distinctive personalities. "I had no idea that this, according to the ethological discipline of the early 1960s, was inappropriate," she later wrote.

Jane's approach, so different from the narrower and more generalized focus of her predecessors, represented a critical change in orientation, a shift from an emphasis on counting and measuring things, to a focus on mapping relationships. Field zoology up until this time, writes Peterson, "typically consisted of specimen collection: shooting wild animals and measuring their remains." Jane was looking for something else: she was searching for connections, the ties that bound together the monkeys as a species, but also as a community of distinct individuals, a society connected by a web of relations and interdependencies. It was an approach akin to that of Jacobs, who argued that generalizations about cities got one nowhere. It was only by observing the unique and particular features of individual blocks, in distinct and individual neighborhoods, and how they connected, that one could possibly get a sense of how the city as a whole worked. Just as Jacobs renounced "the statistical city," which described urbanism as a mathematical abstraction rather than as a living community of people, so Goodall intuitively rejected the notion of approaching the chimps as numbers and statistics. "In the form of statistics, these citizens could be dealt with intellectually like grains of sand, or electrons or billiard balls," Jacobs had written. The same, Goodall would soon argue, was true of animals. They couldn't be abstracted as a species. One had to begin from the bottom up, with the individual and the particular and build from there.

LOUIS FINALLY GAVE THEM THE GO-AHEAD TO PROCEED TO GOMBE IN late June. Jane and Vanne returned to Nairobi at once, loaded up the Land Rover, and then set off on the eight-hundred-mile journey south into Tanganyika Territory. Three days and several punctured tires later, they rolled into the town of Kigoma, a single main street shaded by mango trees, with a red-dirt square, an open market, and a train station, only to encounter their second setback: the dusty port town was in crisis. Violence had flared up in nearby Congo, just twenty-five miles away, on the far side of Lake Tanganyika, following its liberation from colonial rule. Belgian refugees, mostly women and children, were pouring in by the boatload, terrified by tales of machete-wielding hooligans who had turned on their former overlords. Once again, the expedition to Gombe would have to be put on hold. Jane and Vanne joined the volunteer corps that same day, making Spam sandwiches, doling out chocolate and hot soup, helping the displaced Belgians onto trains. In their free time, they wandered the market, where one afternoon they hired a cook for the expedition, a local Kigoma man named Dominic Charles Bandora, whom they came to love. Finally, a little over two weeks in, the game warden, a gentle young Brit named David Anstey, deemed it safe to proceed.

THE BREEZE WAS UP ON THE MORNING THEY LEFT, WHIPPING THE WAter into a meringue of small waves. Anstey had arranged for a launch to ferry them twelve miles up the lake to Gombe Stream Chimpanzee Reserve, tethering the little aluminum dinghy Leakey had sent to the stern. Jane remembered standing on deck, eyeing the little dinghy and having the strange disembodied feeling that she was "living in a dream." The dinghy would soon be their only link to the outside world.

They chugged steadily north, plying their way up the jagged coastline. Deep ravines cleaved the steep, flat-topped cliffs in places. Pockets of forest filled the narrow valleys, through which fast-moving streams spilled to the lake. Occasional fishing villages crouched at the foot of these forested slopes, simple mud-and-grass huts, a few roofed in corrugated tin. And then, as the southern boundary of Gombe appeared, the landscape noticeably changed. Now the mountains were higher and steeper, swaddled in impenetrable-looking tropical forest. A few makeshift fishermen's huts speckled the white beach; on the sand beside them, silvery fish glinted in the sun; otherwise there were no signs of civilization. Later, Vanne would admit that she had been "horrified" by the sheer slopes, the "impenetrable appearance of the valley forests"; Anstey told Jane afterward that he had guessed that she would be "packed up and gone within six weeks." She had looked, he recalled, so "terribly young." Anstey had also been worried that blowback from Congo "would send baddies, vagabonds, hooligans, murderers or what-have-you across the lake and knock her in the head," he confessed.

A small gathering of people awaited them on the shore—a handful of fishermen, two African scouts who lived by permission in the preserve, and an ancient, birdlike man of imperial bearing, ceremoniously attired in a red turban and an equally red European overcoat. The old man was Iddi Matata, they soon learned, the unofficial head of the fishing camp. Matata proceeded to deliver an impassioned welcome speech in Swahili, whereupon Jane and Vanne, as Anstey had counseled, presented the old man with a small gift. All parties apparently satisfied, the onlookers dispersed, and then David Anstey and Matata's six children helped Jane and Vanne set up camp. Dominic's tent was placed just above the beach; Jane and Vanne's was erected on slightly higher ground, in a clearing shaded by oil palm trees. It was, they felt, ideal. Their tent had a raised flap veranda in the front, a separate washroom in the rear, and was roomy enough to accommodate two camp cots set side by side, as well as various tin boxes that held their gear. A "small gurgling stream" tumbled right behind it, with cool, clear water that ran so fast they could wash their feet, or simply soak them on torrid days when the heat stung. Vanne had spotted a small pool deep enough, she hoped, to wash her hair, which, she reported home in a letter, was now "so stiff with dust that it stands up like a halo." In short order then, a deep hole was dug to serve as a latrine, and a baffle of woven palm leaves built to ensure privacy. And finally, a makeshift kitchen—a few poles and a straw roof—was erected near Dominic's tent. That evening, the little party, which still included David Anstey, dined on a splendid meal that Dominic had cooked: soup, followed by a stew with potatoes, then canned oranges and coffee. And then they all trooped off happily to bed.

JANE FELT AN IMMEDIATE KINSHIP WITH THE AFRICANS SHE MET. SHE was startled therefore to learn that her embrace of the locals was not altogether reciprocal. A rumor had spread that she and Vanne were government spies, and that their purpose in coming was to exaggerate the chimpanzee count, which would further buttress the government's case for keeping the forest a protected preserve. Jane was sympathetic. The locals had once cut wood for their fishing boats in this forest; many believed they would be reborn there, after death, as chimpanzees. They still dreamed of reclaiming the thirty square miles of rich black earth for themselves. It was a delicate situation and David Anstey, familiar with local ways, negotiated it with enormous aplomb, meeting with an agitated crowd of some twenty fishermen and villagers the day after her arrival. They talked it through, agreeing finally on a compromise: Jane would do her chimpanzee watching in the company of a few "hired assistants," who would monitor her counts, making sure she wasn't inflating them. One of her minders, it was decided, would be the "good-humored" Adolf

Siwezi, a game scout already employed at the reserve; the other, the "tall and lean and silent" Rashidi Kikwale, who would serve as her guide and porter. It was also understood that she would hire the son of the chief of Mwamgongo village, a fishing settlement just north of the preserve. Jane was upset by the news. She had imagined that she would set off through the forest unencumbered and alone, she told her family. That night, she went to bed feeling "depressed and miserable." But by the next morning, she'd resolved she would make the best of an imperfect situation. "I do hope it's all going to work," she added.

Gombe was an Eden of sorts, more stunning, even, than Jane had imagined. The forest was festooned with flowers; the plunging valleys were lush with green. "I wish you could be here—even for a day," she wrote her family soon after arriving. "It is so beautiful, with the crystal clear blue lake, the tiny white pebbles on the beach, the sparkling ice cold mountain stream, the palm nut trees." Wildlife abounded. There were hippos and herds of buffalo, bushbucks and bushpigs, a few "reclusive leopards" and even the occasional hyena. Birds swooped across the forest canopy in great numbers. Crocodiles lurked along the lakeside. The primate population was copious, including olive baboons; red colobus monkeys; blue, red-tailed, and vervet monkeys; and an astonishing number of chimpanzees. There were smaller species of course too: civets and elephant shrews; genets and mongooses; chameleons, geckos, and skinks. And snakes: giant pythons, some a terrifying sixteen feet long, with back-curving teeth as big as a dog's; slumberous, six-foot-long Nile monitor lizards; and naturally, a number of poisonous varieties too (night adders, spitting cobras, black mambas, vine snakes, bush vipers). Gombe was also home to all manner of biting insects: venomous giant millipedes, scorpions, and assassin bugs; spiders, tsetse flies, and safari ants—this in addition to an army of malarial mosquitoes, all to be avoided.

Jane had arrived anticipating these perils, secure in the knowledge that she was where at last she was meant to be: living amidst the feral splendor of a beautiful and fecund, if at times violent, Africa. She knew, for instance, that chimpanzees were dangerous wild animals, capable of ripping off a person's arm; that leopards were solitary hunters more threatening to humans, many said, than lions; that a black mamba's bite would paralyze you in less than an hour, with only a few more left before you died. Still, it came as a shock when, during her first week, two fishermen rushed up excitedly and led Adolf, Rashidi, and her to a tree near the lake. Its bark was gashed in a hundred spots, the result, apparently, of a lone bull buffalo that had charged one of the fishermen the night before. The terrified man had managed to clamber up into the tree, where he clung to an upper branch for more than an hour, while the buffalo rammed the tree repeatedly, trying to shake his quarry loose.

Jane took the fishermen's warning to heart, later describing her own eerie encounter with a buffalo, which she nearly bumped into one morning as it lay in the dark before dawn, no more than six yards away. "Fortunately the wind was strong, its sound covering the small noise I made, and it was blowing from him to me," she wrote. "I was able to creep away undetected." Another time, while camping alone in the hills, she heard the "strange sawing call" of a hunting leopard close by in the dark. She was terrified, having what she called "an ingrained illogical fear" of leopards. She put a blanket over her head, hoping for the best, and was lucky. Another evening, this much later, she was walking back to camp along the lakeshore, wading in the water to avoid a huge rock, when she saw the slippery black body of a snake. She instantly froze. It was six feet long, and from the slight hood and dark bands striping its neck, she knew it was a Storm's water cobra—a lethal snake with a bite for which, at the time, there was no antivenom. It moved toward her, riding an oncoming wave, and then part of its body actually settled on one of her feet. Jane stared down at it, disbelieving, barely breathing until the wave rolled back into the lake, sweeping the snake with it. Heart "hammering," she leaped out of the water then. These incidents, however, were exceptions. In general, as she would later write, her fears of being hurt by a wild animal were "almost nonexistent" at the time. She truly believed that the animals she was living amidst would sense that she meant them no harm, and would thus leave her alone.

Louis tended to share this belief, while also insisting that she have "a reasonable understanding" of how to behave if she encountered an animal unexpectedly. She knew, for example, that the most dangerous thing one could do was to get between a mother and her young, or to come upon an animal that had been wounded and couldn't run, or one that had, for some reason, "learned to hate man." But these were perils, she added, "no more dangerous—and probably even less dangerous—than those that could beset one in any city, and I was not concerned." Indeed, unlike the few researchers before her, Jane was inclined to identify and even empathize with the animals she moved among, rather than to fear them. She possessed a fundamentally different conception of her own place in the scheme of things. Instinctively, she neither envisioned nor experienced the natural world as a hierarchy in which mankind stood at the top, separate and superior. She felt herself to be as one among a complex web of creatures: part of, and not separate from, the animal kingdom.

Curiously, this difference in her fear response (beyond Jane's obvious bravery) has some biological basis, it turns out. Recently, UCLA researchers discovered that, biologically, men and women respond differently to stress, explains Nina Simons, codirector of the Collective Heritage Institute. "Men tend to react with a fight, flight or freeze response. But when women are stressed, our bodies release a hormone called oxytocin, which is also released in childbirth. Women tend to respond with a desire to connect with others—we socialize. Instead of 'fight or flight,' we 'tend and befriend.' The very act of connecting with other people calms the bodies' response to the stress." It was this deep-felt desire to "connect" with her study subjects that most separated Jane from her predecessors. She was unafraid, in effect, to identify with them. "There is a way of looking at chimps which is an absorption of them and not a projection of you," Jane explains.

HER FIRST DAYS OUT IN THE FIELD SEEMED PROMISING. ADOLF, HER game scout, had reported seeing chimps feeding in a giant msulula tree in a valley near the northern border of the preserve the morning after she arrived. The next day, she, Adolf, and Rashidi set off in the aluminum dinghy, disembarking at the mouth of a river. From there, they started up the forested valley, following "the fast running stream," she remembered. It was beautiful and cool, with "thick vegetation," the forest canopy high above their heads, filtering the sunlight. They saw buffalo prints and bushpigs, brilliant red and white flowers, a kingfisher flash by. After twenty minutes, the climbing grew more difficult, the undergrowth thicker, more tangled with vines. At times they had to crawl. And then they heard the cries of a party of chimps in the valley to their north, "low, resonant pant-hoots," a wild chorus that grew louder as they drew closer. Jane was enormously excited. Adolf led them to a grassy clearing directly across the valley from the msulula tree, a good spot, he assured her, for her viewing. Moments later, she saw the first of several chimps clamber up a palm trunk and into the branches of the giant tree.

For the next hour or so, Jane peered across the ravine through her binoculars, trying to sort out what was happening in the leafy msulula. It was frustrating. The foliage was extremely dense. At such a distance, the best she could make out was a few dark shapes, a perturbation in the leaves, the flail of an occasional arm. There seemed to be a lot of comings and goings. She watched a group of chimps file down and vanish into the forest, and then another party arrive. Or was it the first group returning? She couldn't be sure. Nor could she distinguish anything about size or sex. Were they all males? Females and their young? A combination of both? The second group gorged for a while on fruit and then they, too, filed down the palm ladder and disappeared. She was eager to sample the berries herself, keen to see if she could glean anything about the chimps' feeding preferences, so a few minutes later, she, Rashidi, and Adolf edged quietly toward the msulula. "This was a mistake as they were still in the lower trees nearby and we startled them," Jane wrote. "We heard them moving about, cracking twigs." But "only one animal did we see during this time," she added with exasperation.

Day after day, for the next two weeks, Jane watched the feeding ritual in the msulula, noting what she could. Sometimes she saw large groups feeding; other times it was only one or two individuals. Often, when a party left, they divided in two and headed in different directions. Sometimes she heard cries in the forest all around her. Were they signaling to each other, calling out their locations, keeping track of the whereabouts of the whole troop? Or was it incidental, nothing at all? On two separate nights, she insisted on sleeping out in the forest, so she could observe the chimps as they rose in the morning. Adolf and Rashidi lay close to the little campfire they built, sulky and unhappy to be out. Jane slept farther back, swaddled in a blanket.

And then, at the end of July, the msulula abruptly stopped fruiting, and just as abruptly, the chimps disappeared too. Now Jane's days were spent scouring the forest valleys from daybreak until dark, tracking elusive creatures that rarely, if ever, appeared. It was an impossible situation. The chimps, if she glimpsed them at all, seemed to vanish in an instant, dark and fleeting apparitions. For a while she chalked it up to there being three of them. She tried to leave Adolf and Rashidi on an overlook, from which they could still track her progress, while she pressed on alone. But the apes were no less skittish. Even from across a ravine, they seemed to sense her approach and flee. "How can I ever see any behavior?" she lamented in her journal, admitting that a "mood, a depression" had settled in. Time was passing; soon enough her funding would run out. If she couldn't produce results, Leakey would be unable to generate any further support. She tried not to feel despondent, but it was hard.

Eight dispiriting weeks followed, and Jane's successes, when they occurred at all, were short-lived. The threesome combed Gombe's valleys as best they could. Sometimes the dense undergrowth made passage impossible. Other times, the slopes were too slippery to ascend, or the ravines too sheer to climb either up or down. They followed the streambeds deep into the mountains, searching for signs of feeding. But they found no other fruiting trees.

Added to these challenges was the more delicate human problem of dealing with her "minders," as Jane had begun to call them. Both Adolf and Rashidi were lovely men, able and well intentioned. But neither was particularly inclined to keep up with her pace, having neither the stamina nor the will to do so. She couldn't completely blame them, she wrote home. What madness it must have seemed, from their point of view, to push up and down the densely forested slopes, relentlessly, day after day, torn and scratched by brush and thorns, bitten by bugs, clawed at by branches, searching for creatures far more agile than they. She suspected both men thought her unreasonable; she could see it in their faces sometimes: dismay when she insisted on beginning work well before dawn; distress when she decided to stay out all night, sleeping on the damp forest floor; confusion when she forgot to break for lunch. "This is the trouble with having to be accompanied on my observations," she complained in a letter home. The two Africans, she went on to explain, had been famished that day, pressing for a return to camp for lunch. "People do need food and things & I must try to remember," she admitted. She didn't mean to be a "slave-driver," she added. She knew it was hard and exhausting work, a job for them. This she must remember too.

Vanne, meanwhile, was engaged in some gentle diplomacy of her own, aware that some of the locals still regarded Jane and her with suspicion. Before leaving England, Leakey had told Vanne that the surest way to win the locals' hearts was to come bearing medicine. Vanne had taken him at his word (despite having no medical training), packing great quantities of aspirin, cough medicine, bandages, Epsom salts, and other home remedies. As soon as they were settled, she'd let it be known that she would offer simple medical help to anyone who needed it. Her first patient was an ancient, emaciated man with oozing ulcers on his leg, "the most ghastly, livid swelling on his ankles that I have ever seen," Vanne wrote, "the bad part . . . all red and yellow." Vanne was worried he might lose his foot and urged him to go to the hospital in Kigoma, but the old man refused. So, for the next three weeks, she washed and soaked the leg in hot water and antiseptic. And then one morning, to everyone's amazement, the ulcers began to drain. A week later the sores were clean, the swelling gone.

And so began Vanne's clinic: four poles and a thatched roof. Word of course spread quickly. Soon people were lined up for treatment—thirty on the first official day, sixty by the following week. Nothing did more to dispel any lingering suspicions about the two peculiar women who had traveled so far to live in Gombe's forest than Vanne's clinic, Jane would later say.

AUGUST WAS SWELTERING. "SUFFOCATING BY 9 AM," VANNE WROTE home. Their hair, she added, was "permanently wet & hot." Vanne started to run a fever in the middle of the month; within days Jane was seriously feverish too.

On the morning it hit, Jane was miles from camp. She and Rashidi had been sitting quietly, in a high spot with a clear view of an open glade, when they saw two chimps pass below them. Then Jane heard "a measured tread." Down the hill, heading straight for her, came an older, white-bearded chimp. Male, "palish face, long black shining hair." He got to within ten yards, and suddenly saw her, his expression stunned. He stopped. Stared. Tilted his head from one side to the other, as if mildly quizzical. Then he turned and galloped off into the thicker underbrush. But he didn't completely disappear. Though he was out of view, she could hear him circling around until he was below her. He climbed up a tree, his head just visible through the foliage, and watched her for several minutes. Then, his curiosity apparently sated, he clambered back down and resumed his journey down the ravine. It was the beginning of a change in her luck, though she didn't yet know it.

Vanne's temperature spiked to 105; Jane's hovered at 104. For days the two women lay side by side on their cots, too weak to do anything but reach occasionally for the thermometer. Dom begged them to see a doctor in Kigoma, but both felt too sick to make the trip. So he tended them as best he could, feeding them tea and hot broth, urging them to eat. Things went from bad to worse then. One night Dom found Vanne "collapsed and unconscious outside the tent" and he helped her back to bed. They had been told there was no danger of malarial fever, as it didn't exist in those parts, but this turned out not to be true and it was likely this is what had struck them. At the urging of friends, they began taking antimalarial pills.

Jane by now was feeling "frantic" about the time she was losing. Three months had passed and "I felt I had learned nothing," she wrote. During the last week of August, her fever now beginning to ebb, she forced herself up and out, setting off alone for the mountain directly above camp. It was cool as she left, still early, dawn a pale flush at the horizon. A part of her didn't want Rashidi and Adolf to see her in her weakened state. But she was also tired of "coddling" the two men. She knew that she was risking "official displeasure." But at that moment she didn't care. Free of her minders, she would be able to move at her own pace. Still, the going was slow. Several times she had to stop, unsure if she could make it. "Earth kept vanishing & head throbbed like an engine," she wrote.

She pressed on, climbing what seemed an agonizing slope, steep and terrifically slippery in places, until she reached a rocky overlook halfway up the ridge. "Ogre" she would call that ghastly hill. She knew at once that it was an ideal viewing spot. From where she stood, she could peer down into the thickly forested valley just below her, but also to the ridges beyond.

Too weak to climb farther, she decided to sit for a while, scan the valley for signs of chimpanzees. After some minutes, she thought she saw movement on the black, charred slope just beyond her. Lowering her binoculars, she turned. There, standing and staring at her, stood three chimps. She was sure they would flee, for they were no more than eighty yards away. Except for the white-bearded chimp, who had stumbled upon her quite by accident, she had never gotten closer than five hundred yards—and even then, they had always vanished before she could observe anything especially illuminating. But the three chimps continued to stand and then after a moment calmly moved on, slipping into the leafy vegetation below. Here was affirmation of what she'd felt all along: the apes would be less frightened if she was by herself.

Later that morning, still perched in the same spot, she was rewarded again: peering through her binoculars, she sighted a party of chimps heading down the opposite slope, moving toward some fig trees growing along the stream that cut through the valley. They were followed shortly after by a second troupe, this one traversing the bare, burnt slope where earlier she'd seen the three. She was sure this group also saw her, as "I was very conspicuous on the rocky peak." But while they all "stopped and stared and then hastened their steps slightly as they moved on again," they didn't run in panic. After "some violent swaying of branches," they joined the first group and then fed together in the fig tree. Later, she followed their progress as they moved off, this time as one enormous group. She spied two small infants "perched like jockeys on their mothers' backs." It was a turning point, the first sliver of light after three dismal months of no sightings. "I've discovered more—since my fever, in about five days, than in all the dreary weeks before," Jane wrote home excitedly. And so began a new chapter.

By the middle of September, Jane had lugged a small tin trunk up to the "Peak," as she'd dubbed the overlook, stocking it with coffee and a kettle, a blanket and a sweater, as well as a few tins of baked beans. When the chimps slept close to the overlook, she often stayed there too, to save herself the awful trudge up Ogre in the morning. By now, Adolf and Rashidi had been replaced by two professional game trackers, Soko and Wilbert, courtesy of Derrick Dunn, a white hunter and an old friend of Brian Herne's. Dunn, a huge, "square-faced" fellow, had met Jane in Nairobi and instantly fallen in love. Having learned of the trials she was facing, he had sent two star trackers from his safari business to help—and, perhaps, also to advance his romantic prospects.

Soko, the more temperamental of the two, didn't last long. He was quickly replaced by a third scout, a compact little man nicknamed Short. Wilbert, Jane reported, was enormously tall, and "always looked immaculate even after scrambling on his belly along a pig trail." But his tenure too was short-lived. He couldn't adjust to the food in camp.

By the middle of October, Jane and Short had settled into a workable routine: each would set off in a different direction, with the understanding that they would meet up at various times to exchange information. Once a chimp was sighted, even if they were together, it was understood that Short would hang back so that Jane could work her way closer alone. On the nights she decided to stay at the Peak, she would let Vanne know by sending a message down with Short, who always swung by to check on her late in the day.

Life fell into "a rhythm." Every morning, Jane rose at 5:30 A.M., dressed, ate a slice of bread, hastily made coffee, and then trudged up the mountain in the dark, emerging at the Peak just as dawn was breaking. For the rest of the day she wandered the mountains, searching the rugged terrain alone: walking, pausing, thinking, looking, remembering—seldom returning until well after dark. Dinner was always late. Sitting in the gentle glow of their campfire, she and Vanne quietly shared stories. Then, perched on her flimsy cot, Jane would write up the day's field notes under the halo of her camp lantern, swatting at mosquitoes as she wrote, often still at it until well after eleven.

Dogged, seemingly tireless, Jane continued at this pace throughout the fall, her energy rarely flagging. She was learning a new kind of slowness: how to watch and wait, alert to the slightest quiver in a tree, the snap of a branch, a faint cry rising from somewhere in the forest. She knew by now that the chimps often moved in small groups, at times even alone, almost always in absolute silence. She knew they were edgy, excitable creatures, easily agitated, acutely sensitive to the tread of an alien in their mazy green world. She had learned that sometimes it was necessary to wait for hours and hours, even for the privilege of a single sighting.

Still, bit by bit, she was beginning to "piece together" something of the chimps' patterns: the way they groomed and mated, moved and slept; their food sources and facial expressions, group sizes and repertoire of calls. As the weeks slid by, time collapsing, rarely a day passed now when she didn't have at least one useful sighting. She was coming to know the terrain: the open woods and twelve steep-sided valleys; the sheer ravines and grassy ridges; the rugged slopes and craggy overlooks; the dense green forest pockets, where the chimps spent most of their time.

And by now she had established her methods. Just as she had done at Lolui Island, Jane approached the chimps openly, but always with extreme deference. She gauged her distance by reading their gestures, moving in as close as they seemed to allow. She dressed blandly, shunning colors and "unnatural" patterns, trying her best to blend in. If the apes seemed upset or distracted by her nearness, she tried to feign indifference, sometimes making a show of scratching herself, other times pawing the ground for food, hoping to appear as just another primate foraging for its supper. "The first chimp went close to the trunk and hid in the leaves, watching me," she reported. "I began to look for insects, digging in the ground with my hand & pretending to eat." Though she could never *really* be sure, there were times she felt the ploy worked. Of course often it didn't: "I thought we could get closer to the 5 in the pocket & possibly see them nest building, but this was a failure," she wrote. "As we got round they climbed down & disappeared."

Occasionally though, the chimps seemed to tolerate her presence, provided she was seated and remained perfectly still, and the apes were "in fairly thick forest," as happened in mid-September: "The first old boy sat with his knees up, his arms folded across them and his chin on his arms," she noted with surprise. "Only the other large one did not look quite at ease. The other two both sat very comfortably, and they all stared at us." Soon after this encounter, she reported a second, this one with two male chimps, one an "old & grizzled male with white beard," who had sat "only about 15 yards away" and quietly watched her: "Absolutely no fear. Scratched one shoulder & then the other. Rubbed his chin." She was astonished. She had never been closer than eighty yards.

Jane recorded every aspect of what she observed, from the obvious to the obscure. Like Carson, she approached her work with a kind of blinkered intensity, drawing upon all her senses, filling her notebooks with close, physical descriptions, precise and meticulous detail. "This is the first opportunity I had of seeing the male genital organs closely. The testicles of the old man were enormous, hanging down like a great bag," she wrote. "The penis, which was not erected, was a dull pink."

She labored to capture the tenor of their cries, to weigh their character and mood, even to guess at their meaning. Writing in September, she described hearing a spell of prolonged screaming that erupted every ten to fifteen minutes. "Each occasion was a series of high pitched, fairly short, loud screams. Pain or fear?" she wondered. "Pretty sure it was the same animal."

She noted how they moved: "They all went rather fast, using a movement that was almost like a 'crutch walk' but consisted of moving both arms forward together & bringing the legs forward altogether," she wrote of one group.

She studied what they ate, even going so far as to sample the nuts and sour fruits she saw them consume. One "acorn-like nut" tasted "rather like palm nuts." It was "oily," she added, with "the most unpleasant, bitter pungent taste—I could not get rid of it for a long time." Another, a berry, was "round & purple with a small stone." She collected feces and probed them for dietary clues. "Dry, dark brown, and fibrous," she wrote of one sample.

She watched and recorded how they made their arboreal nests: "It squatted in a leafy tree, near the top. It then rapidly pulled small leafy branches towards it . . . treading on them to hold them in place. It then sat down . . . stood up & pulled off a branch from higher up which it incorporated into the nest. This it did 4 times . . . It then lay down, hardly visible, and later picked a very small bunch of leaves, which it appeared to place under its head. Then it stretched right out so that its feet projected beyond the structure of the nest." Later, once the nest was vacated, Jane crept up and tried it out for herself. "Very comfortable & springy indeed," she pronounced.

"WE INTERACT WITH THE ENVIRONMENT FIRST AND FOREMOST THROUGH our bodies," Diane Ackerman observes in *A Natural History of the Senses*. "There is no way to understand the world without first detecting it through the radar-net of our senses." Like Carson and Jacobs, Jane began her work here: with her senses, the lived and felt and observed, unbound by ideology or preconceptions. She not only "saw things differently, she saw different things," to paraphrase Sally Helgesen, who has written at length about the distinction between men's and women's styles of seeing. Men, Helgesen observes in *The Female Advantage*, tend to have "a bottom line, sharply focused, linear way of thinking" that often excludes any role for emotion or empathy. Women's style of observation, by comparison, is often more "broad and wide-ranging." Women are continually perusing the situation "for more information," while men tend to focus more narrowly on an immediate goal, restricting information to make actions more efficient. The origins of these differences, Helgesen suggests, may go back to hunter-gatherer societies. Men went out for an occasional big hunt, a specific event with a definite climax; women foraged and planted, tasks requiring repetition, perhaps explaining something of their "process orientation."

Nissen, with his "guns and grassfires" and club-wielding "helpers," saw no place for empathy or emotion. He cared little about process. In his clumsy and ultimately reductive approach, he couldn't imagine the chimps as a pool of individuals with distinct and separate personalities—or see them as sentient. Instead, he approached them in an abstract and mechanistic way, conceiving of the species as a "biological monolith," as Jane's biographer notes, an unvarying and homogenous corps about which "simple and definite laws of behavior" could be easily deduced, and then uniformly applied to all. He saw, in other words, a sharp line between us and them.

Jane, by contrast, was drawn to the small dramas and affections that passed between particular chimps. It was precisely the individual and the emotive that interested her: "I saw one female, newly arrived in a group, hurry up to a big male and hold her hand toward him," she wrote. "Almost regally he reached out, clasped her hand in his, drew it toward him, and kissed it with his lips." Another time, she watched "two adult males embrace each other in greeting."

She was fascinated by the "spectacle" of the youngsters at play, their visible joy and obvious imagination. She saw "youngsters having wild games through the treetops, chasing each other or jumping again and again, one after the other, from a branch to a springy bough below." She watched "small infants dangling happily by themselves for minutes on end, patting at their toes with one hand, rotating gently from side to side." Once, "two tiny infants pulled on opposite ends of a twig in a gentle tug-of-war."

She was struck too by the palpably close and nurturing relationships between mothers and their young. "After eating two fruits . . . the mother reached out—right arm, and picked up her child," Jane noted. "She held it to her breast—in exactly human fashion— right hand behind its shoulders, left cradling it, & for 5 minutes it sucked." She was intrigued by the apparent friendships between adults as well. Often, during the heat of midday or after a long spell of feeding, she saw "two or more adults grooming each other, carefully looking through the hair of their companions." These were more than incidental encounters, she sensed; they indicated relationships, bonds, and connections. They suggested sentience, feeling, perhaps minds.

While many details of their social interactions were still obscured by distance and the poor quality of her binoculars, Jane was beginning to recognize certain individual chimps. As an aide-mémoire, she tried to record whatever distinguishing features she could see. "Dark mark on left side of face, behind and below eye," she wrote of one female. The infant with her had a "large white rump patch," among other salient traits. Just as she had at Lolui Island, intuitively Jane sensed that the riddle of the species "as a collective" could best be cracked by observing individuals. And, just as before, she began giving those chimps she could recognize names. "I now know some of them by sight," she wrote home excitedly. "I know the hideous Sophie with her son, Sophocles. I know the bearded grizzled old

Claud, and an almost bald old lady who, I think, must be Annie." It was a clear departure from Nissen's shallow and generalized focus.

Yet seeing things differently didn't automatically translate into seeing them better. Though Jane had now spent more days than Nissen actively tracking chimps in the field, and was slowly and quietly habituating them to her presence, she was also painfully aware that she had yet to add in any significant way to his general observations. Nissen had noted that the chimps slept in arboreal nests and "wandered" a lot; that they liked company and were early risers; that they "ate fruits and berries and other vegetable matter." Jane had corroborated these general conclusions, certainly, but she still hadn't pushed beyond them. She still hadn't found what Louis was looking for: some fundamental behavior that both modern humans and modern apes shared. And then, sometime in October, George Schaller appeared.

George Schaller was a zoologist of some renown. Warm and approachable, uncommonly perceptive, he and his wife, Kay, had just completed a groundbreaking study of mountain gorillas in the misty volcanic forests of the Belgian Congo, pushing far beyond what Rosalie Osborn and her successor had seen in their gorilla study in Uganda. Curious, naturally, to hear what Schaller had learned, Leakey had contacted the couple, inviting them to the Coryndon; after a morning of animated conversation, he had urged them to go to Gombe to visit Jane.

Jane liked the Schallers immediately. For two days, she and George searched the valleys together, visiting all her "usual haunts." Unfortunately, their luck wasn't good: there were no chimp sightings. But their conversations both evenings were tremendously illuminating. In his months of wandering the forest, Schaller had adopted a method much like Jane's, he told her: moving openly, trying to appear "boring," never carrying a weapon, avoiding direct eye contact with his subjects. Like Jane, Schaller believed that the apes were "sentient creatures with humanlike emotions" and that the species could best be understood by studying individuals. It was "really nice to talk to someone who really understood what I was doing & why, & who didn't think I was completely crazy," Jane wrote her family. She had shared her fears of failure that evening, she added, her sense that time was running out, and George, listening closely, had offered his advice; "George said he thought that if I could see chimps eating meat, or using a tool, a whole year's work would be justified," she reported.

George Schaller, like Nissen before him, had concluded that wild apes were vegetarians. Gorillas, from all he had observed, seemed to enjoy a diet of leafy green plants, a slight variation on the fruits, nuts, and berries Nissen described as the menu of choice for chimpanzees. Apes were neither hunters nor meat eaters, it was generally assumed. And since neither man's fieldwork seemed to contradict this notion, the idea still held. Though of course, as Schaller provocatively suggested, no one *actually* knew for sure.

The idea now planted that her chimps could possibly be carnivorous, Jane was perhaps more curious than usual when one morning, late in October, she spotted a wild ruckus in the trees. She peered through her binoculars. It was hard to see exactly what was happening. The foliage was especially dense at the spot where she saw the frantic blur of motion. She heard a few "angry little screams" and then glimpsed three chimps, one of whom was holding "something which looked pink." She stared. After a moment or so, the chimp—a big white-bearded male—seemed to be chewing away at the pink object. "Suspected meat," she scrawled in her notebook excitedly. The limp, pink something—"no hair or fur . . . a baby of some sort. No head"—was indeed meat, Jane soon confirmed. "But *impossible* to know what it was," she added. She watched the big male chimp clutch the thing to his chest and then "he lifted it to his mouth & seemed to 'suck' rather than to bite, at the limp end." He moved then to some lower branches, only to be followed by a female who "beseechingly" put out her hand and touched his, repeating her entreaties again and again. "No response," Jane observed. Then the female "presented her bottom to him" (a sexual and submissive gesture)—one further bid, it seemed, for a taste of the prize. Still no response, despite the fact that she continued to stare longingly at the male, or was it at the tantalizing pink flesh he continued to hoard?

The "unidentified victim," Jane would later conclude, was an infant bushpig, a fact she would soon confirm in subsequent sightings, all variations of the same fascinating, albeit gruesome scene. It was a momentous discovery, the first "eyewitness" account of meat eating among wild chimps. And, in less than a week, it would be joined by a second, even more remarkable discovery, this one life changing: she got a clear sighting of chimps using tools.

The incident occurred in early November 1960. Exhausted after a morning of bellying through the underbrush, Jane had set off along the mountain path just above camp, drawn by the pant-hoots of apes somewhere up the ridge. The trail was considerably more open than the one she'd just battled; she knew it well. She was approaching within a hundred yards of a termite mound when she stopped. There was "a black object" in front of the mound and she couldn't remember having seen a tree stump there before. It was a chimp, she now realized, dropping down quietly. She ducked behind some greenery, hoping she could peer unseen through the scrim of leaves. The chimp seemed to be "picking up things" from the termite mound and putting them in his mouth.

"Very deliberately he pulled a thick grass stalk towards him & broke off a piece about 18" long," she noted. Then he poked the grass stem into a hole, and withdrew it. Unfortunately his back was toward her, obscuring a lot of her view, but she could see he was chewing. She watched him climb onto the top of the mound, his back still toward her. "Then he got down—after peering hard in my direction, & vanished down the hill."

As Jane would soon learn, in the world of chimps, bugs of all stripes—ants, crickets, wasps, beetle grubs—are vital sources of protein. But of all the insect foods, the mound-building termites are the most coveted. Other creatures—baboons, birds, monkeys, and even humans—catch the winged members of the colony on the fly, pouncing upon the "plump" termite specimens as they emerge from their mounds and take wing. But only chimpanzees have invented a means of extracting the *nonwinged* members—the "soldiers"—from their tunneled homes, fashioning a tool to fish them out. The chimps choose a long stem of grass or twig, adjust it "to create a long, smooth, and flexible probe," and then poke the tool strategically into one of the sinuous exit tunnels of the mound. The soldier termites, provoked by the incursion of this strange object, clamp on with their mandibles and hold fast, only to be drawn out as the chimp extracts his tool. He then slides the termite-clad straw through his lips, sweeping them one by one into his mouth, and happily chews them up. This is what Jane had just witnessed.

Jane's first glimpse of the "termite-fishing" chimp was followed two days later by a longer and more satisfying sighting. This time there were two male chimps, one quite visible, the other more skittish, moving quickly out of sight. Quietly, Jane approached through the tall grass. The bolder chimp appeared to be aware of her. But after a pause, in which he seemed to look around cautiously, he returned to his labors.

> *After a few minutes . . . he looked in my direction, peered, got up, climbed to the top of the [termite] hill, and gazed directly at me. Then he got down, resumed his original position, & continued eating termites. I could see a little better the use of the piece of straw. It was held in the left hand, poked into the ground, and then removed coated with termites. The straw was then raised to the mouth & the insects picked off with his lips, along the length of the straw, starting in the middle . . . He chewed each mouthful. Occasionally sat with his lips open as he poked up a new load.*

Jane realized it was the same male she had sighted two days earlier, the white-bearded one. Actually his beard was more gray than white, she now decided. "Grey beard, fingers looked greyish, dark face, only a little bald. Very handsom," she scrawled. She dubbed him "David Greybeard" and watched for another forty-five minutes, at which point he rose and left, following behind the second chimp, who had briefly reappeared. Jane waited fifteen minutes and then moved in to "examine the scene of the repast." But the instant she got there, she heard "low hoots" and then noisy screams. "So—they had been watching me had they! I pretended to eat termites—which must have infuriated them!—& then quietly moved away & sat down."

Greybeard was, she was now sure, the same chimp she had watched "termite fishing" for the first time. He was also the same male she had witnessed a week earlier with a pink slab of meat in his hand. Jane didn't see Greybeard again at the termite mound, though she would see others feeding there. But she did have another astonishing encounter. She was walking one day and saw Greybeard just below her path. They both stopped. She sat, and then he did too, pulling up his knees and facing her directly as he calmly watched. "I was able to observe him excellently," she wrote in her journal that evening. He groomed his wrist and his knees, and then looked up again, reaching to scratch his back. "He then spent about 5 mins stroking his beard with his left hand (like a man thinking), & rubbing his thumb along upper lip. During this he occasionally glanced casually at me. He is a very nice chimp."

A few days later she came upon him again, this time sitting in a tree. "He had his back to me, but as I sat down he turned to face me. He was *perfectly* aware of my presence—as he had been all along." While the other chimp in the tree appeared more anxious, he also didn't flee. And then "a large male baboon walked past below them . . . He glanced at me casually & went on. I felt that it was the proudest moment of my whole life—all 3 accepted me."

David was, as Jane would reflect two and a half decades later, "the first chimpanzee I saw eating meat, the first to demonstrate the use of tools, and the first to permit my close approach in the forest." Almost from the start, he was less fearful than the others, more tranquil and subdued. Perhaps he had more curiosity? His "quiet, almost thoughtful" acceptance of her "alien, ghostly, ponytailed" presence in the woods seemed to "calm" the fears of the larger community. The chimps were clearly tied to one another in mysterious and fundamental ways. "Because he lost his fear of me so soon, he helped me to gain the trust of the others," Jane would write.

LEAKEY WAS STUNNED BY THE NEWS. IN HIS WILDEST MUSINGS, HE hadn't imagined a breakthrough of this caliber or import. Jane had been mailing him carbon copies of her field notes on a weekly basis, and after her second sighting at the termite mound, sure by then of what she'd witnessed, she had sent a telegram with the news. He could barely contain his excitement. Having devoted his working life to searching for the tools of ancestral humans, Louis believed—as did most of his colleagues—that making and using tools effectively defined "human." That Jane had witnessed chimps not only using, but also shaping tools to their own ends came then as culture changing, a revelation with profound and sweeping implications. Cabling her back that same afternoon, he offered his now-legendary words: "Now we must redefine 'tool,' redefine 'man,' or accept chimpanzees as humans."

Jane's plans were to leave Gombe on December 1, 1960, at the official end of the five-month study. But with the meat-eating and tool-using observations of October and November, Louis was now determined that the work continue. His instincts about Jane's energy, stamina, and determination had been correct. However unorthodox and untutored her methods, they had yielded impressive results. Now the challenge was to legitimize this young, scientifically unschooled woman in the eyes of the world. He would begin lobbying for more funds at once, he told her. More importantly, Jane would need academic credentials now. Otherwise, the scientific community would dismiss her work as naive, the fruits of an eager if impressionable amateur; without a degree, she would be openly challenged, an easy object of derision.

As always, with a well-placed mix of gentle pressure and persuasive charm, Louis worked his magic. Jane was soon enrolled in a doctoral program in ethology, the science of animal behavior, under the direction of Professor Robert Hinde at Cambridge University, no mean feat given that she lacked the usual prerequisite—an undergraduate degree. She would begin her studies in a year. Leakey's appeal to *National Geographic* for additional funds went equally well. Still dazzled by the great man's drawing power after his celebrated article about the world's earliest man, the *National Geographic* agreed to a second grant for $1,400—enough to support Jane's work at Gombe for another year. It was bracing news.

Vanne returned to England in the middle of November. Jane left for Nairobi soon after, sorely in need of a break, spending Christmas with friends before her return to Gombe in mid-January 1961. This time she was alone, however, and her social world was completely African. The stipulation that she be accompanied by another European had been conveniently overlooked, and her ties to the community, once so tentative, had by now deepened, as had her affection for her small camp staff, who increasingly felt like family. Even her Swahili was coming along.

Jane's love life, by contrast, had grown complicated again. Derrick Dunn was sending her "unrequited love letters," his interest rekindled after having seen her in Nairobi. Her engagement to Robert Young still hung in the balance. Recently she had sent him a letter declaring that "everything was rather hopeless" and that she "didn't see how it could work." In his return letter, Robert had mentioned a rumor he heard that she was "living with Dr. Leakey, who had now left his wife."

Camp life, meanwhile, was presenting its own small calamities. A giant rat had taken up residence in her tent. It was eating her insect nets and lately had moved on to her blankets and sheets. Though it was captured in a trap, a second rat was soon picking up the slack, its appetite equally ravenous. Happily, it too was caught. One afternoon, she was vaguely amused to discover a scorpion on the sleeve of her sweater. A few days later, she saw a leopard stalking shiveringly close to camp. For weeks, she'd suspected the beautiful dark creature lived somewhere nearby. She had heard an unearthly sound, "unlike any sound I can describe" and then seen a "tail above the grass—held up & over his back a little," she said. As always, despite the obvious dangers, she took these sightings in stride.

By now she'd shed all but Saulo David, her one remaining field assistant. On most days, she moved through the mountains alone, meeting him at some appointed time. George Schaller had left her with a polythene sheet before his departure. Now, on rainy days when she encountered wet vegetation, she could peel off her pants and tie them around her waist, keeping them dry beneath the waterproof sheet—yet another advantage of being alone. Sometimes it was so wet she moved shirtless through the tall grasses too.

The months that followed were unlike any she had ever known. It was an interlude of great rapture and solitude: immersive, piercingly beautiful, at times hard. Living alone in the forest, there were moments when she felt a mystical union with the elements, tied to the rain and the winds, the sun and the moon, the chimps that moved about her in the great primeval woods. The world seemed intensely alive. "Inanimate objects developed their own identities," she wrote. "Like my favorite saint, Francis of Assisi, I named them and greeted them as friends. 'Good morning, Peak,' I would say as I arrived there each morning; 'Hello, Stream' when I collected my water; 'Oh Wind, for Heaven's sake, calm down' as it howled overhead." She loved to sit amidst the giant forest trees when it was raining, "to hear the pattering of the drops on the leaves and feel utterly enclosed in a dim twilight world of greens and browns and soft grey air." As with Carson's experience of the sea, and Jacobs's the systems of the city, Jane felt the forest as a "living, breathing entity," an organism that pulsed with life. She became "intensely aware of the being-ness of trees," she wrote. Touching the "rough, sun-warmed bark of an ancient forest giant," she felt "a strange, intuitive sense of the sap as it was sucked up by unseen roots and drawn up to the very tips of the branches." Though she missed Vanne terribly, Jane had always been at home with her aloneness. Sleeping by herself on the forest floor, listening to the rustle of the ancient forest trees, watching the moon over the "soft sighing lake," there were times when "a powerful, almost mystical knowledge of . . . eternity" swept through her.

"I was getting closer to animals and nature, and as a result, closer to myself," she wrote. "The longer I spent on my own, the more I became one with the magic forest world that was now my home."

It was during this time that Jane would have her most memorable sighting, a drama more startling and indelible, she wrote her family, than any behavior she had yet seen. It was the end of January. She had passed the morning watching a circle of chimps romping and wrestling in the trees. And then the rain had begun. It was a drenching downpour. Within minutes she was soaked to the skin, water streaming through her hair, down her back. She imagined the chimps would take shelter under one of the forest giants. But they hadn't. Instead the chimps divided into two neat groups, one led by a large male she called Paleface, arranging themselves into two parallel rows, leaving about fifty yards between them. "It was most organized," she reflected. Both lines began moving slowly up the grassy slope and then, as each neared the top, one by one, in rapid succession, each chimp left his troop and hurtled back down, charging diagonally across the hill, running and twirling, arms swinging and flailing like scythes. Sometimes they leapt at low-hanging branches as they passed, grabbing at them, snapping off sections. Other times they dragged the branches behind them, or waved them wildly in the air, leaping and charging as they ran, all this "in a pouring rain with thunder rolling above, & vivid lightning flashes," she wrote. On and on they danced, wild and ecstatic, the scene increasingly dreamlike: "primitive hairy men, huge and black against the skyline, flinging themselves across the ground." For it was only the males, she noted. And so it continued for thirty minutes, a scene of such mysterious intensity she struggled to describe it. And then "the rain dance," as she was calling it, ceased. The chimps climbed into the trees, seeming to gaze now quietly in her direction. "I felt all the time that it was for my benefit. I wonder?" Jane scribbled in her journal. Then slowly they descended, making their way quietly to the "brow" of the hill. "Silhouetted on the skyline," several climbed a tree for a moment, waiting for the others to catch up. Paleface, who had been the ringleader, was the last to go. "He stood

up, holding a sapling in his left hand, looking at me. His giant silhouette against the grey sky was impressive . . . The actor taking his
curtain call."

It was another turning point, not only revealing new depths of
emotion and play, but also, more importantly, the presence of ritual
in the social life of primates—organized, perhaps even symbolic behavior shared and understood by all. The sharp line dividing human
from nonhuman had further blurred. Later, Jane would take these
ideas further, seeing in these rain displays the seeds of early human
spirituality, the animistic worship of water, sun, and things that primitive man didn't understand. Though she couldn't yet name it, what
she had witnessed, in effect, was the primal beginnings of religion.

The rain dance was one of Jane's most stunning observations
of that first year: an incident of great beauty and incandescence.
Yet for all its seeming magic, it hadn't occurred by accident, nor
had it come quickly or without significant struggle. Jane's openness
and great powers of intuition; her insistence on moving alone and
always without a weapon; her reliance on qualities that, for want of
a better word, are generally associated with the feminine—empathy,
vulnerability, self-effacement—what Dale Peterson has called her
"revolutionary" approach to "sterile old masculine science," had
something to do with it certainly. Yet much of her achievement was
also the consequence of qualities that were "neither especially masculine nor feminine but rather neutral and sexless," Peterson adds.
Jane's pioneering breakthroughs in that early period, her fresh insights into the animals she moved among, were also the result of
fiendishly hard work: grueling days of false starts and dead ends;
weeks of exhausting, often fruitless, tracking; months of physical
hardship and inclement weather. It took grit and perseverance, an
almost superhuman level of physical stamina, to accomplish what
she did.

JANUARY BROUGHT LASHING RAINS, UNRELENTING HEAT, HUMIDITY
that was choking. The grass "shot up" until it was over twelve feet

high in places, making slow, drenching business of Jane's daily slogs. There were days when nothing would dry, when the trails were perilously slick, when a rogue wind kicked up from out of nowhere. Sometimes, unable to see her way through the vegetation, she had to climb a tree to get her bearings. And there were other challenges. The dampness made a swamp of her bedding. It rotted her clothes and spread mildew through her books and papers like a galloping pox. Unassuming scratches turned into oozing wounds in such high-voltage heat. Skin ulcers erupted on her legs. At times she was beset with inexplicable fevers, headaches that hammered at her skull, strange bouts of insomnia she couldn't explain. Lately, she wrote her family, a mysterious "white & fungus-y" thing had taken up residence between her toes. She assumed it was from being in wet sneakers "all day long, for 4 months." Now "suddenly it's gone *under* the nails—all the cuticles. . . . will my toes drop off. *What* can I do?" she added with a note of dark humor.

Still, even with all this, she soldiered on, her spirits bright, never slowing her pace, or scaling back her efforts, spending long, solitary days slogging through the mountains, rising before dawn, returning to camp after dark, just as before. And so the weeks passed.

By early February she was getting at least one good sighting a day, some that lasted for hours at a stretch, a welcome change from before Christmas. Distance, however, still remained a challenge. "100 yards," she wrote home, was still "not far away." And with the foliage, it wasn't always possible to identify individuals, even at close range. Added to these obstacles, in late February she hit a rough patch. It was odd. Weeks passed and she wasn't able to recognize a single chimp she had seen before, despite the meticulous notes she kept. David Greybeard, whom she had watched at such close quarters in November, had seemingly vanished.

But then things began to improve again in March when, to her surprise, a few chimps began to show up at camp. At first it was only one or two. Several palm trees near her tent had started to produce nuts. And there were new black seeds on a yellow tree nearby. The chimps, she realized, were coming to feed. At first, she didn't make too much of these camp sightings. They were occasional and always brief. One or two chimps would arrive, feed, and then vanish, with rarely any variation in the pattern.

Her forays into the mountains, meanwhile, were getting better. The chimps seemed less and less afraid; she was finally getting close enough to identify individuals again, amassing more and more detail about their patterns and social habits, their facial expressions and individual personalities. Lately she'd felt sure enough about six new chimps to give them names. She could now recognize Mike ("large black faced male, not a very conspicuous beard") and William ("round chin, a long upper lip that wobbled"), Wilhemina and Lucy, Lord Dracula and two-year-old Fifi, "who rode everywhere on her mother's back." She felt she was finally beginning to accomplish something. Waiting at camp for an occasional sighting now seemed wasteful by comparison.

And then, in early May, David Greybeard made his first cameo appearance in camp.

She had been in her tent that morning when she'd heard "a rustle" and then seen a dark male chimp streak up a tree, where it quietly began feeding. She'd known somehow it was Greybeard, even before confirming it with her binoculars. Unperturbed, he'd remained in the palm, feeding calmly as she walked around the tree with her camera, finally standing directly beneath him. "Can it be true? I have just been under the tree talking in a loud voice to him," she wrote her family. "He didn't even look. Isn't it too ridiculous for words? Better than George's [Schaller] 'Junior' who sat 25 yards away. This is not 25 feet." That night, dragging her camp bed and its mosquito net out of the tent, she planted herself under the tree, only to experience "the strangest awakening I've ever had!" When Greybeard appeared the next morning, he paused on his way up the palm to peek under the mosquito net. Then, apparently satisfied, he calmly resumed his climb, gorging on palm nuts for the rest of the morning as Jane watched him from her bed below: "chimp watching in bed!!!"

Soon Jane was leaving David Greybeard ripe bananas, which

he usually "demolished—skin and all" on the spot. One afternoon, after passing him on one of the trails, she stepped aside, deciding to head back toward camp, only to discover that Greybeard was following her, though at some distance. Every time she stopped, so would he. She reached camp before him, in time to toss a few bananas near the palm tree and grab her camera. Sitting at the open front of her tent, she watched him amble into camp ten minutes later. Barely glancing in her direction, he grabbed the bananas and sat, no more than six yards away, facing her as he polished off the fruit. She could scarcely believe he was comfortable letting her sit so near. It was the beginning of a deeper level of habituation for the chimps.

JANE LEFT FOR A BRIEF HOLIDAY TO REST AND RECUPERATE IN NAIROBI at the end of May. Gaunt and exhausted, for the first time in months she slept in a proper bed, ate real meals, and saw old friends, including Clo, who to Jane's dismay was now trapped in a violent marriage. When she returned to Gombe two weeks later, the rains had let up, as had the oppressive humidity. She could finally admit how tough the conditions had been. "There were times, during my last session here, when I wondered if I could possibly exist through those 4 months," she wrote Bernard Verdcourt. She had been "full of foreboding" upon her return. Now, however, she was pleasantly surprised. The chimps were being "utterly charming & friendly," she told him. They seemed "less anxious, easier to approach." Although they had been "reasonably tame before" they were now "suddenly more so." Recently, she had sat "for ages" only ten yards away from a large male. "At times he came even closer to reach for a fruit," she added.

This, in the end, was the triumph of Jane's first year at Gombe, as Peterson writes: that she'd successfully habituated these normally skittish and secretive wild creatures to her presence, slowly and patiently earning their fragile trust. Beyond her discovery of chimps eating meat, making and using tools, and dancing ecstatically in the rain, Jane had found a way to move openly among them, getting closer to wild apes, and seeing more of their social lives and their behavior than any human ever had. No one, with the exception, perhaps, of George Schaller, had entered their world to such an extent.

BEFORE JANE GOODALL'S TIME AT GOMBE, ZOOLOGICAL STUDIES IN THE wild, if they occurred at all, were still relatively short in duration. Those few researchers who ventured out tended to keep comfortable distances, watching from protected blinds. The accepted paradigm was still scientist as "manipulator and voyeur: elevated, protected, distanced." Typical was the Dutch primatologist Adriaan Kortlandt, who, following scientific convention, built a sort of outdoor lab on a papaya plantation in Belgian Congo, where he then staged a series of controlled experiments, watching for results from crow's nests in the trees. In one of his best-known experiments, Kortlandt placed a "realistic," mechanically operated stuffed leopard in areas where wild chimps would see it, "to test the apes' anti-predator defense technique." (Never mind that the chimps might not have found the mechanical leopard all that realistic looking.) To guess at their food preferences, he put out "egg-laden nests," chameleons tied to nets, and a small, freshly slaughtered forest antelope "made to look . . . alive." In each case, the relationship was always the same: observer as spin doctor, observed as object, separate and subordinate.

Jane's alternative approach put "observer and observed in the same field," facing each other on the same terms, equals. It was a shift in stance that carried stunning implications. By sharing the same footing as her study subjects, moving freely and openly among them, "on foot, unarmed, lightly clothed, often alone, always unprotected," Jane Goodall had crossed into unknown scientific territory. By envisioning her apes as living "in a parallel universe," separated, as her biographer has written, "not by an uncrossable gap in feeling and perceiving," but by "a partially reversible failure of communication and understanding," Jane Goodall had opened the door to the previously unthinkable: that chimpanzees were dynamic and willful beings with personalities and a repertoire of similar, perhaps identical emotions and perceptions to humans. "This young, scientifically naive woman," observes Peterson, "had chosen to sail right off the edge of the map." The results—unexpected, astonishing, culture bending—would ultimately spark a revolution in the way humans see animals and, in turn, themselves.

NATIONAL GEOGRAPHIC WAS A GENEROUS PATRON, BUT THEIR CONTINU-ing support came with conditions. In exchange for underwriting the research at Gombe, the staff was expecting an eye-popping story about Jane and her wild ape project, complete with the sort of elegant, high-resolution photographs for which they were famed. And so, by necessity, Jane's focus was now to change. The challenge going forward was how to capture these elusive, highly acrobatic creatures in some acceptable way on film.

Louis, as always, was overly optimistic. Without fully understanding the complications, he had misrepresented how close Jane was getting to the chimps, suggesting to the magazine's editors that a professional photographer could "possibly" go to Gombe by July 1961. That person, however, would have to be a woman, he added—for Jane's comfort, but also for appearances, although "only if Miss Goodall agrees and things are going well."

Jane didn't agree, and things weren't going terribly well. A second person, she worried, would shatter the chimps' all-too-tenuous trust, which she had worked so hard in the last year to win. Although the apes weren't frightened of her, they didn't like to be watched, she insisted. Even alone, it was often not possible to approach closer than fifty to a hundred feet. She couldn't imagine how they'd respond to an additional person with "a big glassy eye." Privately, she also had deeper qualms. She was wary of sharing her achievement with someone else, especially a woman whom she didn't know. And beyond this, there were technical challenges she wasn't sure that anyone, Leakey included, understood. The heavy, cumbersome equipment would have to be lugged up treacherous slopes, often through thorn-bushes and extreme thicket, sometimes over slippery rocks, flooding rivers. The heat and humidity would wreak havoc on cameras and lenses, no matter how well covered. The forest was dim and shadowy, the apes quick and kinetic, reducing the likelihood of adequate light or clear shots. If an outside photographer ultimately proved "abso-lutely" necessary, she preferred to have someone she already knew, she told Leakey in a letter. In the meantime, "I want to do my own photos—or have a jolly good try first."

National Geographic was dubious. But at Leakey's urging, they initially relented. Perhaps, they wrote, sending their own female photographer was not such a good plan, since admittedly the woman they had in mind had no experience with wild animals. But what about hiring Leakey's son Richard instead? wondered Melville Bell Grosvenor, president of the Geographic. Perhaps he would be more acceptable to Jane, "since he is no stranger to her." Richard knew a lot about photography, and also "knows the animals and would not be unduly alarmed about being alone with them in the jungle." Regardless, the National Geographic Society was air-expressing "a wide-angle Rolleiflex camera" and a good supply of high-speed film, he told Leakey, which would be arriving any day. He was hoping for twelve to fifteen pages of "top quality" pictures to accompany Jane's article.

Louis's answer to Grosvenor's proposition was an emphatic no. Richard couldn't possibly upend his plans and go to Gombe at such short notice. Furthermore, Jane, who happened to be in Nairobi that week, had tried the camera the National Geographic sent and found it "far too heavy and clumsy," and "will not take it with her into the field," he explained. "What would you like me to do with it?" In an expression of solidarity with Jane's plight, he had bought her a lighter, "more straightforward" camera with a telephoto lens she could handle, he said. It was small enough to serve as what he called "a camera round her neck." He hoped, however, that the Geographic would send out a second camera with better telephoto lenses and a tripod.

National Geographic responded with dispatch, immediately posting the second camera, three stronger telephoto lenses, batteries, a tripod, a light meter, film, caption cards, and special cartons to simplify the return shipments of exposed film. Also included in the package was a note from Joseph Roberts, the magazine's assistant director of photography, which he called "the shortest correspondence course in photography ever written."

Short or not, the level of technical detail it contained immediately gave Jane "a raging headache." The situation was laughable, she wrote her family. "If only they realized the conditions . . . For instance, yesterday I was inching my way up a mountain side, clinging onto rocks & roots. And I bumped into some chimps. Well, there they sat, yelled at me a bit, ate a bit . . . & went on their way. Wonderful view. Super photos—if I could have taken them. But as I really hadn't enough hands & feet to secure me to the mountain's surface, *how* could I take out my one simple little camera? Let alone if I'd had millions of lenses & tripods stuck all over me!"

Five weeks later, a penitent Jane was writing back to Roberts, apologizing for not having touched the second camera, which remained in Nairobi and confessing to "not having previous camera experience, I'm afraid." She explained that she didn't think she'd use the camera until later anyway, "as I feel that a determined effort at close-up photography will very definitely upset the animals." Finally, she admitted that she'd exposed only a single test roll of film—taken from "the neck-hanging" camera—in two months. She was enclosing it with the letter.

The roll, once developed at the magazine's photo lab, proved even more dismal than the editors expected. "Of 37 exposures on this roll," reported the illustrations editor, Robert E. Gilka, in an internal memo, "16 are so underexposed as to be unreadable, 10 are spoiled because of camera motion, and 6 are not useable for other reasons." There was one shot the magazine "might possibly use, a shot of Miss Goodall sitting on a hillside scanning the area for chimps."

It was clear that a second person would *absolutely* be necessary to man the big camera. Louis and Jane now floated the idea of sending out Jane's sister, Judy. At the time Judy was working at the British Museum. Perhaps she could take a short leave from her job, suggested Leakey, if the Geographic could pay her expenses to Africa and back? Judy looked a little like Jane, which meant the chimps might be more accepting of her. And she had "some experience of color photography," he assured Grosvenor. The response, sent by National Geographic's vice president, Frederick G. Vosburgh, was decidedly clipped: "The production of satisfactory photographs . . . requires considerable experience as well as aptitude, and this assignment would tax the skill of even a professional photographer." The answer was an unequivocal no.

Unmoved by Vosburgh's objections, Louis pressed on with the plan anyway. If National Geographic was unwilling to be more accommodating, if they couldn't see their way to spending a little bit more for such a worthy story, then he would seek support elsewhere, since "it has got to be done." In an uncharacteristically cool letter back, he told Vosburgh that, as they were caught in an unfortunate "deadlock," he saw no alternative but to sell first rights to the story and pictures to another British publication. The matter was left there.

Judy arrived in Gombe that September, startled to see how "skeletal" her sister had become. Jane was glad for the company, and Judy, as always, was game, cheerful in the face of grim skies and peculiar rations, the standout being "crisped termites." For the first week or so, the weather was good. The chimps were present in significant numbers, and both sisters had opportunities for promising shots. But then quite abruptly the rains descended, earlier that season than usual, heavier than anyone remembered. The blanketing clouds dimmed the light; the humidity gummed up the cameras, the mist-shrouded forests made it impossible to see. Periodically, Judy now had to take one or the other camera to Kigoma, losing time while she waited for frozen shutters or stuck gears to be repaired. "I can't remember when I last wrote," Jane wrote Louis in early November. "Had I been ill with a weird fever that left spots all over my face?" By now, the weeks of failed photography were not only weighing on Jane's spirits but also her health. She had developed shingles and was feeling increasingly hopeless. "I hate to write such depressing news, but you'd better be warned—it has made me feel that my entire work has been a failure. This, I suppose, is not true, but I just can't help feeling miserable about everything." In truth, there had only been one really good day of photography during that entire fall, she admitted.

The year limped at last to a close. Jane and Judy both left Gombe in December 1961, reaching Bournemouth in time for Christmas. Their twenty-four rolls of exposed film—one in black-and-white and the rest in color—were immediately airmailed to Washington, D.C., to be developed at the National Geographic offices. They waited nervously for some word.

When the letter arrived on January 2, the news was not good. Robert Gilka noted that except for one shot, the photos were "not exciting." Due to "a lack of good pictures of the animals in their native habitat," he wrote, the magazine had decided not to publish the story of Jane and her chimp research project. "I wish that we were in a position to publish an article on your subject because it is an unusual and fascinating [one]," he added patronizingly. However, "this shortcoming is so serious as to preclude attempting to illustrate a story." The project was effectively dead. Deflated, both Jane and Judy left Bournemouth a few days later. Jane was headed to Cambridge to begin her first semester, Judy to London to reclaim her job at the British Museum.

SHE MISSED GOMBE IMMEDIATELY. CAMBRIDGE IN WINTER WAS GRAY and socked in; it seemed cold and colorless after Africa, meager and claustrophobic. And she was worried about almost everything. Leakey had persuaded the higher-ups at Cambridge to accept her field journals from Gombe as the basic research for her doctoral thesis. She would be working, he told her, under the wing of Professor William Thorpe, who would guide her readings. The plan was to spend half her time at Gombe, collecting more data, and the rest at Cambridge, writing up her results in an "appropriate format."

Her first interview with Thorpe, however, didn't go well. The eminent professor seemed stiff and humorless, vaguely distracted, and she was upset by his myna bird, who after swooping about the room, landed on her and began pecking, eventually drawing blood. She was relieved when she got word that Thorpe had decided to reassign her to one of his colleagues, Robert A. Hinde, the resident monkey expert.

A brilliant scholar, Professor Hinde was by 1962 a legend at Cambridge. "Incredibly handsome," as a former student remembered, with "piercing blue eyes," a "craggy" face, and "silvery gray" hair, he was proud, certain, serious, and on occasion, volatile—a whip-smart critic who was known to suffer no fools. "He was not aggressive," recalled another former student, "but he was very smart—and he could make you feel completely stupid." He was reputed to have reduced many a young female student to tears. "He would ask a penetrating question and look at you with those penetrating eyes." Yet Hinde was also one of the few professors who readily took on female students. Ultimately, he was a mentor to women, a rare champion where champions were few.

Jane was "terribly in awe" of the great professor in those first tentative months. During their weekly tutorials, which took place in his apartment at St. John's College, she would say little, she remembered, as he "pointed out the flawed reasoning behind some attempt to describe and quantify a portion of the data, or explained just why it was that certain words were not acceptable in the scientific circles of the time." He wasn't stiff or formal, she added; he was "youthful," and a touch odd. He would "sprawl" on the floor in front of the fire, as he read and critiqued her work. There were times when he would tell her that "I'd better go and do a lot of reading before I continued to make a fool of myself (not that he put it quite in those terms, but his meaning was clear)." She often trudged home "filled with frustration and sometimes despair. Back in my digs I would hurl everything into the corner of the room: page after page, written so carefully, now marked all over with Robert's comments and criticisms. How desperately I longed to give it all up and go back to the chimpanzees and the forests."

But anger can be a potent spur. Proud and a touch stubborn, tenacious by nature, Jane didn't give it all up. And often, by the next day, even when she didn't agree with everything Hinde had said, she did understand *why* he had said it. As time passed and they grew more familiar, when Jane continued to disagree, instead of holding back, she returned for another round of heated debate. The distinguished professor and defiant pupil grew accustomed to locking horns.

Hinde's task, as he saw it, was to teach Jane the language of "ethology," his first challenge being to help her convert her handwritten Gombe journals into an acceptable scientific form. This meant a language, as he put it, that was cool, "quantitative," and scoured of sentiment, which is to say, standardized. Instinctively, Jane had set down her raw data as a narrative, a series of anecdotes and stories, complete with an elaborate cast of characters. This, she learned, was naive and unscientific, unacceptably subjective. Such an approach supposed humanlike individuality, feelings, and sentience. From an ethological point of view, "ascribing personalities to the different chimps" was a cardinal sin, she wrote. "Only humans have personalities, I was told. Nor should I have been talking about the chimpanzee mind—only humans, said the scientists, were capable of rational thought. Talking of chimpanzee emotions was the very worst of my anthropomorphic sins."

She wondered, at times, if she'd been mistaken in coming for her Ph.D. The only things people at Cambridge seemed to appreciate, she lamented in a letter home, were "graphs & statistics." It got, she added, "rather depressing."

"I had no undergraduate degree. I had not been to college, and there were many things about animal behavior that I did not know. For one thing, I did not think of the chimps as 'study subjects' but as individuals, each with his or her own personality. I was learning *from* them, not only *about* them."

And, as Jane was also learning, there was another, even more fundamental problem with her method. According to the language of ethology, her focus was supposed to be on the species as a *collective* and not on chimps as individuals at all—even as numbers; it was best to concentrate on behaviors that were "typical," she was told, those that were innate and predictable, fixed and demonstrably true. The atypical was deemed less important, if not irrelevant.

This was hugely problematic from her point of view, only a part of the story. She didn't object to writing about typical behaviors—chimpanzee nest-making practices, mating rituals, standard feeding preferences. These were elaborate and fascinating behaviors—worthy, certainly, of attention. She wasn't blind to the importance of knowing and understanding the typical. But how could she discount the atypical behaviors, the mysterious exceptions to the rules? The unusual friendliness of David Greybeard? The singular mothering style of Ollie and Flo? The inexplicably close kinship between David Greybeard and his friend Goliath? She had seen David and Goliath holding hands, tickling, and chasing each other in circles, wrestling and laughing in seeming delight. These were anomalous behaviors, but they were no less revealing than those that were common to the species. It could be argued, in fact, that they were *more* revealing. They indicated emotional complexity, depths of expression beyond the automatic and instinctual, the unnerving possibility that chimpanzees might, like humans, have individual personalities, emotions, and even minds.

She and Hinde continued to spar over these questions. Hinde was wary of soft science, mistrustful of anything that couldn't be absolutely verified. Like most primatologists at the time, he hewed to a traditional view of ethology, advocating a stripped-down, mechanistic approach, science made immaculate and predictable. It was important, he felt, to scrub the data of the subjective. The goal was to describe animal behavior in measurable ways, to parse out the irrefutably true, while avoiding the messy and subjective—the "possibly true."

In truth, their philosophical differences went to the heart of a fault line that was beginning to appear not only in science, but also in the larger culture. In Jane's empathy and respect for her study subjects, she was challenging the very notion of a value-free neutrality, countering the view that the world could be parsed out in wholly objective, mechanical ways. Like Carson and Jacobs, she was refuting the idea that in nature there existed a simple, mechanistic template that could be applied to all (whether in the service of eliminating city slums or bothersome insects) or that elements in nature could be examined—or understood—stripped of their context. This meant factoring in the social fabric, recognizing the fragile web of connections and interactions that constituted the larger system. Her work, like theirs, left room for intuition and the efficacy of the individual, the anomalous and the contradictory, personal testimony and the common sense of direct observation. It was a shift in emphasis that in the early sixties was quietly beginning to build in reaction to the increasingly specialized and autocratic direction of the culture. "This is an era of specialists, each of whom sees his own problem and is unaware of or intolerant of the larger frame in which it fits," Carson had written.

Though she didn't necessarily intend it, like Carson and Jacobs, Jane stood at the crossroads of the interplay between intuition, personal experience, and social change. In drawing attention to the being-ness of animals, Goodall was speaking directly to a cultural shift that was just beginning, a gathering moral commitment to social justice and the humane treatment of all races, and, by extension, all living things—from animals to the earth.

IN THE SPRING OF 1962, AS CARSON WAS EDITING *SILENT SPRING*, GOOdall received an invitation to speak at a scientific conference. It would take place in London, in mid-April, and was sponsored by the Zoological Society of London. Louis, of course, insisted she attend.

The conference, in retrospect, would prove something of a turning point. Even at the presymposium party, remembered one of the participants, there was "a sense of occasion. There really had not been a primate conference before. It was the feeling that a field might be opening up." Among the attendees expected were the leading lights of the primatology world, including Desmond Morris, Jane's old boss at Granada TV and film unit; Irven DeVore, who had studied baboons in Kenya; Adriaan Kortlandt (of stuffed mechanical leopard fame); and Solly Zuckerman, the formidable secretary of the Zoological Society and the chair of that day's events.

Solly Zuckerman was by 1962 a force to be reckoned with. Big and self-important, with a barbed wit, a "red face" and a "perfectly parted" mop of wavy white hair, he was a man at home with power: smart, ambitious, charming when he needed to be. Unapologetic about his appetites, a shrewd collector of people, he was partial to fine suits, good whiskeys, and well-connected friends, a political animal to his teeth. As a young scientist, he had been a research anatomist at the London Zoo, his specialty the social behavior of baboons, a position he had parleyed into an academic post at Oxford and a well-received book, *The Social Life of Monkeys and Apes*, for a time the bible of pre–WWII primatology. These notches in his belt were soon followed by an appointment to the Ministry of Defense as chief scientific adviser, and, eventually, a knightship, in recognition of his services during the war. Officially "Sir Solly" now, a titan in the primatology world with an entourage of rich and powerful friends, he was not a character to contradict or cross.

On the first day of the symposium, Jane nervously took the floor, the fourth of eleven speakers, following directly after her two most significant predecessors, Adriaan Kortlandt and Rosalie Osborn, Louis's former secretary and lover.

Standing before the sea of mostly male faces, Jane touched briefly on her discovery of chimps making and using tools, and then almost immediately moved on to speak about chimpanzee feeding behaviors: the amount of time chimps spent eating, the methods they used to retrieve foods, the wealth of foodstuffs they consumed. "Sixty-one different vegetable" varieties, she told the audience, "four different insect species," *and* the meat of mammals. It was the first such mention of apes eating meat, and was followed that day by several other speakers who discussed the same, including an upbeat Irven DeVore, who showed a grisly film clip of baboons devouring the bloody flesh of some unrecognizable animal.

The protocol at such conferences was still quite buttoned down in 1962. Men wore suits and ties; women—those few present—long skirts, white gloves, and the occasional pillbox hat with a veil. If a member of the audience wished to ask the speaker a question, it was directed first to the chair, who in this case was Solly Zuckerman, who would then turn to the speaker and restate the question. Only then was the speaker allowed to answer.

That day, recalled Desmond Morris, Solly Zuckerman was "very hostile" toward Jane, disparaging her behind the scenes as one of the "amateurs."

During his brief baboon study at the London Zoo thirty years earlier, Solly Zuckerman had observed adult baboon males herding females to form harems. Based on this somewhat dubious bit of proof, given that the baboons were living in artificial confinement and *not* in the wild, he had decided that a harem system was "characteristic of all primate groups." After Jane's talk, curious to hear her views on the matter, Desmond Morris rose to ask if she had observed harems among the chimps. Solly ignored the question, whereupon Desmond asked it a second time. Again it was ignored by the chair. At this point Desmond, breaking with protocol, turned to Jane and asked her directly. Jane responded. No, she explained, the male chimpanzees she had observed did not acquire stable harems.

This was clearly too much for Solly, who, having written at length on the subject, remained deeply invested in his own conclusions. At the close of the day's proceedings, during his official summary, he opened with the cutting comment that "there are those who are here and who prefer anecdote—and what I must confess I regard as sometimes unbounded speculation" in their study of primate behavior. He wasn't at all sure, he added smugly, that this constituted "a real contribution to science."

Solly Zuckerman now proceeded to grandstand. On and on he droned, touching on "dominance relations" and "primate sex ratios," territorial issues and chimpanzee feeding patterns, circling back repeatedly to the "over-riding importance" of his signature book of thirty years before. While it might be true, he asserted, that some baboons had been seen eating meat on occasion, these sightings were to be discounted, since the baboon was clearly "a non-carnivorous animal in most places where it exists." As for Miss Goodall's most irregular account of chimps eating meat, it would be "a useful point to remember," he continued, his voice oozing with condescension, "that in scientific work it is far safer to base one's major conclusions and generalizations on a concordant and large body of data than on a few contradictory and isolated observations, the explanation of which leaves a little to be desired."

The matter, in other words, was settled. In spite of what Jane Goodall or anyone else might have claimed that day, Solly Zuckerman could "assure" the assembled that all the monkeys and apes of the world were vegetarians and not carnivorous. One needed only to refer to his book.

Desmond Morris remembered receiving a letter of appeasement from Sir Solly soon after the conference. He hoped Desmond would forgive him for having avoided the question about chimp harems, he said. "I realize you were only trying to provoke discussion," Solly offered, "in the same way," he continued, that he was sure Desmond could appreciate his "anxiety" that a subject of such scientific import "should continue in the unscientific shadows because of glamour."

Solly Zuckerman's glib putdown of Jane's presentation, his efforts to trivialize, if not undercut, her work, calling it essentially "glamour" masquerading as science, would be a foretaste of a certain kind of resistance Jane would encounter on occasion through much of her distinguished career, writes Peterson: the idea "that her legs were too nice, her hair too blond, her face too fine, her manner too feminine for anything she said or wrote to be taken all that seriously."

But primatology, which had pretty much always been a boys' club until then, was changing. It was Sir Solly's contention—and he was not alone—that the secret to all primate behavior could be unlocked with a single key. That key, he further insisted, pivoted on two grand but simple themes: sex and violence.

Yet even by the 1960s, this view was beginning to be challenged. Instead of "great simplicity," a fixed pattern of behavior that applied to all members of the species, researchers were discovering great variety and complexity in the social lives of many primate species. Yes, there were stable harems, as Sir Solly claimed. But there were also "monogamous pairs" and solitary outliers, promiscuous couplings and shifting alliances within stable groups. Beyond the basic hormone-driven "sex and dominance" behaviors of Solly Zuckerman's muscular vision were activities and actions that spoke of other, more nuanced impulses: friendship, "choice, kinship, learning, maternity, paternity, planning and politics."

Indeed with time, despite the scorn of the Sir Sollys of the world (and there were others), Jane's own view of chimpanzees as autonomous and emotional creatures who are self-aware and self-directed would increasingly become the mainstream view. Though Jane would be unjustly dismissed as an "amateur" and not a professional, just as Carson and Jacobs both were; though she would be assailed, like them, for framing her arguments using "anecdotes," rather than the dry jargon of science—and scorned, also like them, for writing for a popular audience and not to the elite of her field, she would also find herself on the right side of history. (In 1967, before her Ph.D., when her first bestselling book, *My Friends the Wild Chimpanzees,* was published, her Cambridge mentor is said to have gasped in horror, "It's—it's—it's—for the general public!") Solly Zuckerman would find himself a footnote in the long slow march of science, while Jane Goodall, in time, would become one of the most acclaimed scientists of the twentieth century, her methodology used for recording the social lives of animals from whales to ants. But not, of course, before she had absorbed her share of flak.

BACK IN WASHINGTON, NATIONAL GEOGRAPHIC'S MELVILLE BELL GROS-
venor sat stewing in his office. Two months had passed since rejecting the Jane Goodall piece and he was having second thoughts—not
about the photographs, he still believed they were unacceptable—
but about letting go of the story, which he now believed was a big
mistake. Recently, a stunning letter from Leakey had arrived, endors-
ing Jane's work and updating him on her discoveries—including the
meat-eating and tool-using sightings, about which Grosvenor had
been unaware. Grosvenor had always believed in the Goodall story.
It was strong and exceptionally dramatic. But now he also realized it
was scientifically significant. The National Geographic "must make
every effort to get it," he told David Boyer, the staff photographer
he wanted for the project. If, that was, Leakey hadn't already sold
it to another publication. He wrote now to Louis in Nairobi, and
Jane at Cambridge, suggesting that they pay Jane a sizeable sum to
leave Cambridge for three weeks in February and fly to Gombe with
Boyer to get the pictures they needed. He hoped, he added, that Jane
would also be willing to write an in-depth article telling the story
of her discoveries. Jane wrote back immediately. It was impossible,
she explained. She was presently at work on her Ph.D. Additionally,
given the travel time involved, it made no sense to go for so few
days, especially in February, which was often the worst of the rainy
season. The chimps would be hard to find, extremely challenging
to photograph. Leakey's return letter was even more emphatic. The
answer was no.

But the National Geographic Society wasn't so easily spurned.
That March, in an effort to convince Leakey of their seriousness
(and perhaps to woo both Jane and him back into the fold), they
voted to award Jane the Franklin L. Burr Prize for Contribution to
Science, which came with a cash stipend. The society hoped, they
wrote, that Jane would fly to Washington in April to present her
findings. In a separate letter to Leakey, they also agreed to fund her
research at Gombe for another year. Louis was sufficiently seduced.

Back in Nairobi, Leakey returned now to the photography question. He had recently met a brilliant wildlife photographer, he told

Jane, a young Dutchman named Hugo van Lawick, who he thought would be ideal for the job. The young man, he went on to inform the *National Geographic* editors, was "familiar with wild animal behavior." He "lives in Nairobi, so could get to the chimp site within a few days." As for the delicate business of his being male, if the Geographic would be willing to pay for Jane's mother's expenses as well as Jane's, Vanne could again serve as her "chaperone," handily solving any questions of propriety.

Baron Hugo van Lawick was young, gifted, and dashingly handsome. He was also painfully shy. Reasonably sure he'd glimpsed Jane Goodall a year earlier, in June 1961, at the home of his then-employers, the filmmaking couple Armand and Michaela Denis, he was feeling apprehensive about his new posting. The Denises were friends of the Leakeys and Jane had stopped by briefly for photographic advice. Hugo doubted she'd noticed him that day, though he remembered her beauty and poise quite distinctly. How, he must have wondered, would the two of them fare, living in a tiny, remote camp in the forest, far from creature comforts or the usual human distractions? Jane's mother was due to arrive in several weeks, but until then, it would be just Jane and him. With some hesitancy, he cabled to say he would be arriving on August 15.

Hugo needn't have worried. Jane returned to Gombe that July, having left Cambridge at the close of the term, and when Hugo appeared a month later, they took to each other immediately. Beyond Jane's seeming ability to live without eating, which Hugo, who was rather partial to food, found perplexing, and Hugo's occasional smoking, which required adjustments on Jane's part, they got on famously. Hugo, like Jane, was ardent about his work. He shared her love of nature, her sense of kinship with the Africans, her identification with the pure, primitive beauty of the landscape. He felt, as she did, that he belonged to Africa.

Twenty-five years old at the time, slight, with a muscular build, thick dark hair, and "a face burnt red-brown by the African sun," Hugo had mostly lived a nomad's life. Born in Indonesia, at age four he had lost his father, a Dutch pilot, to a plane crash. His mother, overnight a widow with two small sons to support, had moved the family to Australia and shortly after to England, before finally settling in Holland at the close of the war. Hugo, who had been happily ensconced in an English boarding school until then, elected to stay in Devon, an arrangement that lasted until his mother could no longer pay the tuition. Though a titled family, the van Lawicks were not a wealthy one. At ten, when the family coffers tapped out, young Hugo was forced to leave England to finish his schooling in Holland.

He was drawn to animals and wild nature, even as a boy, and by age fourteen, he had added photography to these passions, hoping he might find a way to combine them to make a life. The family had a long and esteemed history of military service, so a career in photography ran against this tradition. But his mother encouraged him to follow his heart, however tenuous his plans appeared to her. And so, after a short stint in the Dutch army, Hugo had joined a film company as an assistant cameraman, and then worked for a time as a still photographer before finally decamping for Nairobi in 1959, where he was hired as a cameraman for the Denises, who were making a name for themselves producing the first nature films for television. It was while working for the Denises that Hugo met Leakey, who happened to be looking for someone to make a background film for one of his *National Geographic* lectures. Struck by Hugo's work, Leakey hired Hugo for the film, and, extremely impressed with the results, eventually recommended him to Grosvenor, who proceeded to hire the young Dutchman, which is how he came to be at Gombe that summer of 1962.

"Hugo is charming and we get on very well," Jane was soon writing to her friend Bernard Verdcourt at the Coryndon. "We are a *very* happy family." They quickly established their system. Jane would find a spot where she felt the chimps would return, and the following day, the two of them would haul up Hugo's heavy equipment—a bulky wooden tripod, several lenses, metal storage boxes. At the "crack of dawn" they would then build a rough hide, where she would leave him with "her blessings" and then continue with her own observations.

The chimps, she was pleased to see, seemed to remember her from six months before, which was enormously cheering; despite her worries, she hadn't lost ground during her time away. David Greybeard was becoming so comfortable in her presence, in fact, so accustomed to visiting camp, that one morning she awakened to the sight of him sitting quietly beside her bed, finishing off a banana he'd cadged from a storage box. "He IS a devil!" she wrote home. "I shooed him off & closed my eyes." But the day's most astonishing development, she added, had come later, when once again Greybeard ambled into camp. It was "the happiest: the proudest, of my whole life to date," she wrote excitedly. "David G—yes—he has TAKEN BANANAS FROM MY HAND. So gently. No snatching. The first time I held one out he stood up & hooted, swayed from one foot to the other, banged the tree, & sat down. So I threw it to him. The next one he came & took."

Even more encouraging, Greybeard's increasing tameness seemed to extend to a tolerance for Hugo too. The chimps "have accepted the presence of Hugo van Lawick with his tripods and lenses in the most wonderful way," she wrote to the Geographic Society. "The method I have always followed—never hiding from the chimpanzees, never following them when they have moved away from me, and never appearing particularly interested in them—has, at long last, paid dividends."

Those dividends soon included roll after roll of photographs, which both Hugo and Jane continued to take throughout that fall. Robert Gilka, increasingly pleased with the images he was seeing, now began to ask for specific shots. He wondered, he wrote Jane, if Hugo could possibly get a photo of her standing in a stream washing her hair? Though it is possible, he added, "Hugo might be too embarrassed to ask you to do this for him." Jane sent back a letter in late September: "You will by now have stills of me washing my hair—though Hugo is going to take some more in a more forested part. You needn't worry about him being embarrassed. I haven't noticed it yet!"

Hugo, as Jane would later write, was "just the right person for the job." As they sat together by the campfire each night, serenaded by the whir of insects, the larger world outside seemed "so far and remote," she wrote her family, "and our conversation is mostly chimp-chimp-and more chimp. Hugo loves them as much as I do, and we have got some simply wonderful film as a result."

By the beginning of December, *National Geographic* was writing to say that they had enough strong photographs "to make a fine layout" and were now ready to move forward with the chimpanzee article. William Graves, a member of the editorial staff, reiterated the *National Geographic's* preference for what he called "a first person, anecdotal type of article which includes not only the remarkable and hitherto-unknown scientific data you have gathered on chimpanzees but also a little of your dramatic personal experience while making the discoveries. In fact, not a little of the personal drama but a lot." He urged Jane to begin writing at once, telling her they needed the manuscript "as soon as you can possibly send it." What he didn't yet know was that Jane had been working on the article for some time. In a return letter dated December 19, 1962, she noted that she was just putting the finishing touches on her first draft and would send it on immediately.

DAVID GREYBEARD'S BOLD ACCEPTANCE OF THE BANANA JANE HANDED him stayed with her. Both Jane and Hugo longed to get closer to the apes, not only for better still photographs and film footage—Hugo was simultaneously shooting a movie—but also for more thorough and precise observations. Despite the frequency of Jane's encounters, they were still a matter of happenstance, random and serendipitous, as they both knew. Jane's gut instinct had always been to reduce the distance between her world and theirs, to effectively "tame" the wild apes, so she could move more freely among them.

The idea of luring the chimps into camp on a regular basis—

effectively "provisioning" her study subjects—thus held obvious appeal. It wasn't so different, she and Hugo reasoned, from having a fig tree in camp that was perpetually in fruit. And so, however accidentally it started, the practice of setting out bananas in camp began.

The chimps were quick to catch on. By the summer of 1963, they were sauntering into camp with breathtaking regularity, sometimes twenty apes in a single day. David Greybeard and his friend Goliath, an alpha male, were the first to appear. But other males soon followed, becoming bolder and more approachable by the day. And finally came the females, adolescents, and youngsters, opening a thrilling new window onto mother-child relations, as well as the ritual particulars of mating.

This randy topic came into especially sharp focus that summer, when the soon-to-be-famous Flo, "the most hideous old bag in Chimpland," as Jane put it, but also the most sexually "popular" with the males, began to make regular visits, "quite won over to the idea of popping in for an odd banana." Flo's appearance in camp drew "millions of males along here with her," Jane wrote Leakey—particularly that August, when she "developed her first pink swelling," providing opportunities for observations Jane hadn't yet seen. (During estrus the female chimp's genitals redden, signaling sexual availability.) David, Jane told Bernard Verdcourt in a letter, sometimes greets Flo "in the most fabulous way—did I tell you how he once kissed one of her nipples and pinched the other! The men have a naughty habit of poking a finger up their lady friend's vagina when greeting her!"

But if the quality and depth of Jane's sightings were multiplying, so too were the dangers. Provisioning was increasingly becoming a point of vulnerability. The bananas drew baboons to camp as well as the chimps, sparking violent skirmishes between the two species. Additionally, the chimps often sparred among themselves in their contest for fruit, adding to the general aggression in camp. One day an agitated chimp "rushed about with all his hair out," Jane wrote her family, and then hurled a two-pound rock almost twenty feet.

"It was quite fantastic," she added, downplaying the danger. On another occasion, Hugo had thrown a rock at Goliath, fearing he was about to chase Jane. Ever since then, she reported, "Goliath has borne Hugo a grudge." He "has chased him 4 times."

Both Leakey and Melvin Payne, the Geographic Society's executive vice president, wrote to express their concerns. "I know you have complete confidence in your friends, but I have a continuing apprehension that they may suffer a momentary lapse and forget their friendship with you," Payne cautioned. Louis was similarly alarmed. A weak chimp could easily rip the arm off a strong human, he reminded them. Not only were she and Hugo at risk, he warned, they were also endangering the chimps, since if any human at all was injured, including one of the local fishermen, "some interfering officer" would no doubt shoot a chimp. Jane was angered by Leakey's letter, which she found mildly patronizing if not unfair. He wasn't there, after all; he had no idea. As a precautionary measure, Jane and Hugo had by now had a steel cage built, which would provide them with a retreat in case of emergency. The cage was "foolproof," she assured both men. But what really raised her hackles was the implication that she was jeopardizing the chimps. "Louis, can you really think that I honestly haven't thought and thought about the safety of my chimps?" she wrote. She and Hugo were also experimenting at this point with ways to regulate the number of bananas that any single chimp could take. Eventually they would move the feeding area some distance from camp, as well as dispense bananas from boxes by remote control, to reduce the fighting. But that first summer they were still finding their way.

Jane continued to insist that the provisioning was worth the risks. "We now have 21 regular visitors to camp," she wrote her landlady at Cambridge. "This means that for the first time I can get continuity in my observations."

"To be able to follow the interrelationships from DAY TO DAY, instead of simply seeing the same two animals together once a week or even once a month—well, I can really say now, that I know chimps," she insisted in a letter to Leakey.

Jane's practice of provisioning her chimps would eventually come under fire. Her critics would claim that the banana feeding made the apes "more aggressive than they would otherwise have been." Whether this was true or not is debatable. Certainly the practice was very much in keeping with the "manipulative traditions" of European ethology. In time Jane would distance herself from the practice, just as she would later discourage the researchers to whom she passed the baton at Gombe from having direct physical contact with the chimps—most immediately, to reduce the risk of anyone getting hurt, since chimps were four times stronger than humans. But also because of concerns over spreading infectious diseases to the apes, which would later become a problem. Yet in the summer of 1963, provisioning would prove spectacularly successful in lifting the veil on the mysterious and endlessly fascinating lives of the chimps at Gombe, allowing for a deeper level of habituation than Jane had yet achieved.

IN AUGUST 1963, LESS THAN TWO YEARS AFTER THE PUBLICATION OF Jacobs's *Death and Life*, and a year after Carson's *Silent Spring*, the National Geographic Society sent out to its subscribers three million copies of its popular magazine. In it, readers encountered an exotic story by an extremely pretty and unknown young British woman who had been living alone with wild chimpanzees in a remote African forest for more than a year. Jane Goodall's thirty-seven-page-long feature, "My Life Among Wild Chimpanzees," captured the public's imagination for myriad reasons. With its charming, first-person narrative and intimate tone, its personal anecdotes and unapologetic expression of empathy for the animal world, it was clearly written by an amateur and not a credentialed scientist. And yet, its content indicated that its obviously fearless twenty-nine-year-old author had made significant scientific breakthroughs. In photo after photo, Goodall—slight, ponytailed, sometimes even shoeless—was pictured standing astonishingly close to wild animals known to be dangerous and unpredictable. In one, she was handing a chimpanzee a banana, as if the creature were tame. In another, a chimp was letting her touch him. There were shots of chimpanzees carrying meat, proving they were carnivorous; chimps shaping twigs and using them to fish for termites, evidence of tool use; chimps with the author in the forest, conveying her open embrace of the natural world and her courage in the face of danger. The article opened with a moody shot of Goodall sitting, swaddled in a blanket, in front of a flickering campfire, her face in silhouette against a vast lake below her and the night sky above. To paint the scene further, there were photos of local fishermen by the lake, an image of the author's mother handing out medicines under a thatched roof with poles, and, in a nod to the Spartan conditions of camp life, a fetching shot of Goodall standing in a rushing stream washing her hair.

Readers were enchanted. Within days, Jane was buried in messages from friends, letters from readers from around the globe, solicitations from publishers and journalists. It was the beginnings of a celebrity that Jane could neither have imagined nor been prepared for, a degree of visibility that, if at first exhilarating, was soon overwhelming. For in addition to the acclaim her article brought came all manner of intrusions and assumptions, notes from naysayers and weirdos as well as fans, all of whom wanted to weigh in, if not get close to the young, blond-haired beauty who had traveled to Africa and lived with wild animals in an isolated rain forest miles from any village and unreachable by road.

A Connecticut woman sent a letter pleading for more information. She had "been conducting cancer research on my own body," she said. She was now "fighting for survival to complete my work." Her "only hope in this battle" was to know the precise names of the "eighty-one different kinds of chimpanzee foods" mentioned in the article.

A kook from California, upset by the spiritual ramifications of chimps making tools, wrote "to question the state of Jane's soul": "I thought to myself, gee here is a beautiful young lady, brains, ability, plenty God given talents, HOW MUCH DOES SHE GIVE 'GOD' IN RETURN??? . . . You are working hard for your Doctor of Philosophy, how hard are you working for YOUR CROWN IN HEAVEN?"

"Whenever I think of Africa and Apes," pronounced a young male from Pennsylvania, "a funny feeling charges up and down my spine." In the woods near his house, he said, he enjoyed diving "through the trees free as an Ape." He often camped out in "high platforms" he had made in the trees, and was now determined to go to Africa after high school graduation to "live among the apes, just as Jane was doing." He felt confident that he wouldn't "have much trouble following them because I am built like one. After I get in contact with them I could get cameras and other equipment and record their habits and daily life."

There were of course letters from well-wishers too: Malcolm MacDonald, a former British MP and cabinet minister, sent a note saying that he was so impressed by the article that he was having it "bound in leather" so it could live in a prominent place in his bookshelf. Sir Julian Huxley, the eminent zoologist and grandson of the renowned Victorian biologist T. H. Huxley, wrote to ask if there was any chance of joining Jane on her next trip. A celebrated German zoologist sent a letter proposing a film collaboration about her work. Hugo, who had been pushing hard to finish his own Gombe film, was understandably threatened. "Your fascinating account of life among the chimpanzees leads me to believe that the subject might serve as a basis for a popular book," wrote more than one publisher, including Paul Brooks of Houghton Mifflin. Brooks (who had just published *Silent Spring*) would later become the American publisher of Goodall's 1971 bestseller, *In the Shadow of Man*.

Jane tried her best to rise above the uproar, gently discouraging the solicitations from strangers, the proposed visits by journalists, the suggestions for film collaborations and related stories. Even so, the human world seemed increasingly to intrude. The shadow side of her instant celebrity would be a loss of the great, rapturous solitude she had enjoyed in the forest with her chimps. That fall of 1963 would be the last she and Hugo would spend at Gombe alone.

JANE AND HUGO LEFT THAT DECEMBER, HUGO DECAMPING FOR NAIROBI, Jane to Bournemouth for Christmas, and then on to Cambridge for her third term. By now, however, they were romantically involved and were thus feeling particularly unhappy at the prospect of what was to be a four-month separation. Hugo decided to take matters into his own hands. On the day after Christmas, Boxing Day in Britain, he sent a telegram to the Birches. "WILL YOU MARRY ME LOVE STOP HUGO."

Jane accepted by return cable. "You can't imagine the cables . . . that went whizzing . . . between here and Africa, Africa and Holland, Holland and here," she wrote Sally Cary. "My goodness. 7 letters all at once from Hugo . . . did I like emeralds!! So back went a cable— 'love emeralds love you'!!! Back came a cable—what size was my finger? . . . Back went an answer . . . And in the middle of all this frantic cables from Africa to Washington, Washington to Africa and to England . . . about files and tapes and more films."

The wedding would take place three months later, on March 28, 1964, Easter Sunday, at Chelsea Old Church, London. Jane wore white and carried daffodils and arum lilies. A clay model of David Greybeard stood atop the wedding cake and bubbly wine was served instead of Champagne. (Champagne, Jane said, was "just a snob value waste of money.") Louis Leakey, unable to attend, sent a tape with his congratulations, after which a telegram was read aloud from the National Geographic Society, announcing that the bride had been awarded a second Franklin L. Burr Prize for Contribution to Science. Giddy with happiness, the Baron and Baroness Hugo and Jane van Lawick returned almost immediately to Gombe, stopping for a brief reception in Nairobi, followed by a quick side trip to the Leakeys', before pushing on to rejoin their beloved chimps.

AS EARLY AS THE SUMMER BEFORE, JANE HAD DECIDED THAT SHE wanted to establish a permanent research center at Gombe so the work might continue long term. She had already realized, as she put it years later, that "there was no way that one person, no matter how dedicated," could make a truly "comprehensive study of the Gombe chimpanzees." That fall, having secured Leakey's blessings, she approached National Geographic for funds, and by the end of 1964, plans had been approved for a series of primitive, semipermanent buildings. By March 1965, two prefab aluminum units sheathed in bamboo were in place, the smaller of which—dubbed Lawick Lodge—a single room that would become Jane and Hugo's sleeping quarters. Two assistants had also been hired, one to do clerical tasks, the other to continue the research while Jane and Hugo were away. Jane was due back in Cambridge in March 1965, to work on her doctoral dissertation; Hugo had an assignment in East Africa, to film wild animals for *National Geographic*.

Grudgingly, Jane returned to England as planned. But her focus was increasingly divided. Beyond the punishing work on her thesis, she now had other pressing obligations. She had promised *National Geographic* a second chimpanzee article (which she sent them in February), as well as a draft of her first popular book, *My Friends the Wild Chimpanzees*. (The latter, when it was completed a year and a half later, would make her an even bigger star among the public.) Additionally, she'd agreed to give a lecture in September in Vienna, for the prestigious Wenner-Gren Foundation, and another in October at the Royal Institute in London, both of which required preparation. By November, she was working "flat out on the final stages" of her dissertation, she reported, writing herself ragged and living on "Nescafe and the occasional apple," rarely getting enough sleep. She "surfaced" briefly in December, she told a friend, long enough to register the groundswell of excitement over the December issue of *National Geographic*, the cover of which carried her picture, and the CBS television special *Miss Goodall and the Wild Chimpanzees*, which aired two weeks later, on the twenty-second, raising her profile even further. (The audience that evening was estimated at twenty million.) Feeling the wind at her back, she caught a flight for Nairobi, arriving in time for Christmas with Hugo, and then spent "a delightful month on the Serengeti," only to have to rush back to England in early February for "horrible oral exams," followed by a series of lectures in the U.S.—one in Washington, D.C., before thirty-five hundred people in the grand auditorium of Constitution Hall. By the spring of 1966, having received formal notification that she had passed her doctorate exams, Jane was finally back at Gombe, after more than a year's absence. But by now she was a much-lauded, much-courted figure, increasingly visible and in demand. Though she didn't yet know it, she had entered the next, more public and peripatetic phase of her life; the seeds of her future activism were already in the wind.

JANE'S LIFE OVER THE NEXT DECADE WOULD BE AN ITINERANT ONE. Hugo's photography work would take the couple to the Serengeti, then north to the national parks in Uganda and back to Ngorongoro Crater, where for weeks, sometimes months at a stretch, they watched, tracked, photographed, and wrote about wildlife, most immediately hyenas, wild dogs, and jackals for a book they collaborated on called *Innocent Killers*. Despite these hiatuses, the research at Gombe went on, aided by a changing cast of international research assistants, which grew from a handful to two dozen or more with time.

The highs and the lows of the work would continue too. In 1969, in a heartbreaking tragedy, one of the researchers, Ruth Davis, fell to her death. A polio epidemic struck in 1966, paralyzing and then killing some of the chimp families that congregated at the feeding site. At Jane's urging, an emergency supply of oral vaccine was flown in and administered to both the people and primates at Gombe, but not before more chimps had succumbed. Other sadnesses followed: David Greybeard died of natural causes in 1968, and then ancient Flo in 1972, a year after the publication of Jane's bestselling book, *In the Shadow of Man*. Flo's son Flint, "hollow-eyed, gaunt and utterly depressed," died soon after of heartbreak.

The research, meanwhile, continued to deepen. By the mid-1970s, the first shocking observations of chimp warfare and chimp cannibalism had been made, revealing the darker side of the chimpanzee psyche. Human violence ratcheted up too. In May 1975, rebels from neighboring Zaire kidnapped four researchers from camp, including three Stanford students, nearly shutting down the Gombe operation. Stanford, which had been funding the project since the early 1970s, pulled its support, and for the next few years, American students stopped coming, leaving it to the Tanzanian staff members to continue the data gathering, which they did.

And then, quite abruptly, the trajectory of Jane Goodall's life changed. It began in 1986. Jane had been writing and doing a bit of teaching at Stanford. Her seminal book on chimpanzee behavior, *The Chimpanzees of Gombe: Patterns of Behavior,* had just been published to considerable acclaim. She and Hugo had raised a son, Hugo Eric Louis van Lawick—nicknamed Grub—mostly at Gombe, while continuing with the research. "I was in a dream world," she says. "I was out there with these amazing chimpanzees. I was in the forests I dreamed about as a child." And then in November 1986, Goodall attended a primate conference in Chicago. For the first time, all the chimp people across Africa were brought together, she explains, as well as people studying chimps in noninvasive ways in the lab, and in captivity. There was a session on conservation. "That's when the shock hit me," she says. "Every place in Africa where chimps were being studied, their numbers were plummeting. Forests were disappearing; human populations were growing; habitats were being destroyed. Chimps were being killed by the beginnings of the bush meat trade; there was still some live animal trade too, shooting the mothers to get the babies, which utterly shocked me. So actually, as far as I know, I didn't make any decision. I went as a scientist—I had my Ph.D. by then—and left as an activist. Just like that. I couldn't go back to that old beautiful life."

Goodall had no set plan. "I don't know what the hell I thought I could do," she recalls. "That's what was absurd." She set off across Africa, searching for partners and financial sponsorship, talking to whomever she could, determined to make a difference. She found eventual support from Conoco in the Congo, and from America's James Baker, who sent letters to heads of state in advance of her visits. But it wasn't until a few years later that she saw clearly what needed to happen.

It was the early 1990s and she was in a small plane, flying over Gombe. "I knew there was deforestation," she remembers. Throughout the years, in her trips across the lake, she had noticed bald pockets along the park's borders. "But I wasn't prepared for what I saw from above, which was this little island of forest surrounded by miles and miles of bare hills; more people than the land could support; people too poor to buy food, land over-farmed and unfertile." That's when the lines of connection between the fate of humans and the fate of the earth's endangered species and ecosystems hit home, she says: all were intertwined.

"I realized that until we addressed the poverty of the people who were chopping down the forest for firewood and farmland, we couldn't even begin to try to save the chimps, and this is true all over, not just for the conservation of chimps, but the conservation of any species."

Goodall teamed up with a partner, and together they launched a local conservation program, which they called TACARE. It was important from the very beginning, she explains, that their approach be respectful, that it not be a "bunch of arrogant white people telling the locals what to do to make their lives better." So they pulled together a group of amazingly committed Tanzanians, local people who went from village to village talking to the residents and asking them what *they* thought could be done to improve their lives.

The approach worked, creating a natural opening. The people wanted better health facilities, better education for their children, better techniques for growing food, "so we tackled those things first," Goodall remembers. "We had very little money, a tiny grant from the EU for twelve villages around Gombe. But we started anyway." The team's first challenge was to restore fertility to the land. It had to be without using chemicals, not only because they couldn't afford them, but also because "I already knew how evil they were," she says. (Goodall had read *Silent Spring*, and is a great admirer.) They developed innovative ways of capturing the soil, which had washed away when the trees were cut. Free nurseries were set up, providing fast-growing tree species for building and firewood. Water projects were introduced, and then women's health, education, and microcredit programs.

That was perhaps the most important part of the initiative, Goodall believes. As the women's education improves, family size begins to drop, and the standard of living rises for everyone. "You do this by keeping the girls in school during and after puberty," she explains. "Mostly they quit because there are no proper toilets, no sanitary supplies, no privacy, so you realize you've got to have proper lavatories."

The microcredit loans were tremendously important too. "The women choose their own projects, and whatever they do, it has to be environmentally sustainable, so they start little chicken farms, tree nurseries so they could sell seedlings, little pineapple farms." They begin small, but their projects grow, as did the program.

TACARE proved to be a magnificent success. "The real encouragement," Goodall says, "is that as soon as the people's lives began to improve, they began to allow trees to come back." As a result, they have set aside the land the government requires them to put into conservation in such a way as to make a buffer between the Gombe chimps and the villages. "The villagers understand the watershed. They understand that you can't destroy the trees along the edge of a stream or the water level will decrease. They've seen it happen. So they completely understand the problem. The trees and the water and the environment and their future wealth and happiness are all mixed together, all interconnected."

As for the chimps' long-term prospects, this is more difficult to say. Today TACARE is in fifty-two villages, spreading beyond Gombe to the south, so the chimps in the greater Gombe area now have three times more forest area than they did ten years ago. They have an opportunity to interact with other known chimp groups, which is critical to maintaining genetic diversity, their only real hope for survival. It's a start in offsetting the sharp population declines they have suffered in recent decades, but only a start.

Goodall feels that time is running out. Her worries these days extend beyond the fate of the chimps, to the fate of endangered species across the globe. She's now on the road three hundred days a year, advocating for sustainable development and worldwide protection of habitat. She's committed to working on a global scale, to lobbying for changes that will have the greatest impact. "We are now in the sixth great extinction," she says, a note of despair edging her voice. "Our human impact on the planet, our greenhouse gas emissions, our reckless damage to the natural world, all these have been devastating."

If a bug disappears, one might think it doesn't matter. "And it might not," she adds.

"But that little bug that looks so insignificant could be the main food source of a fish, and that fish could be the main food source of a bird, and that bird, you know, could be the main creature that's distributing the seeds of a certain kind of plant, and the seeds that grow into that certain kind of plant could be the main food source for some larger creature, and on it goes. We now know that whole ecosystems can collapse because of the loss of one little piece. In other words, it's the web of life. I think it's unfortunate that we talk about 'biodiversity' because people, ordinary people say, what the hell is biodiversity? But the web of life, this makes sense."

What doesn't make sense, Goodall reflects, is the fact that our intellect is so hugely developed, and yet we persist in destroying the planet. "How is it possible that the most intelligent creature to ever walk the planet is destroying its own home?" she asks. "There seems to be a disconnect between the clever brain and the human heart." Instead of saying "how does this decision affect my family, our people years ahead, we're saying how does this affect me, my pocketbook, the next shareholder's meeting." We're not borrowing from the future, she insists, "we're stealing from it. We need to change that."

To this end, Goodall has lobbied senators and trade group leaders, World Bank officials and policy makers, timber company CEOs and ambassadors from the U.S., France, Tanzania, Burundi, and other countries, often with success. She recently convinced a consortium of timber-company chiefs to make protecting wildlife a part of their business code. Her push to have chimpanzees recognized as endangered led the U.S. Fish and Wildlife Service to recommend that chimps in captivity be granted the same protections. "I think a lot of what I do is just talk to people," Goodall says. "And I am not aggressive." But it's more than this. Like Carson and Jacobs, Goodall knows how to inspire people. Her ability to win hearts and minds, to move an auditorium packed with people, or a wealthy donor she's seated beside at a gala dinner, gives her a persuasiveness that few can claim. Her manner may be gentle, but Goodall's passion and conviction are clear. She's an extremely compelling character.

Goodall is eighty-three now, though she hardly seems it. Her gaze is direct, her gait strong, her energy unflagging despite a schedule that would be grueling for someone half her age. If her spiritual home remains Gombe, her actual home is the road. She hasn't slept in the same bed for three consecutive weeks in more than twenty years, she recently told Paul Tullis of the *New York Times*. "It never ceases to amaze me that there's this person who travels around and does all these things," she adds. "And it's me. It doesn't seem like me at all."

Driving her on, she says, is the need to raise awareness. "I have to raise money," she admits, "but my goal is awareness." Which is where children come in. In a world beset by increasingly dire news about the environment, Goodall finds great hope in the promise of young people. "If you get children out into nature, in the right setting, they just thrive," she says. "You don't have to teach them. You just have to allow them to be. And yet so many never get that opportunity." Her first priority, therefore, is growing the youth program she started in affiliation with the Jane Goodall Institute, the conservation NGO she founded in the 1970s, "because if the young people of today lose hope, we might as well give up."

Roots & Shoots, as the program is called, began with twelve students in Tanzania in 1991. Today it's in 114 countries, with more than 100,000 active youth groups, each initiating projects to improve the world and protect the environment. Its growth has been swift and dramatic. At COP21, the United Nations conference on climate change in Paris, Goodall recently unveiled a giant interactive map showing every Roots & Shoots chapter across the world. "We're calling it the tapestry of hope," Goodall says. "You just click on a certain place and the projects there come up." There are chapters in North America and parts of Latin America; in Europe and Asia, including China; in Africa and the Middle East, mainly Abu Dhabi. "It's completely magical, and it's not even finished. There are some groups that we haven't even heard from yet, because they have only occasional Internet connections."

Equally inspiring is the range of projects the kids undertake. There are kids planting trees, kids doing urban gardening, kids growing food or butterfly gardens in schoolyards. There are children volunteering to work in dog shelters, groups going out on weekends to clear invasive species from a prairie area in Texas, others doing the same in wetlands in Taiwan. There are kids raising money to help victims of the 2015 earthquake in Nepal, groups working with street children, groups starting Roots & Shoots programs in refugee camps. In China children are going to visit other children who are long-term patients in hospitals, to cheer them up.

The important thing, says Goodall, is that the kids generate the projects themselves. "Our job is to listen, to hear their voices. We don't tell them what to do; we let them find their own passions, empowering them to take action." What's exciting is "they are all growing up, moving into the adult world, taking that philosophy with them."

The larger message, Goodall insists, is that individuals can make a difference. "I think the main reason that people do nothing, that people go along in a state of denial about what is happening, is that they look around the world at the problems and they feel helpless and hopeless, and they think, well, I am just one person, what can I do? So the thing is that as one person, we can't do anything. But if we start making considered choices, choices in our everyday actions, the little things—what we buy, what we wear; if we think carefully about the consequences of those choices—how was it made, where did it come from, was it child slave labor, was it cruelty to animals, et cetera, then we start making *different* choices. Small choices. *But* multiply those small choices by a hundred, a thousand, a million and then a billion and then you start to see a different kind of world."

Cultivating "a romance of limits," to use the author and environmentalist Bill McKibben's words, would seem to be a difficult sell in this age of rampant materialism. But to Goodall it comes quite naturally. She likes to tell the story of a lecture she gave several years ago. Afterward, she received letters from two separate people who had attended: one was from Hong Kong, the other from Holland. Both writers loved sports cars, they said, and both were about to buy the sports car of their dreams. "In my lecture I had repeated Gandhi's quote 'The world can provide for human need, but not human greed.' So when we go to buy something, we should say to ourselves, 'Do I really need it?' Both of these men, from two completely different cultures, were going off to get their sports cars and then they remembered what I had said, and they said, '*Damn Jane.*' One of them actually gave the money he would have spent on the car to the Jane Goodall Institute. I've got many stories like that. One man sold his house. He said, 'I realized I didn't need such a big house.' You never know who it's going to be, or how they will react," she adds. "It's why I keep doing what I do."

Nearly fifty-six years have passed since Goodall first set foot onto Gombe's pebbled shores. Yet in some respects, she is not so changed. She still requires minimal food and little sleep. She still exudes accessibility, despite near constant intrusions by flocks of admiring strangers wherever she travels. She still favors the same simple uniform—khakis, a neutral-colored shirt, sneakers—whenever she returns to Gombe. Twice a year, on her brief trips back, she still stays in the same cinder block house where she, Hugo, and Grub lived in those early pioneering years, and which she later shared with her second husband, Derek Bryceson, a principal in Tanzania's first democratically elected government and now deceased. The house, reports a recent visitor, is filled with old magazines and animal skulls.

It has a proper door, "but the windows are chicken wire." The idea is to keep animals out, but to let in "as much of Gombe's atmosphere as possible." Which is just as Goodall wants it. Five and a half decades after her first consequential visit, her gray hair pulled back in the same low ponytail, Goodall still requires few creature comforts.

Her days are crowded, she admits, even at Gombe. Yet shoehorned between meetings in Kigoma to discuss land-use policies, visits to local Roots & Shoots chapters, side trips to Nairobi and Burundi, conversations with ambassadors and officials, Goodall on occasion still walks the same forest paths that she first explored so many years ago, her senses attuned now, as then, to every stir in the trees, every rustle and quiver in the tall grass. She still finds solace in the peace of the deep green forest where it all began.

"People often ask me, 'Why aren't you still there?'" she says. "I wish I could be," she admits. "I yearn for the days that are gone. I have a wonderful team there and I still love being out in the forest on my own. But Gombe is not like it was, and we have to save those chimps, which also means saving the earth. As long as I can, I have to keep trying."

chapter FOUR

....

ALICE WATERS

Alice Waters stepped into the little stone house in Brittany, trailed by her friend Sara Flanders. It was the spring of 1965 and she was just twenty-one, a footloose Berkeley student on a semester abroad in France. An elfin creature with expressive gray eyes, she was small boned and slender, barely five two in height, and there was a faint air of dreaminess about her. She wore a little antique hat that lay close to her head and a delicate flea market dress. But her gaze was intent, keenly focused on every detail: the ancient stone house, the stairs leading up to the dining room, the pink cloth-covered tables—there were no more than twelve in the entire room. The décor was rustic and unassuming, the plaster walls a soft ivory hue, the floors old and planked, yet the room still felt elegant, generous. The cutlery gleamed; the chairs seemed to invite sitting, lingering and talking, losing track of time. For all its simplicity, there was a sense of occasion about the place, an understated beauty. Alice noted the breeze wafting in through the windows as they sat, the burbling stream just outside, the emerald green garden in the back. Most of all she noted the menu—or the lack of one. She'd never experienced this before. The chef, a woman, stepped out from the kitchen and announced the night's fare: cured ham and melon, trout with almonds, raspberry tart. The meal was simple and exquisite, wondrously fresh. The trout had just been pulled from the stream, the raspberries picked in the garden. It epitomized everything Alice was coming to feel a meal should be: down-to-earth, straightforward, steeped in the perfume of its place. "Elsewhere, even when I found the food to be wonderful," she would write years later, the French "would say only that it was 'all right.' But after the meal in this tiny restaurant, they applauded the chef and cried, 'C'est fantastique!' I've remembered this dinner a thousand times."

For three months now, Alice and Sara had been eating: in tiny corner bistros with lace curtains and blackboard menus outside; in bustling cafés, where the air smelled of licorice and green bottles gleamed from behind the bar; in brasseries on grand boulevards; in antique dining rooms with high ceilings and gilt mirrors and long, luxurious banquets; in patisseries with little marble tables.

Alice was especially partial to the simple, working-class places—storied haunts like Au Pied de Cochon, where one stood at the bar, and had oysters out front, and then moved on to the "downstairs part," where for four francs fifty, "you could eat at the bar and have a *blanquette de veau* and a glass of wine and a basket of great bread."

She already loved the classics: coq au vin, cassoulet, steak frites, pot-au-feu, lentils and salmon, hard-boiled eggs and homemade mayonnaise. There was something about the rituals of French dining, the timeworn traditions, the care taken over every detail of the table, that moved and thrilled her: the cheese plate that arrived at the end of the meal, the little cup of dark espresso served with a curl of lemon peel on the saucer, the square of chocolate that accompanied the check. Even at the humblest places, small felicities mattered: the pretty cloth napkins, the bouquet of flowers on the bar, the soft lighting, the late-night glass of cognac on the house.

Paris was intoxicating, an ongoing revelation of taste and sensation. Everywhere she turned there was beauty and food, pleasures for the eye and for the palette: neat trays of bonbons in little candy shops, fresh fruit tarts beckoning from patisserie windows, bustling outdoor markets crowded with carts of fresh produce and cut flowers, tables laden with country pâtés and pyramids of fresh cheeses, butchers' trucks with ropes of homemade sausage.

Alice loved Paris in the early morning, when the city was still wrapped in a chill; she loved the great chestnut trees that lined the avenues, the elegance of the Luxembourg Gardens and the grand cafés. "Everything in Paris was magical to me. You'd walk past a church and you'd hear music, and you'd walk in and sit down and they'd be playing Bach at lunch. Concerts—we went to so many concerts. And those magnificent museums." But most of all Alice loved the food.

Before coming to Paris, Alice hadn't really thought much about food. It was Sara who had been looking forward to the wonders of French cuisine, who had suggested they take a semester at the Sorbonne. Alice had been looking forward to improving her French— and to getting "some experience with French men." A major in French studies, she assumed she had license to do both.

Alice was extremely fond of men. Born in Chatham, New Jersey, in 1944, she had grown up in a household of girls, the second of four daughters—the source, perhaps, of this predilection. It was a typical suburban family, modest and conforming. Her father, Charles Patrick Waters, was a staunch Republican, a management consultant and a rising star in the Prudential Life Insurance Company; her left-leaning mother, Margaret Hickman Waters, a stay-at-home mom. Like most women of her time, Margaret did all the cooking. Years later, Alice would recall her applesauce and rhubarb compote, but also that "my mother didn't really know how to cook; she was only interested in our good health, always feeding us vitamins and heavy brown bread—an unfortunate combination." In matters of food, what Alice would remember most was the garden out back, its sprawl of fresh greens her father's great pride, its exquisite cut flowers her mother's domain. Poor eyesight had kept Pat Waters from active service in World War II; to do his part, he had planted a "victory garden," as it was then called, a practice encouraged by the government as a hedge against food rationing. One Fourth of July, as if in anticipation of Alice's future career, her mother costumed her as queen of the garden. Red peppers wrapped her ankles, a necklace woven of long-stemmed strawberries wreathed her neck, and she wore a skirt of fresh asparagus. "The skirt was kind of itchy," she remembered; "it was summer, so the big feathery tops had bolted."

She was a child of her senses, even at an early age, acutely sensitive to extremes of hot and cold, the scents of the garden and the seasons: the musty smell of leaves in the fall, the spring air sweet with lilac and apple blooms, the dry rush of the wind in the trees.

"We played outside," Alice remembered. "I never wanted to come in. We had to come in for dinner at seven. But otherwise we played outside until we fell over."

Alice shared a bedroom with her older sister, Ellen. On summer nights, she often lay awake for hours, too hot to sleep. In winter, she complained of the cold. Sometimes her mother would take her to the basement, zip her into her snowsuit, and "stand her shivering in front of the furnace," she recalled. There was a burbling stream behind their house, where she liked to wade and play, wild raspberry bushes, and a great wood. "We climbed the trees and played with the rocks in the stream. We rode our bikes. No one was keeping track of us. We were trusted to take care of ourselves. I was out in the woods all day."

Her mother was passionate about flowers and trees. She taught Alice their names, helped her know their leaves, their shapes and distinctive hues. "I could name all the trees," she remembered. "They became my friends. I am in the willow tree, I am in the elm tree, I know the lily of the valley, the myrtle, the bearded iris, the kinds of roses my mother grew." In the fall, the family went on drives to look at the turning leaves, the golds and russets and flaming reds. "I loved those rides," she reflects. "We did them in spring too. I am convinced that it was the outdoors, the victory garden and the experience of nature that shaped me, touched me in unforgettable ways."

There was an ancient mulberry tree in the yard, she recalls. "The mulberries seemed very sour to me at the time." But the branches dipped all the way to the ground, "which always made a great little house to play in." Later, after she started the restaurant Chez Panisse, and they went out searching for the best fruits, Alice found a man with a similar mulberry tree, giant and prolific. Mulberries are incredibly delicate, but with the man's blessings, they brought them back to the restaurant. Alice began making what would become one of her signature dishes, mulberry ice cream.

Chatham, New Jersey, was a quiet, tree-shaded town, one of the growing commuter suburbs just outside the city. Alice turned six in 1950 and the economy was booming. Americans were on the go, speeding along on freshly built highways; moving into spanking new suburbs; buying TVs and washing machines and sleek new chrome-accented cars, all for the first time. Farmland was disappearing, gobbled up by hungry developers near every major urban center, a lot of it in places with the most fertile soil. The spread of "man-made ugliness," as Rachel Carson had called it, and the "trend toward a perilously artificial world" had begun. But most Americans were moving too fast to notice: they were having babies and buying on credit and enjoying a level of affluence few had ever seen, mesmerized by the promise and conveniences made possible by mechanization—everything from assembly-line cars, to mass-produced houses, to the new fast food. In 1948, the McDonald brothers had opened their first fast food joint. What was new, beyond the cut-rate prices—fifteen cents for a burger, four cents more for a slice of cheese—was that everything in the kitchen was automated. A machine turned out standard-size patties; another whipped up milk shakes. Paper bags, wrappers, and disposable cups replaced silverware and plates, saving the labor of dishwashing. Now families could eat quickly and carry out their food; they could even dine in their cars. The new ethos was that faster and cheaper was better. The more standardized the product, the more quickly it could be moved. The company would therefore choose the condiments: ketchup, mustard, onions, two razor-thin pickles. Condiment stations meant delays, inefficiency, disorder. "Our whole concept was based on speed, lower prices and volume," Dick McDonald said.

"Here was the perfect restaurant for a new America," writes David Halberstam. Or so some thought.

Pat Waters was by dint of his age and position a part of this new, revved-up world. A good company man—smart, affable, well organized—he rose quickly up the corporate ladder, moving his family with each promotion, in 1959 from New Jersey to Illinois, and then in 1961 to Southern California, for Alice's last year of high school.

At Van Nuys High, Alice was intensely social. Though she ran with the bookish crowd, the cool, more intellectual kids heading for college, there were always boyfriends, she remembered, mostly older guys and "ones from the other side of the tracks." In those days, there was time to burn, a lot of cruising up and down the main street in town, a lot of riding around in "big Bonneville convertibles." Though no one smoked pot yet, there was also a lot of drinking, Alice said.

She was a diligent student, and her parents hoped that she would go to Berkeley, the most rigorous of the California schools. But Alice chose UC Santa Barbara, where she spent her freshman year in a debauch of parties and drinking, a fog of casual dalliances with Ken-doll-cute frat boys. "That was a dark period of my life," she told Thomas McNamee, author of *Alice Waters and Chez Panisse*. "I was just moving from one party to another. There were various people I was involved with, but I don't remember a single person's name. I never got into pot smoking, but alcohol affects your memory too. Especially when you drink enough to pass out."

She was bright and beginning to feel uneasy, aware that she was squandering her time. One of her sorority sisters, a gentle, soft-spoken beauty named Eleanor Bertino, had been talking about transferring to UC Berkeley, where the atmosphere would be more serious, the students more intellectually engaged. When she suggested that Alice and another friend from Alpha Phi, Sara Flanders, might join her, Alice liked the idea. In the fall of 1963, a year after *Silent Spring*, and just months after Jane Goodall's *National Geographic* article appeared, she, Sara, and Eleanor submitted their applications for transfer to Berkeley and were accepted.

Heading north that January, the erstwhile sorority sisters were soon swept into the turmoil rocking Berkeley's campus. Politics hung in the air, fanned by the assassination of President John F. Kennedy two months before; by film footage of Sheriff Bull Connor unleashing his dogs on civil rights protesters in Birmingham, Alabama, the previous summer; by photographs of four black girls killed in a church; by edginess over the escalating arms race, which seemed despairingly close. Nuclear warheads were being designed just blocks away, at the Lawrence Berkeley National Laboratory. That summer of 1964, Ku Klux Klansmen in Mississippi murdered the Freedom Riders James Chaney, Andrew Goodman, and Michael Schwerner. Three dozen black churches in Mississippi were torched. Congress voted to give President Lyndon B. Johnson full war powers, allowing him to dispatch sixteen thousand "military advisers" to South Vietnam. By the following May, he had sent in thirty thousand more troops, beginning a war that would rapidly escalate. In San Francisco, at the Republican Party convention that year, the party's candidate, Barry Goldwater, called for an expansion of the Vietnam War, with the possibility of deploying nuclear weapons. He argued that federal intervention in the civil rights struggle should be scaled back.

Years later, when asked what sparked her political activism, Alice would cite the Free Speech Movement (FSM). "That changed everything," Alice said. "It inculcated the idea that politics meant more than voting, that political activism was how you lived your life, what you physically fought for."

By the fall of 1964, Berkeley was boiling. Thousands of students were protesting a university edict that barred all political expression on campus. Many of the student leaders had traveled south with the Freedom Riders that summer, working to register voters in Mississippi. Many had been organizing picket lines against racial discrimination in hiring practices in the San Francisco Bay Area. They were deeply committed to the civil rights movement; outraged by the administration's interference with their rights to free speech; unnerved by the university's close ties to California business interests, which included some of the country's leading defense contractors. The university, they charged, had fallen short of its core mission, which was to encourage knowledge for its own sake. They had come to Berkeley to learn, to grow, to question, not to become cogs in the wheels of American industry, nor to be muzzled in the face of social injustice. The administration seemed more intent on business than on ethical values, they charged.

That October, ignoring the prohibition on political organization, a Berkeley agitator named Jack Weinberg set up an information table and was arrested by the campus police. Instantly, a surge of students moved in to surround the squad car, preventing it from carting Weinberg away. For the next thirty-two hours, the squad car remained—Weinberg still inside—as more and more students massed around it. By midday, some three hundred students had gathered, and the roof of the fortressed car had become a speakers' podium, where a continuous stream of speakers held forth for the next thirty hours, including a telegenic undergrad firebrand named Mario Savio. The next morning, photos of Savio atop the car were splashed across the nation's papers. And so was born the Free Speech Movement, which coalesced that October.

On December 2, Savio led thousands of free speech advocates in an occupation of Sproul Hall, Berkeley's main administration building. The sit-in, while peaceful, would prove profoundly unnerving to university officials. As Joan Baez cheered on protesters with Bob Dylan's civil rights anthem, "The Times They Are A-Changin'," Berkeley's president, now frantic, was working the phones. A call went out to Deputy District Attorney Edwin Meese III, who promptly dialed up the governor, seeking consent to proceed with a mass arrest. Nothing like this had ever transpired on a college campus and no one knew what to do.

By midnight, as Savio beckoned to the crowd from the steps of Sproul Hall, urging his fellow students to resist the administration, hundreds of policemen had begun assembling. "There is a time when the operation of the machine becomes so odious, makes you so sick at heart, that you can't take part, you can't even tacitly take part. And you've got to put your bodies upon the gears and upon the wheels, upon the levers, all the apparatus, and you've got to make it stop," Savio cried.

"This is part of a growing understanding . . . that history has not ended, that a better society is possible, and that it is worth dying for," he continued.

Finally, sometime after 2 A.M., the police moved in. Close to eight hundred students were hauled off to jail, the most sweeping mass arrest of students in the nation's history. A month later, when the university pressed charges against the leaders of the occupation, an even larger campus protest ensued. It was the beginning of a surge of student activism that would convulse college campuses across the country, eventually spilling into the streets—and the world.

Alice had taken part in some of the demonstrations that fall. Like a lot of people she knew, she believed in what the FSM stood for. But she was also shaken by what had happened, the overreaction by the police, the violence and untethered fury. People had been unreachable that night, tempers white-hot. It seemed as if something worse, something more random and violent, might be next—the kind of bloodshed and beatings they had seen at civil rights rallies in the South. When Sara suggested they take a semester in Paris, it felt like a good time to get some perspective.

THEY SLEPT THROUGH THEIR FIRST DAY IN PARIS. ALICE'S RICH AUNT had given her the name of a place to stay. It was in the First Arrondissement, right behind the Place Vendôme, a hotel that was much more expensive than they could afford. But it was dark when their flight landed in Luxembourg, and they were tired; the bus to Paris had taken another five hours; it seemed too late to look for something else. Alice had never been to Paris, never been anywhere in Europe. All she and Sara could think about was sleep. They closed the curtains—great heavy blue curtains, Alice remembered—and fell into bed. When they woke up, they discovered they had missed the entire next day, slept right through it, which was especially painful given how expensive the hotel was. To get their bearings, they went down to the dining room and ordered lunch. Alice was shy about her French, intimidated by the waiters. Unsure what to order, they both asked for the soup; it was the cheapest thing on the menu, and they could pronounce it: "*soupe de légumes.*" It wasn't pureed, she remembered, "it was just finely chopped up, and it was so delicious. I felt like I had never eaten before. And everything that went with it—those big, old, thick curtains and a bed that was made with those sheets that had rolled-up cushions at the head of the bed—it was a sensibility that was not part of my life, had never been."

They left and found a cheaper hotel, a seventh-floor walk-up on the Rue des Écoles. Wandering the cobbled side streets in those first thrilling days, they eyed every bistro they passed. Initially they were afraid to go in, still self-conscious about their poor French. "That was the time when no matter what we said in French, in a restaurant or to anyone, they said, '*Comment?*' They just refused to understand us. It was so demoralizing, and so painful. '*Comment?*'" Instead, they took their meals at the Self Service Latin-Cluny, where you didn't have to speak French to a soul. "They had pretty decent pâté, with cornichons, and oeufs mayonnaise, two hard-boiled eggs coated with mayonnaise. We'd have a glass of wine and a glass of mineral water and a yogurt. I can still taste that yogurt."

With help from the Sorbonne, they settled into more permanent lodgings, a studio apartment on the Place des Gobelins, with a bathroom down the hall. Alice was determined to make true café au lait, with steamed milk, which they drank out of bowls. Every morning, they went out to procure hot croissants and *pains au chocolat* from their favorite patisserie, and then carried them back to their flat. "We would just sit there in that room," Alice remembered, "and eat this little treat, pastries and café au lait. Then we missed our first class, and it kind of went on like that."

Their attendance at the Sorbonne continued to be occasional.

Now and again they showed up at class, but mostly they scouted out places to eat. It hadn't taken them long to surmount their skittishness about restaurants, and once that door to pleasure opened, there was no turning back. Soon, they were dining out multiple times a day, their passion for French food bordering on "obsession." They'd get up and have breakfast and immediately begin searching for where to lunch. As soon as lunch was finished, they would lay out their plans for where to go for dinner. "It was a wonderful time to be in France. It was real and genuine and honest," Alice remembered. "And really affordable. We were able to go to restaurants twice a day for five francs."

Alice was extravagant by nature, far less concerned about prices than Sara. What worried her was the idea of doing something gauche, she told her biographer. "I was still very intimidated everywhere we went," she says. "I didn't want to order the red wine with the wrong course or the white with the wrong course, so I just ordered rosé. And I always ordered too much food. The waiters must have all been laughing hysterically in the back room. That's always how we felt, like no matter what we did, we were doing the wrong thing. But sometimes you learn well when you're on your knees, you know?"

Alice did learn well. She paid close attention to every nuance: "What the fruit bowl looked like, how the cheese was presented, how it was put on the shelves, how the baguettes twisted. The shapes, the colors, the styles." She was enamored of the whole French aesthetic, from the crisp linens and exquisite table, to the small pleasantries—the "Bon appétit!" at the beginning of the meal, to the seriously good food—the wholesale celebration of sensual pleasures. An oyster here, a touch of pâté, a glass of wine.

Paris was "a complete seduction," she would later say, "a whole different way of life. Sitting in cafés, stopping in the afternoon to have coffee or tea, the whole aspect of eating was an important part of living, a sacred part of the day. The social part was important. You either went home and ate with your family, or you ate at a little restaurant with friends. You never would skip that. Nothing had to be so fast that you would skip a meal. People went to the market twice a day. They took two hours for lunch. These were people who thought of good food as an indispensable part of life." The table, Alice saw, was where community began, where one found friendship and good conversation, connections to the larger world. "I had a couple of French friends, and politics was always a part of the conversations," she remembers. "We spent a long time eating. A three-hour dinner was typical. It wasn't a half hour. We really had time to get to know each other."

When spring break arrived, Sara and Alice went to Barcelona and then on to the South of France for three weeks. They drove through ancient stone towns with trickling fountains; past walled terraces planted with olive trees; by simple farmhouses and lavender fields, great allées of umbrella pines. Everywhere they stopped, Alice was amazed by the food, even at the humblest places. It was bold and seasonal, straight from the land. The herbs were aromatic and just picked, the olive oil unlike any she'd ever tasted, sharp and green—what the French called "first-pressed." "I'd hardly even heard of olive oil at all, much less 'extra virgin,'" she remembered. Alice couldn't forget any of it—the open-air markets, the elemental flavors, the sublime freshness of everything they tasted.

When they returned to Paris, they met two young Frenchmen, Amboise and Jean-Didier, who took them out into the countryside again. "That," she remembered, "was our real introduction to the food." Amboise was from Brittany and had laid out an itinerary, a little chart with all the places they should go (including the little restaurant with the pink tablecloths Alice would always recall). They stopped in Honfleur, Alice remembered: "I'll never forget those mussels. The fishermen brought them in from the boat, rinsed them off for a minute, threw them into this big cauldron, and then scooped them out."

She loved the *crêpes Grand Marnier* they found in Brittany. "I'd have a couple dozen oysters for lunch and a couple of crepes, savory and sweet, and a bottle of apple cider. Oh, that cider." The first time she and Sara tried it, they hadn't known it was hard cider. After lunch, they were "walking along the road getting sleepier and sleepier, and finally we just lay down in the grass and drifted off to sleep."

"And the *fraises des bois*! I didn't know what they were. Strawberries of the wood . . . We had them with crème fraîche and a sugar shaker. The waiter would come to the table with this big bowl, and you help yourself. Unbelievable."

Alice had fallen in love with the sensuous good food of France and she had fallen hard, enchanted by what the charmingly irreverent M. F. K. Fisher had called "the old gods" of France—food, good living, and sensual pleasure. In sly and seductive books like *Serve It Forth* and *Consider the Oyster*, M. F. K. had "opened a door to pleasure," writes Luke Barr in *Provence, 1970*, paying serious attention "to everything from shellfish to freshly picked green beans to the pre-departure glass of Champagne at the train station café."

> *The atmospherics of desire and betrayal, the seductive
> pleasures of a shared glass of marc, the fleeting ripeness of
> peaches and zucchini flowers: the human appetite for food
> and love were one and the same in her writing . . . This was
> a philosophical joining, an alchemy, that could only have
> happened in France, where an anti-puritanical attitude about
> both prevailed.*

Twenty-five years later, Alice had walked through that same door, without even knowing what it portended.

"When I got back from France," Alice recalls, "I wanted hot baguettes in the morning, and apricot jam, and café au lait in bowls, and I wanted a café to hang out in, in the afternoon, and I wanted civilized meals, and I wanted to wear French clothes. The cultural experience, that aesthetic, that paying attention to every little detail—I wanted to live my life like that."

It was harder to do than Alice imagined. While she and Sara were away, fire and riots had erupted in Watts, a black neighborhood of Los Angeles. The FSM students arrested during the December sit-in had gone to trial and been convicted. The first U.S. Marines had shipped to Vietnam, sparking a wave of antiwar protests from Berkeley to Harvard. There were whispers of revolution everywhere, acts of civil disobedience, draft card burnings as thousands defied conscription. Friends were registering as conscientious objectors; others were leaving for Canada. By the close of the year, some two hundred thousand American soldiers were already in Southeast Asia. The mood on the UC Berkeley campus was explosive.

"God, it was a wild time," Alice would remember. "A terrible time in many ways. But it all felt so important. History was being made in Berkeley, and we all felt that we were part of it. I didn't lose my French aesthetics . . . but what was going on at Cal seemed bigger, more important."

It was a rich and combustible moment to be coming of age, a heady time to be young and unfettered. The Bay Area was awash in music, radical politics, art, and alternative lifestyles. Relations between the sexes were open and unbinding, sex joyously flaunted, communal living considered de rigueur. Materialism was shunned, spontaneity embraced, ecstatic experience—whether drug induced, or rooted in some mystical Eastern practice—actively courted. In a Berkeley attic on any given night, one might find a nude dance performance paired with kinetic art; live music with someone reading aloud about the "horrors of Babylon" from the Bible; a teach-in on left-handed Tantric sex. In San Francisco's Golden Gate Park, hippies bedecked in beads and tattered jeans gathered daily to salute the rising and setting of the sun. On soundstages from Berkeley to Newport, Rhode Island, rock bards from Janis Joplin to Country Joe & the Fish rued the war, and mocked the commercialism of the larger culture. "Oh Lord, won't you buy me a Mercedes Benz?" Janis crooned. A group that called themselves the Diggers set up a store in San Francisco where everything was free. Called Trip Without a Ticket, people would come in and take what they wanted. The Diggers also gave out free food and staged "living" theater pieces at Free Speech Movement rallies, combining wiggy street theater with anarchic politics. In one, they had a guard marching away a prisoner and beating him. When the crowd started screaming, "Don't do that!" the Diggers jumped into their VW bus (named the Yellow Submarine) and drove off.

Many among the radical young were truly disillusioned with a society they found bigoted, materialistic, and war hungry; others were simply along for the ride. On the front page of the *Berkeley Barb*, the Bay Area's leading underground paper, one might find a cogent, left-wing slam of official Washington; in the back, a piece about cosmic awareness, or an interview with Timothy Leary about tripping on LSD, or a "crazy mandala for the local yogis." No antiwar demonstration is without "a hirsute, be-cowbelled contingent of holy men, bearing joss sticks and intoning the Hare Krishna," wrote Theodore Roszak in *The Making of the Counter Culture*. And it was not a criticism. The bread and wine and flowers and festivity, the kisses and candy and costumed hippies, the delegation of self-proclaimed holy men added levity to what were otherwise sober, life-and-death issues. The humor and playful bonhomie drew people in. To many, the carnivalesque elements seemed no more crazy and irrational than the idea of blowing up the world with a nuclear bomb. Less so, in fact. The one issue that everyone had in common was his or her opposition to the war.

"What kind of America is it whose response to poverty and oppression in South Vietnam is napalm and defoliation, while its response to poverty and oppression in Mississippi is . . . silence?" went the official SDS call.

Alice partook of a little of all of it: the sex, the drugs, the rock and roll, but also the earnest political discourse. She was deeply committed to the antiwar movement and to the democratic ideals of the FSM. Her sister Ellen Waters Pisor remembers going with Alice to an antiwar discussion. The speaker was the conservative columnist William F. Buckley, founder of the right-wing magazine *National Review*, and afterward Buckley took questions from the floor. Alice immediately stood up, Ellen remembered. "She was outraged by everything Buckley had said, and she said, 'What are you planning to do about the genocide in Vietnam?' Well, genocide, that was a pretty heady word for 1966."

That spring, Alice volunteered to work on a political campaign for Robert Scheer, the editor of the radical magazine *Ramparts*. Scheer had decided to run in the Democratic congressional primary, on an antiwar platform, not because he thought he could win—he wasn't really a politician—but because he hoped he could force the incumbent to come out against the war. "Alice was only twenty-one," Scheer remembered. "But I put her in charge of our press liaison, and she was so good, so unflappable, so passionate, so focused."

Alice's friends had noticed her focus too. Returning to Berkeley that fall, she'd seemed more energetic, more committed, more self-directed than they'd ever seen her. "I kept drinking my café au lait from a bowl," Alice remembers, "but I was also seeing the world beyond that now a little more clearly."

Her major was French cultural history and she had decided to concentrate on the years between 1750 and 1850—"in other words, the French Revolution," she says. "In part, I suppose, because it felt like the moment we were living in."

"She was never a shy person, exactly," her friend Eleanor remembered, "but she wasn't very articulate. Then, in 1965, when she came back from France, all this intellectual stuff and the arts gave her a means of expressing herself. She was super-energetic. Not a moment of self-doubt."

Alice was discovering another means of expressing herself. She was beginning to cook—to replicate as best she could the food she had loved in France. She, Sara, Eleanor, and another friend had found an apartment with a good kitchen, and a willing teacher who lived downstairs, a Frenchman with a flair for *la vraie cuisine*. She had discovered the British kitchen doyenne Elizabeth David and was cooking her way through David's recipes. She was also following Julia Child's weekly cooking series on TV, cheered on by Child's self-deprecating good humor and "plummy patrician accent," her unflappability in the face of certain culinary disasters—the fallen soufflés and flubbed sauces. ("Never apologize—nobody knows what you're aiming at, so just bring it to the table.") It wasn't long before the apartment was filled with any number of smart, attractive young men who were dropping by for the company as well as the food.

"The door was always open," says Eleanor, "and we always had men who were just friends, which was something a little bit new. Things were very open in Berkeley. The sixties started in Berkeley before anywhere else, you know. You could be a nice girl and still sleep with your boyfriend. Everything was kind of expansive."

It was an auspicious moment to be an aspiring cook. Spurred by Rachel Carson, who'd been the first to shine a light on the perils of agricultural chemicals, food had entered the consciousness of counterculture Berkeley as surely as had calls to end the war, or ban the bomb, giving rise to the organic food movement, and prompting a new hostility toward big business and the industrialization of nature. Marin County communards were planting vegetable gardens and baking their own bread. Hippies were opening up food co-ops and turning to "health food nuts" once thought extreme. J. I. Rodale's sober, quarter-century-old gardening series, *Organic Farming and Gardening*, now occupied space on every hippie's bookshelf, as much a generational touchstone as *The Whole Earth Catalogue* or *The Greening of America*. Euell Gibbons's loopy books on natural foods and foraging (*Stalking the Wild Asparagus, Stalking the Blue-Eyed Scallop*) were enjoying a new vogue. (Rodale, according to the writer David Kamp, was the first to use the term *organic* in an agricultural context.) After reports that the amount of DDT in the breast milk of U.S. mothers surpassed even the government's own limits for DDT in milk sales, the underground press ran a cartoon picturing a woman squirting a noisome fly with breast milk. No additional commentary was needed; everyone was already in on the joke. In the eyes of the counterculture, "organic's rejection of agricultural chemicals was also a rejection of the war machine," notes Michael Pollan, "since the same corporations—Dow, Monsanto—that manufactured pesticides also made napalm and Agent Orange, the herbicide with which the U.S. military was waging war against nature in Southeast Asia."

Being a foodie was also a mark of worldliness. For those with a little time, vagabonding across Europe or India had become a generational rite of passage. Travel was cheap, and exposure to far-flung places was changing and refining tastes, adding a new veneer of openness and sophistication to young American palates. Food and culinary choices were lining up as yet another iteration of the counterculture's disdain for all things artificial, inauthentic, and mass-produced. How and what one ate profoundly mattered.

As the writer Warren Belasco notes in his sly and incisive work of culinary anthropology *Appetite for Change: How the Counterculture Took On the Food Industry*, "White versus brown was a central contrast. White meant Wonder Bread, White Tower, Cool Whip, Minute Rice, instant mashed potatoes, peeled apples, White Tornadoes, white coasts, white collar, whitewash, White House, white racism. Brown meant whole-wheat bread, unhulled rice, turbinado sugar, wildflower honey, unsulfured molasses, soy sauce, peasant yams, 'black is beautiful.'"

Alice and her friends were certain there were better ways to join the world than as corporate strivers amidst the "white food" status quo. They would change the world without compromising their ideals, starting close to home, by living what they believed.

In the spring of '66, Eleanor recalls, "I remember sitting around together and saying, 'You know, we don't have to get married.' And Sara said, 'It's kind of exciting—we can do whatever we want; there are no rules, there's no structure. We have to make them up ourselves, and that's sort of frightening.'" They decided they wanted to start a restaurant, which they would call The Four Muses. "We'd get this great big old house, and Sara would make clothes there, and Betsy Danch, our other roommate, would have an art gallery, and I would have the bookstore, and Alice would do the restaurant." Alice envisioned a restaurant where a different person would cook each night, but where they would all go to eat. "We were starting to think about a more community-oriented, communal way of life—creating a family in that way instead of the traditional mommy, daddy, baby."

That same spring, Alice met David Goines, who soon became her live-in boyfriend. Goines was a well-spoken, bespectacled kid with street cred to spare. He was working at the time at the Berkeley Free Press, where Alice regularly dropped off campaign materials for printing. Tall and lanky, with a helmet of wavy hair, and an elegant, rather courtly manner, he had been a classics major, studying Greek and Latin, before being expelled. A prime leader of the FSM, he had gone to jail in the summer of 1965, serving a month for his role in the Sproul Hall sit-in of December 2, earning him considerable kudos around Berkeley. In the summer of 1967, he would serve another thirty days. Goines, in the meantime, was becoming a skilled printer. Alice liked his droll humor and quiet reserve, his artistic flair and heartfelt political convictions.

David and Alice both loved to cook. And they were fearless, Goines remembers. "Alice would try anything." She wasn't afraid of failure, "which is absolutely essential to any kind of learning." What worried them less than culinary catastrophe, Goines said—"truly momentous disasters occurred on a regular basis"—was the dearth of ingredients around; there were so many foodstuffs they just couldn't find. "The whole trend of American family cooking since the 1940s had been toward faster and easier, and things that were already prepared. It was gradually whittling away the very essence of what it meant to cook dinner for your family."

David had been doing print work for a composer and music critic named Charles Shere, the driving force behind the now-legendary radical radio station KPFA, when he discovered that Lindsey, Charles's wife, also loved to cook. Soon the two couples, who by chance also lived on the same block, were regularly making meals together. Like Alice, Lindsey was slight and fine featured, with gentle eyes and a sweet smile. Her specialty was desserts and, like Alice, she was a stickler about the ingredients she used, unwilling to cut corners or scrimp on the quality of anything. They were immediately fast friends.

"Those meals of ours were truly composed, like music," Charles later told Alice's biographer. "I spent years at KPFA making record concerts, and I was always really interested in how you design a program—how this piece of music goes with that piece of music.

That was one of the things that really impressed me about Alice. Every item on each menu had a relation to all the other items on the menu. When you spent a lot of time composing things, then I think you develop a mind that always looks to see what the interesting connections are, what the running threads are. I think that Alice's awareness of interconnection ultimately flowered in her thinking about the whole concept of sustainability."

Alice was already a perfectionist, Goines remembered. She was becoming a really good cook, but she would never settle for anything less than perfection. Unlike Julia Child, she *didn't* just bring it to the table. If she curdled a béarnaise, "she didn't even try to save it, just threw it right in the garbage. I was shocked. She was very demanding, very exacting. Everything had to be, within reasonable limits, perfect, or she wouldn't serve it."

Alice was striving for an ideal, a level of quality and care, an attentiveness to detail that was being lost—if it had ever existed at all—in an America increasingly habituated to the conveniences of fast and processed food. In her own quiet way, she was beginning to integrate food into her political and social life. Alice, remembers Tom Luddy, "was the only one who kept insisting that the way we eat is political."

By the spring of 1967, Alice was waitressing at an awful place called the Quest, what Goines termed a "trying-to-be-French restaurant." She had graduated from UC Berkeley that January and, after casting about for work, settled on the server's job, seeing it as a logical next step, a way to deepen her knowledge of French cooking. This turned out to be a short-lived illusion. Within weeks, she was frustrated. The food was crummy, the boss surly, the customers demanding, and the tips stank. Morale in the kitchen, not surprisingly, was low. Every night felt like "an exercise in humiliation." Not only was she learning nothing—she was already a much better cook than anyone there—she had less and less time to spend with Goines. David worked days, and she worked nights. It was getting hard to find time to cook together.

Eleanor Bertino and Sara Flanders were in Paris now, although

Sara was about to move to London, to begin psychoanalytic training. Alice still mused occasionally about starting a little restaurant, but "the restaurant fantasy," as Goines called it, was still just that: "a fantasy." The one bright spot was the cooking column she and Goines were doing for the *San Francisco Express Times*, a radical newspaper run by friends. They were calling it "Alice's Restaurant," in homage to Arlo Guthrie's much-beloved antiwar song. Every week, one of Alice's recipes would appear in elegant calligraphy, with illustrations by Goines. "There were a lot of artists and writers who were working together on that, and I would be feeding them," Alice remembered. "It was a great way for me to sort of test the waters, if you will. The whole experience of working together on the paper, and the deadlines, was something that we did as a group. It was a very important time." Alice was once again feeling the link between food and community she had discovered in France.

Still, Alice was never completely satisfied. No matter how hard she tried, the food she cooked never seemed quite as good as what she'd had in France. She knew it wasn't just an illusion, a distortion fed by time or nostalgia. She knew it had a lot to do with the ingredients she could find—or rather, that she couldn't find.

In Provence, Alice had watched the way the French housewives shopped. They practiced what was called market cooking, "*la cuisine du marché*." A woman would wander through the morning market and make her choices according to what was in season and what looked most fresh. She worked by intuition, waiting for inspiration to strike, settling on one ingredient and then the next, depending on what she saw, what sort of harmonious complements she could find, touching and sniffing and appraising each possibility as she went. A menu might begin with a beautiful fennel bulb, or a basket of oyster mushrooms, or a sea bass just pulled from the cool blue waters outside some port town. It all depended on the day's catch, or that morning's harvest, or on what the butcher brought. Every meal was an improvisation, open to change and adjustment, variations according to the season and supply.

"Ingredients!" Alice would exclaim. "Sure, you had to know

technique. But if you didn't start with great ingredients, you could never make great food."

WORK AT THE QUEST WAS WEARING THIN; IT WAS A DEAD-END GIG, A grind without recompense. Alice needed something inspiring to do, something she loved as much as cooking. A friend of her sister Ellen's, Barbara Carlitz, was a Montessori teacher and thought Alice might like the work. And it turned out Alice did. In the fall of 1967, she got a job as an assistant teacher at the Berkeley Montessori School and was enchanted. "Montessori went straight to my heart, because it's all about encountering the world through the senses. That's how kids learn best. The hands are the instrument of the mind—that was how Maria Montessori put it."

The genius of Maria Montessori's method was in her respect for a child's natural development, both physical and psychological. She emphasized independence and designed materials that appeal to children through their senses. The idea was that kids learn best by discovery, rather than instruction, and that interacting with the environment is important.

"It's an observation that's not just with your eyes," said Alice. "You take a little broom and try to get up every crumb on the floor. You take your little tray and you put that back on the shelf. There's a place for it. You learn about everything in your environment. You become familiar with it. And you begin to really see what its value is."

Learning by doing—through tactile engagement, using the senses to navigate one's way through the world: it was a perfect fit for Alice. It was how she herself learned, a way of approaching the world that was also intriguingly similar to that of Carson, Jacobs, and Goodall: privileging the senses, trusting to intuition, starting with direct observation, and then building from there.

BY THE SUMMER OF 1968, ALICE WAS LIVING IN LONDON. SHE HAD DEcided to study the Montessori method at the international Montessori Centre in London, putatively the best place anywhere to train. Barbara Carlitz and her husband, Michael, had relocated to London that spring and graciously took Alice in. Though David still had work in Berkeley, he urged her to go. He hoped to join her later that fall.

David appeared that October and moved in with Alice. The Carlitzes happened to be away that weekend, so it came as something of a surprise to find David holed up with Alice when they got home. "Hello," David said, proffering his hand, "I'm your certified commie creep."

They remained there until Christmas; then Alice went out in search of digs of her own. She found a tiny room in the turret of an old Victorian house in Hampstead. There was barely space enough for a single bed, and no central heating. "You fed a space heater with shillings," Alice remembered. "The kitchen was a closet across the hall—a two burner hot plate with a tiny broiler underneath." Alice didn't care. She loved living in a turret. The only rub was the landlady, who eyed her coolly, and then announced that the room was for one woman only. It seemed that London in 1968 was considerably more buttoned down than free-thinking Berkeley. Living together as an unmarried couple simply wasn't done. David found a room near Alice's and from time to time they returned to the Carlitzes' for a night in their guest room.

Alice, meanwhile, continued to cook, even with a closet for a kitchen. She loved the Montessori work, but she still obsessed about food. "I nearly froze to death that winter, but I cooked up a storm," she reminisced. "I remember steaming open mussels in a battered saucepan, and even managing to bake an apple tart in that broiler. When I remind people that they can cook no matter where they are, I know what I'm talking about."

In early summer of 1969, the Montessori program over, Alice and David set out for Paris, determined to eat their way through the next two weeks. David flew back to Berkeley in July. Alice stayed on, joining Judy Johnson, a friend she'd met at the Montessori Centre. The two women bought a beat-up Austin Mini Cooper and started driving east, their sights set on Bulgaria, where Alice imagined they would find "Gypsy music" and "smell rose essence." Instead, they found "kind of a military state." They moved on to Turkey, where they "hooked up with two French guys" in a little Citroën Deux Chevaux and drove into the interior. Each night the foursome would set up tents side by side. "It was a great thing, because two women couldn't get into anyplace without men," Alice remembered.

It was in Turkey, Alice later wrote, that she experienced a moment of simple hospitality she would never forget. They had run out of gas and were hungry. "A shy, big-eyed boy appeared . . . and he mimed that there was no gas to pump. And we counter-mimed that we supposed we would have to wait . . . then, fingers pointing to mouth, where would we get something to eat?" At which point he invited them in. The house was bare-bones, and his parents, it was clear, were away. Birdcages hung from the low ceiling. Two rough benches leaned against a wall, each draped with lovely woven rugs. In the corner, near a baby brother, sat "a brazier made out of an old gas can." As Alice wrote: "The boy builds us a fire out of pinecones, puts on a kettle, and makes us tea. Then he produces a small piece of cheese and painstakingly cuts it into even smaller pieces, which he offers us gravely . . .

"He has given us everything he has, and he has done this with absolutely no expectation of anything in return. A small miracle of trust, and a lesson in hospitality that changed my life."

It was a lesson that would stay with her: the importance of generosity, the magic of "finding grace in the unexpected."

From Turkey, they headed down the coast and into Greece, ending up in Corfu, where they lived on "practically nothing, very simply, watching the sun and moon rising and setting over the harbor." They swam at dawn each morning, ate fish "fresh out of that sunstruck sea." For the first time, "I was unmistakably part of the natural rhythm of a place," Alice remembered. "It was like a garden of Eden. You just went out and picked things off the tree. We tenderized octopus by throwing it against the wall. We cooked every day. We'd have this beautiful feta cheese with ouzo, and beautiful olives and olive oil . . . We never used a watch. The people had these big pig roasts, with everyone dancing. There was a certain poverty, certainly, but it was a beautiful way they lived their lives in spite of it."

By late summer, Alice was out of money. She flew back to Berkeley, and moved back in with Goines. A new couple from the South of France, Claude and Martine Labro, had recently moved to town and some friends introduced them. Claude was a mathematician teaching at UC Berkeley; Martine, a graphic designer, but also a superb cook. Alice liked her immediately. Martine had enormous style, the sort of natural chic Frenchwomen somehow seemed born to: elegant, understated, a touch bohemian. And her eye was infallible. She could move through a flea market and find the perfect quirky piece that set off a room. Martine loved faded lace and antique dresses, sheer and pleated pieces worn in layers, vintage fabrics and antique buttons. She also loved hats. It was she who convinced Alice that she looked stunning in a beret and in "a tightly wound cloche." Alice was soon wearing both. "Martine was very important to my whole aesthetic," Alice later claimed. "She was very definitive. About light—that was one thing I got from her, an obsession about light." Alice also learned about frugality from Martine. "Martine could feed ten people with one small chicken, and everyone ate well. The American way was to buy one chicken for two or three people."

Claude and Martine loved French cinema and were regulars at the Telegraph Repertory Cinema in Berkeley. They often talked film with the manager there, a movie buff named Tom Luddy. One night they introduced him to Alice.

The attraction startled even her. Alice was instantly smitten, not only with Tom Luddy but also with French film. She had soon parted ways amicably with David Goines, and was living with Luddy in his modest bungalow on Dana Street. By day she was teaching at a Montessori school, by night still cooking regularly with Lindsey Shere and Martine and Claude Labro. But now her circle had widened to include Luddy's film world friends. "Film was the art form of the moment," Eleanor Bertino remembers. "And Tom knew so many interesting people. Susan Sontag, Huey Newton, Agnès Varda, Abbie Hoffman, Jean-Luc Godard. If they were in town, they would be invited over to the house. It was, like, this fantastic salon, in this incredibly modest house."

Alice would cook dinner, and Tom would screen a film—"a political documentary or a Bresson film, or some old American film no one has ever seen," Eleanor remembered. "Then Alice would serve Cognac and a little French tart . . . She was completely obsessed with food at that point."

Alice was getting to know Francis and Eleanor Coppola, whose new venture, Zoetrope Studios, was just getting started. In 1969, the Coppolas had converted an old San Francisco warehouse into a workspace and were now making films. She and Luddy traveled to the island of Noirmoutier, France, as guests of the filmmaking couple Agnès Varda and Jacques Demy, the director of *The Umbrellas of Cherbourg*, among other French classics. While Luddy talked shop with Demy and Varda, Alice slipped into the kitchen, where she befriended Demy's mother, who made the sorts of simple, rustic dishes Alice most loved.

It was Luddy who introduced Alice to the films of Marcel Pagnol, whose classic 1930s trilogy, *Marius*, *Fanny*, and *César*, would strike such a chord. Set in the old port of Marseille, the films follow the intersecting fortunes of a group of French provincials who pass their days in a tiny waterfront café, where they flirt, play cards, argue, and, on occasion, profess love. Alice's favorite character in the trilogy, the one she always "grew mistiest over," says Luddy, was Panisse, the gentle, love-struck sailmaker who offers to marry the jilted Fanny, who finds herself pregnant by the bar owner's son, who has left her for a life at sea. When Alice started batting around names for a restaurant, Luddy suggested the noble Panisse, which to Alice seemed just right: an invocation of the generous, good-hearted spirit she envisioned for the place.

It was also through Luddy that Alice met Paul Aratow, the man who would make her restaurant dreams real. An aspiring filmmaker and a junior professor at Berkeley, Aratow was as serious a home cook as Alice, having caught the culinary bug in France, while his wife was on a Fulbright fellowship. He had a little family money and was willing to invest in the venture. It was he, in fact, who found the house on Shattuck Avenue they eventually settled on— "an ugly, squat, two-story Hollywood-type stucco apartment house that I tore apart with four or five hippie carpenters," as he describes it. It looked "like a rundown hippie crash pad that had fallen on bad days," Luddy recalled.

All that spring, summer, and fall of 1970, Alice, Tom, Claude, and Martine scoped out restaurants in the Bay Area, to see what they could learn. "Usually it was learning what not to do," Luddy remembered. Then, in the little coastal town of Bolinas, they stumbled upon a place Alice liked, a converted Victorian farmhouse that had a casual, offhand feel. The chef served fresh and interesting food, some of it picked from the garden out back. The walls were hung with patchwork quilts. The china was funky, all of it mismatched. But the atmosphere felt right. "Nobody cared if you wanted to stay at your table playing poker all evening," Alice remembered. "I loved the idea of a restaurant in a house . . . I loved that you could stay as long as you liked." The house on Shattuck Avenue seemed to hold the same promise.

By now, Alice and Luddy had enlisted several limited partners— Alice's London friends Barbara and Michael Carlitz; the rock critic Greil Marcus and his wife, Jenny; Alice's parents, who agreed to a $10,000 loan (which they raised through a mortgage on their home). Alice also had some "necessary silent partners," notes Mc- Namee. Drug dealers. Though these were not, he adds, "the scary, Glock-wielding gangsters" one thinks of today. They were laid-back fellow hippies who happened to support themselves by selling a lit- tle pot to friends. "Well, of course, they were the only people who had money," Alice says. "The only sort of counterculture people who had money. We couldn't get it from a bank, God knows."

They combed the flea markets all that summer of 1971, picking up mirrors and cutlery, glassware and china as they found it. They didn't care if anything matched, only that it felt good to the touch: that the silverware had a certain weight and girth; the patterns on the antique plates were pleasing. Alice decided she wanted red-and-white-checked oilcloth on the tables rather than white linen. She didn't want the room to feel too stuffy or formal. The seating would be simple: straight-backed oak chairs; tables of all sizes—small and intimate enough for couples, big and sufficiently comfortable for larger groups. The planked floors would be clean and polished, the plaster walls painted a soft antique white. Alice wanted light and air, fresh flowers throughout: a big, overflowing bouquet near the bar, smaller vases on each table. There would be a bank of windows facing west, overlooking San Francisco Bay, through which light would pour: the dazzling white light of late afternoon, the Campari reds of sunset. Every note needed to be just right, she insisted. She wanted diners to feel the breezes rolling off the bay, to inhale the fragrances drifting in from the kitchen, to bask in the rosy glow reflected in the mirrors on the walls. She wanted the room to seduce and delight, play directly to the senses. Most of all she wanted the food to be so good that people paused, slowed, savored each forkful.

As for the menu, there would be no choices; Alice said, "I wanted it to be like going to somebody's house. Nobody gives you a choice about what to eat at a dinner party." There would be a single set menu each night, and every evening's offering would be new. The menus of the week would be posted each Monday.

Later this would prove to be one of the unusual features people most loved. "You ate what was there," Greil Marcus recalls, "and often it was something you had never had, or cooked in a way that you had never imagined. Very quickly Jenny figured out that the wrong way to go to Chez Panisse was to see what the menus were and choose something you thought you would like. The best way was to choose something that you thought you didn't like or you had never heard of. We loved having people come from out of town and taking them to this incredible place that didn't meet anyone's expectation of a good restaurant."

Alice's bibles were by now the cookbooks of Elizabeth David

and Richard Olney, the self-trained American cook who lived in southern France. It was Olney who made a special point about the sequencing of dishes. Each course should flow effortlessly into the next, to create a single, "harmonious whole." It was, in some ways, a Montessori ideal: the notion that the connections between elements in the world were important, that paying attention to context, to how things worked together, mattered.

That August, Alice began assembling the staff. Experience, it seemed, was not a prerequisite. Nor were job descriptions. Alice was no more sure what the work would entail than the next guy. But she was confident that she'd recognize the right people, the people who "got it," as she said, when she saw them. Her first hire was Sharon Jones, an aspiring actress. Alice listened to her meager qualifications, searched her eyes, looked her up and down, and then offered her the waitress position. Sharon smiled. She wondered if she could work part-time; she hoped to continue doing theater auditions. Alice said of course. This was pretty much the pattern. Nearly everyone who applied for a position at Chez Panisse was also pursuing some other line—poetry, filmmaking, music, painting. Many members of the kitchen brigade also held advanced degrees—or esoteric interests. Whatever it was, Alice was open to it. If you could work only one night a week, that was fine. If you needed to go off for a trek in Nepal, or a week by yourself in the woods, or a silent meditation at Green Gulch, that was cool too. Alice was completely flexible, notes her biographer. She liked smart, eccentric, creative people, and the hires she made reflected this. And life in Berkeley at the time was extremely cheap. "A couple could live on three thousand dollars a year," the chef and food writer Ruth Reichl remembers. "Berkeley was filled with people who chose time over money, who were educated and wanted to do honest work that didn't exploit anyone." Alice was very much a part of that.

ON AUGUST 28, 1971, AT PRECISELY 6 P.M., CHEZ PANISSE OPENED ITS doors. It was an evening of pure, mad improvisation. The kitchen still wasn't finished, though for two months running, the place had been crawling with hippie carpenters. At a quarter to six, they were nailing up shelves. The menu was chalked on a blackboard: pâté *en croute* (in homage to Julia Child), a main course of duck with olives, a salad, and a plum tart. Alice wore a vintage lace dress, pale taupe and carefully selected for the occasion. The five waiters wore what they pleased. So it was hard to tell who was a waiter and who a patron. There wasn't enough silverware and the waitstaff kept tripping over one another. Alice didn't yet know that the dining room wasn't really big enough to accommodate more than three on the floor. The chef, Victoria Kroyer, was a novice. Until Alice hired her two weeks earlier, she'd been a grad student in philosophy. None of the other kitchen staff had professional experience either.

"I hired them all, because I liked them. I didn't want professionals. We were just going to figure it out. That's what we were doing, making it up as we went along. It was totally insane," Alice said.

But the ducks were fresh, not frozen. They had found them in Chinatown in San Francisco. And the produce was fresh too; it had come from the Berkeley co-op across the street. The pâté, which had been prepared beforehand, arrived quickly. It was served with Dijon mustard and savory little cornichons. And then an hour passed: a long, long hour. The dining room grew quiet as people waited. Alice looked flushed, then flustered, then anxious; she was in and out of the kitchen, she was pleading with the staff, chatting up the patrons. The problem, it seemed, was the duck. It simmered in great pots on every burner, yet still it wasn't softening to the "melting tenderness" she and Victoria wanted. The diners looked less patient now: fifty-some people waiting at the tables, more waiting at the door to be seated, a long line queued up on the sidewalk.

"I was going out of my mind," Alice remembered. Fortunately, at least a lot of them were friends and family.

And then at last, waiters surged from the kitchen, plates of tender, glistening duck on their trays. People forgot about the time or the wait, even those who had been sitting for hours. For all the chaos, the rank unprofessionalism, by the end of the evening everyone looked happy and sated. The plum tart, which was Lindsey Shere's creation, was perfect, warm and delicate and served just on time. "It was exactly the right historical moment for this to happen in Berkeley," remembers Victoria Kroyer, who had spent two days laboring over the sauce. "The educated clientele, the professors and students who had been to Europe, were waiting for a restaurant like this."

Well after midnight, after the last patron drifted out, Alice and the exhausted staff uncorked some wine and sat in the half-finished upstairs café. One hundred twenty dinners had been served that night. If no one could say exactly how many of those had *actually* been paid for, it hardly mattered. "We were so happy. We were so young," Alice remembers. "We were in love with what we were doing. And we were in the right place at the right time. It was sheer luck, really."

AS THE ELATION OF THE FIRST DAYS SETTLED, ALICE BEGAN TO FRET: She wasn't sure about the light, which seemed too yellow. She wanted it to be white, to illumine the colors of the beautiful food. She worried about the sound of shoes on the hardwood floor. The noise was jarring. Wooden clogs were all the rage in counterculture Berkeley. One of the waitresses, Brigitte, wore them a lot. When a complaint came in from an elderly customer that the lovely Brigitte wasn't wearing a stitch of lingerie, Alice couldn't have cared less. But she did care about the clomp of her shoes.

Alice wanted the dining room to feel intimate, yet warm and alive, to be filled with the buzz of bright conversation, the soft clink of cutlery a gentle backbeat. She didn't want diners to have to shout, or to strain to hear. And she didn't want rugs, which collected food. Chez Panisse, she insisted, should feel generous and welcoming, elegant but informal. It should never have airs, but it should also never scrimp on quality. "No corners cut," Alice told the staff. "Ever."

Alice also worried about ingredients. Food of the quality she

wanted was hard to find. And it was always expensive. "I was looking for food that tasted like the food I'd eaten in France. I was on a
search for that. I remember buying four cases of Kentucky Wonder
beans and just taking the little ones out of the bottom and pretending those were haricots verts. I threw all the rest into the compost."

Later, Alice would pay close attention to limiting waste, but this
was still early and she was searching for the ineffable—an integrity
of taste and ingredients she found wanting in American food, an
integrity that was already being undercut by the efficiencies of industrial agriculture.

Despite insisting on the best, and her tendency to overspend,
Alice was determined that prices remain low. She wanted Chez
Panisse to be a simple place where good friends could gather, the
kind of obscure little restaurant you'd run into by chance in some
small town in France, where the food was both so remarkable and
so affordable that when you got home you told all your friends
about it. The prix fixe for this reason was just $3.95, not only a steal
for a four-course dinner, but wildly unrealistic given the operating expenses—the food, the flowers, the oversize staff and ongoing
construction costs. Chez Panisse was just two weeks old and its
finances were already unraveling. Alice and her partners were out of
money and the upstairs café still wasn't finished. New invoices were
coming in daily, and past construction bills hadn't been paid. Nobody was keeping track of what came in, or what went out, or who,
really, was in charge of the books. It certainly wasn't Alice.

"Those of us who were working hard figured somebody had
a handle on it, that the cash flow was being properly managed—
and we were wrong," says Jerry Budrick, a waiter on opening night
and later Alice's lover. "The whole operation was sliding deeper into
debt."

On September 12, barely two weeks after opening, a memo
went out to the staff announcing that the payroll couldn't be met.
There was $500 to disburse, which amounted to 10 percent of staff
salaries, so only 10 percent of the wages due each employee would
be paid. "Everyone will have to be paid off as soon as we correct

the situation." In the communitarian spirit that was Berkeley at the time, a final note was added: "If anyone wants to return his partial payment, it will be put into a fund and redisbursed to other needy employees . . ."

Alice wasn't sleeping at night. "It was a train out of control, a wreck about to happen," she recalls.

Yet Chez Panisse was somehow finding its way despite its financial woes. The menu was evolving, and plates of food, all of it beautiful, were arriving more or less on time. "We were inventing as we went along," Alice remembered. "There was never any set plan. Every day was an improvisation." The menu was decided by what foods could be found on which days, and where. Alice was perpetually in her car, picking up provisions—fresh duck from Chinatown one day, sweetbreads from "Such-and-such Meats" on another. She lost track of the miles—and the bills—her attention focused exclusively on finding the best ingredients she could.

Paul Aratow had wanted Chez Panisse to be open round the clock, like La Coupole in Paris. This proved unrealistic. Breakfast was dropped—"nobody came before noon," said Alice—and prices for Friday and Saturday nights' dinners were raised. Even so, the hours were still gruelingly long for Alice, which was taking its toll on her health and on her relationship with Luddy. "She was a total workaholic, and I was very impatient and angry with her much of the time," he says. "She had no time for anybody—just driven, driven, driven." Luddy wanted Alice to take Sundays off and be more reasonable about her health. "I remember once coming to the kitchen and seeing her sitting on an upturned pot, and I said, 'What are you doing?' She said that she couldn't see—she was blind. Her system had short-circuited, and she sat there maybe, like, an hour before her sight came back."

Luddy and Alice split up, and Paul Aratow too was falling away, unable to square his desire to turn a profit with Alice's lofty ideals. "We would have heated discussions about putting in more tables," Aratow recalled. "Alice would say, 'No, it has this lovely atmosphere, not too crowded.' And I would say, 'We're losing money! . . . We could serve another twenty meals a night and break even if we put three more tables in!"

But for Alice, Chez Panisse was never about money, says Greil Marcus. It was about "a notion of how to live, an emphasis on certain values," a faith in novelty and living without a script. Yes, it was about aesthetics, a simple love of the sensual and the beautiful. But it was also about giving people something that perhaps they'd never had, an experience of excellence and *plenitude,* showing them just how extraordinary a simple, exquisitely cooked meal could be. It was about creating a sense of place and community, fostering a spirit of generosity and goodwill, a feeling of connectedness between people, offering them something that was pleasurable and honest and true. "It could not have happened without the Civil Rights movement, the Free Speech Movement," says Alice's friend Sharon Jones. "My memories," she told Marcus, "are of a lot of color and a lot of light . . . the light would just float into the rooms . . . I had a smile on my face all the time. And it was true every day. There was nothing routine about working there . . . Out of Alice's excess," Jones adds—having lived through nineteen years of Catholic schooling, college, a stint on a kibbutz in Israel, and then overland travel through Turkey, Iran, Afghanistan, only to emerge almost where she'd started, in a Catholic hospital in India—"I learned to demand the pleasure principle my mother had done her best to *smoosh* out of me. The extravagant part of Alice—the sense of *display,* the beauty—she had a sense of wanting people to have more than enough."

Wanting *more than enough* for people: it was out of this spirit that the restaurant grew. If Alice thus wanted truffles, she bought them, no matter that money was tight. Unhappy with her own floral arrangements, she hired a friend whose sensibility was even more exacting than her own. Carrie Wright's arrangements were gorgeous, wayward combinations of grasses, vines, gnarled branches, and flowers, some with a faintly Addams Family–esque feel. They spilled lavishly from an array of antique buckets—exquisite, unruly, and wildly expensive. Alice wouldn't hear of cutting back. Despite her best intentions, extravagance was in her nature, and she was a perfectionist. What she cared about was community, quality, fairness, generosity, doing one's best.

Alice wouldn't compromise. "She was very stubborn," Aratow said. "Alice had a very pure vision, and she didn't really have the business sense to get the thing off the ground." If Alice felt it was gracious to offer a customer a complimentary glass of Champagne, she poured it; if a friend desperately needed a job, she hired him, even if they were already fully staffed.

In part, it was the times. Berkeley in the late sixties and early seventies was awash in utopian aspiration, a place of grand visions and quixotic dreams. Everyone of a certain mind-set believed that a new, more generous age was at hand, an age based on communitarian goodwill and perfect, instantaneous freedom, authenticity, and a simpler, more "natural" way of living. "Scale down your attachments to modern technology," the Zen poet Gary Snyder advised; "voluntary simplicity would subvert an economy geared to overconsumption."

"'Natural' was a liberated state of mind," writes Belasco, a symbol of opposition to mass production and standardization—the automated assembly-line way of life. It was widely held that the "personal was the political"—that by "revolutionizing" one's own private life, one would also transform the system, rebuild it from the ground up. As the sign on the Diggers free store in San Francisco announced, "No owner, no manager, no employees, and no cash."

If Alice tended to be too generous, so too did her staff. A friend would appear with a bag of perfect yellow beets just pulled from the garden, and rather than money, the chef's return gift would be a meal on the house. Alice gave away desserts, glasses of wine, sometimes whole dinners, and the staff followed suit. It was all part of the generous, openhearted Pagnol-esque spirit she was after. But it was becoming a serious threat to the restaurant's survival.

Aware they were caught in an accelerating death spiral, Alice decided to approach a woman she knew in Berkeley, Gene Opton, who ran a successful cookware shop. She wondered if Opton would help them with bookkeeping.

Opton agreed to join the staff as general manager, pumping in a little money of her own now that she was a partner. "There was a little cottage in back of the house, and in the cottage there was a desk, and it had a drawer," she remembered. "I pulled open the drawer, and in it were stuck all these little bits of paper for their expenses. That was their bookkeeping." No one, according to Opton, had ever sorted them, added them up. Or paid them.

Wine was part of the problem. Everyone working at Chez Panisse was expected to know the list of offerings, which meant tasting them. And once a bottle was opened, well, best to continue drinking rather than letting it go to waste. After the first year, more than $30,000 worth of wine was unaccounted for, most of it drunk on-site. Recreational drugs also played a part in the overall mayhem, though just how much it was hard to say, since marijuana at that moment was such an integral part of Berkeley's culture. "It was quite unremarkable for a waiter lofting a tray to suck back a last-minute toke before plunging through the swinging door to the dining room," writes Alice's biographer. Indeed, it was no less remarkable for customers to arrive "ripped to the gills themselves."

By Opton's calculations, Chez Panisse was serving meals for four dollars that cost at least six. There was no system for keeping track of people's hours. Yet everyone, including Alice, bridled at the idea of a time clock, which seemed punitive. It ran counter, they said, to everything Chez Panisse stood for: its ethos of kindness, family, nurture, trust, goodwill. At one point there were earnest discussions about whether the restaurant should be run as a commune. Opton had a lot on her hands.

"I knew it was going to be hard work," Alice recalled. "What I guess I didn't know was that I wouldn't be able to get control of it somehow. I thought, since we had all these people, surely we'd be okay."

With such fiscal mismanagement, no one quite understood how the restaurant could continue to get better, but somehow it did—

consistently, against all odds. It was baffling even to insiders. Those close in said it had something to do with Alice herself. A natural collaborator, Alice had a certain generosity of spirit, the ability to inspire the best in people, to foster community and a sense of excitement about working together toward a common goal. From the very beginning, remembers Charles Shere, Alice decided that she and her backers should share ownership with the key staff. She gave these people "as big a percentage of the restaurant as she kept." A lot of people "would think that that was a damn fool thing to do, as far as business was concerned. But in fact it was an incredibly important thing to do because that was what energized the restaurant." It wasn't a partnership; "it was really a community—a real community of interests where everyone was there for exactly the same reasons, and in it to exactly the same depth."

Alice had an easy way with people, a gentle, winsome charm. She knew how to draw them out, solicit their ideas, show gratitude when gratitude was due. She wasn't shy, but sometimes she could appear that way: dreamy and a little tentative, more uncertain than she actually was.

Those who knew her weren't fooled. "There has always been this interesting contradiction between the delicate little way Alice looks and how she really is," her friend Eleanor Bertino once observed. "She looked just like a Pre-Raphaelite angel," but in truth she's strong and extremely deliberate. "Once, I had this idea of her as this fragile flower, always on the verge of bursting into tears," says longtime friend and food writer Corby Kummer. "It took me a while to see her will of steel." Alice, friends concur, always focused on the things that mattered to her. And what mattered was that Chez Panisse embody a certain unassuming perfection.

Alice doesn't disagree: "I am uncompromising," she admits. "And I am a purist. I believe in that. I think you're always trying to make things better. Always trying to make things right." It's about "a way of doing work, a way of focusing. Having fine-tuned senses so you can make really right decisions."

A CERTAIN EQUILIBRIUM WAS TAKING HOLD, THANKS IN PART TO GENE Opton's oversight.

But Alice still wasn't feeling satisfied. She continued to fret over the poor quality of the food they were finding. By now they had discovered the Chinatowns of San Francisco and Oakland, which came as a revelation. Here they could find ducks bred on local farms that were still alive or freshly killed, their heads and feet still dangling, just as she had seen in France. They could get chickens in wooden crates that after slaughter didn't taste like chalk, live catfish and lobsters from tanks, wriggling eels from tubs of water on the sidewalk.

But the supermarkets, where they still had to buy most everything else, were discouraging, the vegetables half dead, wilted, and tasteless, the fruit hard as rocks. Too often they had to settle for ingredients that weren't as good as they should be. And there were always items they couldn't find at all.

To fill in the gaps, they began—literally—to "forage," Alice remembered. "We gathered watercress from streams, picked nasturtiums and fennel from roadsides, and gathered blackberries from the Santa Fe tracks in Berkeley. We took herbs like oregano and thyme from the gardens of friends . . . We also relied on friends with rural connections. The mother of one of our cooks planted *fraises des bois* for us in Petaluma . . . Lindsey got her father to grow the perfect fruit she wanted."

"We knocked on the doors of strangers to ask if we could pick the mulberries from their trees."

And so began what Charles Shere has called "the hunter-gatherer culture at Chez Panisse." This was an idea no American restaurant had ever tried, writes McNamee. The food at Chez Panisse was being composed using the flavors of its "place," the distinctive character of its locality. *Terroir* was the term the French used to describe a food's or wine's expression of its origins: the mineral composition of its soil, the quirks of its topography, the tang of its air. What Alice was trying to do was similar: to imbue the meals at Chez Panisse with the essence of its locality. She was approaching cooking as the French chefs did: using fresh, seasonal ingredients that were grown nearby. But the difference, of course, was that there was no American model to follow, no ready infrastructure of purveyors on which she could rely, no distribution system by which fresh indigenous ingredients could find their way from Bay Area farmers to restaurateurs like her.

Alice envied the chefs of France. In Paris, she had wandered the early-morning markets at Les Halles, astonished by the variety and freshness she saw there. To the French the scene seemed unremarkable, simply the way things had always been done. To Alice the abundance of fresh food, the sheer profusion of choices, seemed a miracle. "They always had this local distribution system," she observed. "So much beautiful food came from nearby, less than an hour away."

When the restaurateurs made their rounds to procure the day's provisions, the fish still carried the whiff of the waters from which they'd come, briny Atlantic or "Mediterranean reef," icy mountain lake or alpine stream. Their scales still gleamed and their eyes were clear. The poultry had just been plucked, the eggs collected from the henhouses that morning. There were baskets of cut greens with crumbs of dark soil clinging to their roots; ripe pears picked "one piece at a time and laid gently in straw." There were "blue legged chickens" that had grazed in fine green pastures all their lives; wild mushrooms and little delicate strawberries gathered from the forest floor, picked mindfully by some careful individual; there were savory cheeses from "Burgundy, the Pays d'Oc, Alsace, Normandy, Savoie," each lovingly made by someone on some small local farm.

It was this way throughout all of Europe, and certainly all over Asia—in most of the rest of the world really, where "agribusiness" was unheard of and farms still small. "But not here. Not in Berkeley, not in San Francisco or Chicago or New York," Alice said. "Some good things you could get in some places," she notes. "We were going to stalls in Chinatown. And to ethnic markets that had the *look* of European markets. There were farm stands in the summer. Not a lot, but some. Out in Brentwood, you could get fresh corn and lovely vegetables."

It wasn't that one couldn't eat well in America if one really wanted. Most decent-sized cities had at least one very good restaurant, where discerning palates could enjoy haute cuisine. But it was always at a price, and there were certain caveats. The Dover sole at La Bourgogne in San Francisco, a truly sensational place, might be flown in daily, but the chef, who was usually French, still relied on tinned foie gras. The crayfish and oysters at Galatoire's in New Orleans were deliciously fresh, just culled from local Louisiana waters, but if one ordered escargots, the snails' origins were a can. Truffles might appear at Lutèce in New York, but they were strictly for show, dull as dishwater and just as flavorless. And even if the food was excellent, the atmosphere in these places was often stuffy, the waiters haughty, the menus overly elaborate—as if designed to intimidate. Everything felt too formal, with even bathroom attendants to tip. And the check for such a dinner was always eye-poppingly high—many times the price of a meal at Chez Panisse.

Good ethnic food, if one knew where to go, could still be found in certain cities, savory dishes cooked in the style of the old country. But even those were disappearing, the food losing its distinctive flavors, the cooks adjusting their menus to suit more bland American tastes.

A curious thing had happened in the 1950s: Americans forgot how to eat. Soaring prosperity combined with suburban living—the "Station Wagon Way of Life," as *House Beautiful* put it—had given rise to convenience foods. Grocery stores grew in size to accommodate new aisles of packaged and processed concoctions—premade salad dressings, instant and powdered soups, ready-to-bake cakes in throwaway foil pans—all touted as quick and easy. Why slave over a hot stove, the thinking went, when a meal could be prepared in just minutes? Cookbooks aimed at busy housewives— Poppy Cannon's *The Can-Opener Cookbook*, Betty Crocker's *Picture Cook Book*—flew off the shelves, hugely popular. Just as poison gases developed for WWII had been converted to peacetime uses in the form of pesticides, so too were military rations domesticated and now sold as convenience foods. The TV dinner, chirped the *Washington Post*'s Olga Curtis in April 1957, "can claim the K-ration as an ancestor."

At the same time, Americans of a more worldly bent were beginning to be more interested in food and cooking, thanks to a handful of food writers—James Beard, the wry and confiding M. F. K. Fisher, and most especially Julia Child, whose popular 1961 bestseller, *Mastering the Art of French Cooking*, was followed two years later by her even more popular TV cooking show. Beard, Child, and Fisher had defined for a generation "how to talk and think about food and wine and cooking and life." Like Alice, they had experienced France as an ongoing "revelation of taste." Fisher had been the first, embarking on a serious but playful "literary consideration" of eating and even love in France. Beard and Child followed with clear and accessible cookbooks that sought to demystify French cooking, while also presenting it as an art form—one that was achievable by any ambitious home cook, as long as there were no shortcuts taken, no "skimping" on key ingredients like butter, cream—or time. Inspired by these avatars of good living, all of whom had been changed by their gastronomic experiences in France, serious home cooks were attempting more adventuresome dishes, inviting their friends to elegant dinner parties, delighting in sharing the rituals of good eating.

As Nora Ephron wrote in *New York Magazine* in 1968: "Food acquired a chic, a gloss of snobbery it had hitherto only possessed in certain upper-income groups. Hostesses were expected to know that iceberg lettuce was *déclassé* and tuna-fish casseroles *de trop*. Lancers sparkling rosé and Manischewitz were replaced on the table by Bordeaux."

American men and women were cooking along with Julia Child, Ephron remarked, "subscribing to the Shallot-of-the-Month Club . . . cheeses, herbs, and spices that had formerly been available only in Bloomingdale's delicacy department cropped up around New York."

Ephron was right, of course, but only to a point. In certain zip codes, certainly, specialty food shops catering to this small if growing cohort were sprouting in America's more sophisticated cities.

At Cardullo's Gourmet in Harvard Square, Cambridge, one could find fresh-pressed olive oil, artisanal bread, hunks of Italian Parmesan. Displayed on refrigerated shelves at Bon Appetit Fine Foods, in Princeton, New Jersey, were French triple crème cheeses, imported olives and salamis, caviars and smoked fish. At Balducci's in Greenwich Village, mountains of fresh greens beckoned from the produce aisles. Most of it was out of season, of course, flown in from hot countries half a globe away, and none of it was organic. But there was tremendous variety and the food looked well cared for. Beyond these sparkling emporiums, however, it was slim pickings.

In general, the quality of food in America was seriously declining. Immigrant children raised on wholesome home-cooked meals were embarrassed by their mothers' old-world ways. They wanted what was pictured on TV: frozen prepackaged pizza, canned spaghetti, Pop-Tarts, Rice-A-Roni, Cheez Whiz, fried onions in a pop-top can. Not only were family farms disappearing, "little family restaurants that might have had a few good simple things were being replaced by McDonald's," Alice said.

Even restaurants a notch pricier were now serving processed food that arrived in frozen, individually portioned "heat-and-serve vacuum packets" by way of a refrigerated semi. A lobster Newburg or a veal scaloppini could be slid into a new device called a microwave oven, and then sped out to diners in minutes. Frozen inventory eliminated spoilage; freezers enabled long-term storage; purchasing in bulk ginned up profits, opening the door to this second tier of restaurant chains—places like Red Lobster and Sizzler steak house—that featured more than hamburgers and fried chicken. Advertisers, meanwhile, called this new, gussied-up fast food "progress."

All this processed and packaged food, of course, contained sugar and fat in much higher quantities than home-cooked meals, not to mention a minefield of artificial ingredients. Often a dish purporting to be seafood or chicken didn't actually contain any real seafood or chicken at all. All that mattered was that the texture seemed right. Deep-fried and swaddled in a cocoon of breading, the taste was masked. America was becoming a place of not only *fast* food, but *fake* food. It wouldn't take long before people's health suffered. The slide toward a nation of the overweight had begun. "It was so dispiriting," says Alice. "But mostly I just buried my head. It wasn't till years later that I really saw what was going on."

The truth was, even for home cooks, the quality of the ingredients they could buy was on a slippery downhill slide. There was a deceptive increase in choice, a great expansion in the size of supermarkets. But the additional acreage was mostly taken up by processed foods. Fruits and vegetables once available only in season were now in produce departments year-round. Laced with pesticides, treated with fumigants and preservatives, shipped in from points across the globe, and stacked for months in refrigerated warehouses, such produce lost its nutritional value. All this was a consequence of Big Ag conglomerates that for years had been systematically consolidating operations, until they controlled every step of the global food chain, from growing to marketing to retailing. The bland but blemish-free produce they sold was lucrative, partly because it could be priced lower than anything a small-scale local grower could offer. But it no longer had any flavor, and any health benefits were lost. "If you wanted a peach or a plum that tasted like anything," says Alice, "you pretty much had to grow it yourself."

Alice was keenly aware of the culinary wasteland that most Americans faced. "When I traveled," she remembered, "I had to bring food with me. My life support kit. A bottle of olive oil, a bottle of vinegar, a loaf of bread, a little bag of salad, and some cheese." But at this point, she still wasn't in a position to change the system, or to advocate on "real food's" behalf. "All I knew was that Chez Panisse could be better than it was."

She was beginning, however, to look beyond the usual purveyors. You still couldn't find really good bread in Berkeley or even San Francisco for that matter, but one of the busboys decided he was "going to keep making bread till he got it right," Alice recalls. "We're still serving his bread. We lent him money to start up Acme Bakery" (a wholesale bakery that supplies organic bread to restaurants and groceries in the Bay Area). Driving through the back roads of Marin and Sonoma, Alice found hippies who were raising goats and beginning to learn to make chèvre as delicate and delicious as any she had tasted in France. "They were doing it for themselves, without any thought to being commercial, so they didn't advertise," Alice said. "We were starting to reach outside our own little circle, telling them, 'You can do this.'"

Other changes were afoot too. When seven days a week proved too much, Sunday dinner was scotched. A casual upstairs café was added, serving light fare until midnight. On certain days, the dining room began to serve a modest lunch. And then breakfast reappeared, the offerings strictly French: croissants, café au lait in bowls, *pain au chocolat*. Worried prices had spiked beyond the reach of friends, in the spring of 1972, less than two years after opening, Alice introduced a more reasonable weeknight dinner menu, the tab $3.75 (roughly $16.00 in today's money) rather than the usual $4.75.

Alice was by nature egalitarian. For all her insistence on perfection, she wanted Chez Panisse to be affordable. She also wanted the food to be inventive. She encouraged innovation in the kitchen and was excited by new ideas. If a waiter or even a busboy wanted to try his hand at a dish, Alice was amenable. If it didn't pass muster—and Alice was a tough and exacting critic—it wasn't served. She wanted the Montessori ideal of "learning by doing" to touch every aspect of the restaurant, to be part of its DNA. "I've always believed you can't ask somebody to do a job when you don't know what's involved in it. Say you're asking somebody to wash dishes. You can't know how hard that is or what it's really worth, what people should be paid or how it should be set up, unless you experience it yourself."

Central to this ethos was collaboration, paying attention to everyone's contribution and truly honoring their input. Every afternoon, the staff held a meeting. If the halibut on the menu that night arrived and it didn't look fresh, everyone would brainstorm, tossing out ideas about what to substitute, and where to find it. A cook might suggest a variation on the sauce, different herbs to add as accompaniments. Someone else would point out that some beautiful baby eggplant had just been dropped off. Not enough for a full course, but perhaps they could be added to the zucchini. Then someone would mention that one of the neighbors might have Swiss chard in her garden. A quick call would be made and an hour later, a basket of just-cut chard would appear. "Everyone was switched on and engaged," remembers Kelsie Kerr, a former chef. And so the menu would evolve. Sometimes changes were made even in midevening, "if someone—most often Alice—had a better idea."

The same cooperative approach applied to the cooking. Unlike most high-end restaurants in America, the kitchen at Chez Panisse wasn't organized hierarchically. No one was pigeonholed as a prep cook or a saucier or a grill cook. Everyone could do everything, provided he or she possessed the skills. Similarly, there were no apprentices to sharpen knives, chop onions, sweep up the scraps; no lowly helpers to wash and dry lettuce, prep vegetables. This kept things fresh and interesting. No one had to repeat the same task night after night. Different cooks helped each other, certainly, but in general the person who did the fish was responsible for every step: filleting it, constructing the stock, chopping the herbs, mopping up afterward. It was the same with every dish: the cooks mostly decided among themselves who would do what, and then went about their work. (What this also meant was that there was only one person Alice needed to speak to if she found there was a problem.)

CHEZ PANISSE WAS FINDING ITS OWN RHYTHM. IT WAS BEGINNING TO embody a certain esprit. Everyone who worked there felt personally invested, part of its creation. The kitchen was producing simple delicious food, the service was cordial and exacting, the dining room was consistently warm. The staff felt increasingly familial, and this sense of belonging, of being part of a large and genial tribe, extended beyond the confines of the kitchen. The regular customers and even the suppliers felt it too. They were all a part of the Chez Panisse family, a large and generous circle with a shared ethos: doing one's best, nurturing community, caring about people and the land.

This belongingness didn't end with the workday. Most nights, after the customers left, the staff would hang out and talk, uncork some wine or pull out a joint, crank up the rock 'n' roll and dance. Sometimes, in true Pagnol-esque spirit, they even fell in love. "There was scant division between work life and social life," remembers Patty Curtan." We worked together into the night, and after hours we would eat and drink. On our days off we would gather at one another's apartments . . . dance to 'Bennie and the Jets.'"

In April 1972, barely seven months after opening, Chez Panisse received its first notice, a brief, lukewarm squib in a mimeographed San Francisco newsletter. "There are so many aspects of this new restaurant that are almost touchingly admirable," the writer began, also noting its "talented amateurism" and a pea soup that "had no pea taste."

It was followed a month later by a more flattering review, this one from *Jack Shelton's Private Guide to Restaurants*: "Right now in an unassuming, circa 1900 wood-frame house . . . an exciting experiment in restaurant dining is being carried out . . . Not always meeting with unqualified success, but never anything less than stimulating and often positively exhilarating," Shelton wrote. "Don't lose your marvelous aura of adventuresome experimenta-tion," he added. "Keep striving to improve . . . but don't change, whatever you do!"

But change, it seems, was already in the cards. Victoria Kroyer announced she was leaving. She was moving to Montreal with a new lover, leaving Alice to fill in the breach. "I never wanted to be chef," says Alice, "but there I was." Alice took up her post behind the stoves, producing a flurry of savory dishes, each night some-thing simple and new. But try as she might, she couldn't seem to rein in her prodigal spending, or her "come one come all" approach to staffing. Despite Gene Opton's sensible ministrations, the restaurant was still hemorrhaging money. Hired to put Chez Panisse's house in order, Opton had hit a brick wall. The staff continued to resist her suggestions, disdainful of all talk of money. As Jerry Budrick put it, "She wasn't Gallic. It was a French restaurant, and she was bringing influences that were more of a stern nature, Germanic stuff." Gallic or not, the chemistry wasn't working. Everyone, including Opton herself, agreed it was time for her to go.

By October 1972, Victoria Kroyer was back, her love affair having ended badly. Opton was poised to step down, agreeing to a long-term note to cover the cash she'd invested. Tom Guernsey, gay, well liked, and inspiring to all, was beginning to take over her duties. Life inside the kitchen seemed to be settling in.

And then Victoria Kroyer quit again, this time for good. The restaurant needed a chef and this time Alice wasn't interested. She had decided she was happiest overseeing the food, but not cooking it. She wanted to be in the dining room, fussing over the patrons, drawing inspiration from their responses, and gauging their pleasure.

Chez Panisse had by now become a "destination" in Berkeley, a clubhouse to the film crowd, thanks to Luddy, and as such, a people-watcher's paradise. But it was still uneven, a good place to eat, certainly, but not yet a great one. As the wine writer Robert Finigan put it, it was still "a beef-stew-and-fruit-tart bistro for students and junior faculty."

And then, in February 1973, Jeremiah Tower appeared.

Jeremiah Tower was a thirty-year-old "roué" with two Harvard degrees, a thin work résumé, and an excess of attitude. Rakishly handsome, with chiseled features and a mop of strawberry-blond hair, he had spent most of his life in Australia and England, living in coddled luxury and roaming the world with his parents, wealthy socialites who traveled by ocean liner and stayed in grand hotels. Left to himself in these temples of privilege, Jeremiah had learned early to appreciate the finer points of living: "Escoffier-style *grande cuisine*," expensive wines, the delicacies that could be had in the food halls of Harrods and Fortnum & Mason in London, where he regularly splurged.

Already a culinary boy wonder by the time he got to college (he claimed to have begun cooking at the age of five), Jeremiah liked to say he had found the teachers he needed as he needed them. From his bluestocking Philadelphia aunt, he had learned to love art; from an Australian Aborigine, how "to roast barracuda and wild parrots" on a beach; from an elegant English lesbian, how "to smoke a cigarette in an ivory holder" and enjoy gin; from six years of English boarding school, how to "love boys" and loathe bad food; from his sweet but alcoholic mother, how to cook like an angel.

At Harvard he was known as a gadabout, famous among his louche friends for the phenomenal meals he whipped up, and the even more phenomenal Château d'Yquem (the most hallowed of Sauternes) he served with them—all courtesy of his father's bank account, from which he liberally drew.

His first job after leaving Harvard had been as chef of a pub in Surrey, near his family's former English home. But the shepherd's pie crowd wasn't keen on his haute French cuisine; he was quickly sent packing. Though gay, he lived for a time with a girlfriend. On a visit to her family's private island in Maine, he read Euell Gibbons's *Stalking the Wild Asparagus*. Inspired, he picked mussels in the shallow coves, wild greens from the meadows, newly appreciative of the virtues of fresh, natural ingredients. It was a truth he'd known but temporarily forgotten, having spent many happy hours helping his mother harvest vegetables from her garden in the English countryside, which she'd loved almost as much as her Jean Patou suits and Cartier jewels. He returned to Harvard in 1967, to study architecture at the Graduate School of Design (it was eight years after Jacobs's revolutionary talk). Unable to convince anyone to build his designs (his specialty was outrageous underground habitats), he drifted for a time, living by his wits and making the rounds of the party circuit.

But by January 1973, he was down to his last twenty-five dollars. He'd been crashing in the Bay Area, on the couch of some Harvard friends, one of them the poet Michael Palmer, making pennies working part-time as a gardener. He was tired of drifting, relying on his pretty-boy looks and the kindness of friends. He needed a job. Palmer, by chance, had seen the want ad Alice placed in the *San Francisco Chronicle* and, aware of his friend's prodigious cooking skills, gave him directions to the restaurant. Jeremiah took him up on the tip.

Alice had never encountered anyone quite like Jeremiah Tower. She was dazzled by his roguish good looks, his bad-boy charm, his edge of decadence. But most of all, she was dazzled by his cooking. Jeremiah knew food and was a wizard in the kitchen. "I, of course, immediately fell madly in love with him," Alice remembers. "Yes, he was gay, but that didn't ever stop me from trying. He was incredibly handsome, and he had taste. I was in love with the way he thought about food, the way he handled food, the intellectual approach he had, and the guts."

The two were natural collaborators, and soon thick as thieves, mutually driven by a seemingly unslakable passion for food and great cooking. Like Alice, Jeremiah wanted only the freshest ingredients: live fish, wild fennel, little baby lambs. Whatever Jeremiah dreamed up, Alice would labor to find it. "He was a perfectionist, and so was I, and that's why it worked."

Initially, Jeremiah didn't range too far from Alice's simple bistro menu, producing the requisite pot-au-feu, leek and onion tart, sole meunière. But after a few months, secure now in the job, he began to strut his stuff, aware that his knowledge and culinary skills far surpassed anyone else's there—Alice's included. From the kitchen now came ever more elaborate dishes, all classic French: *crème gratin* and *quenelles à la Lyonnaise, la brioche de ris de veau* (sweetbreads in brioche pastry with Champagne sauce) and *truite jurascienne* (trout cooked in red wine with hollandaise sauce and buttered croutons), chicken stuffed with sausage and roasted with chestnuts, salad of dandelion with bacon. Though it wasn't the rustic provincial fare Alice had always envisioned, the customers seemed to like it. And Alice, at that point, was still in a swoon.

Other changes ensued too. Opera replaced Led Zeppelin in the kitchen. The dining room was dressed up, from oilcloth to white linen. New and even more ambitious dishes appeared.

At first the bearded, "Brillo-haired" prep cook, Willy Bishop, was standoffish, certain that anyone who looked and acted like Jeremiah "could only be an asshole." But even Willy was in due time seduced, won over by Jeremiah's cooking chops. Mordant-witted, emphatically blue collar, a sometime drummer and painter, Willy proved a perfect foil to Jeremiah's peacock strut. When Jeremiah's airs got "too Harvard," Willy, who suffered no fools, was always ready with a raunchy quip.

Jeremiah and Willy prepped and cooked almost everything that left Chez Panisse's kitchen that first year of Jeremiah's term. "Generally, it worked," says Bishop. "And it worked because this guy, Jeremiah, was so manic and insane. If he was gonna make bouillabaisse, he'd go to Chinatown and come back with, like, a six-foot conger eel: 'Look what I got in Chinatown!' I was like, 'What the fuck is it?'"

What was most amazing about their collaboration, says Bishop, is that Jeremiah "wrote out these elaborate, themed menus, a different one for each night of the week, and sent them off to Goines to be rendered in calligraphy, printed and posted for public viewing a few days before they'd be served . . . *without ever having cooked any of the dishes described.*" Which meant that every night was a high-wire act, with no dress rehearsal and no safety net below, absolutely no margin for failure. And the astonishing part was, they were actually pulling it off. Less than two years old, and still running in the red, amateurish on so many levels, Chez Panisse was becoming an important place culinarily, in spite of everything.

"Jeremiah was Escoffier," says Alice, "with the whole extravagant, decadent thing. We used to go out after work and have Champagne and caviar, and he'd order the best, spend all our money."

Wanting the best, of course, also meant ranging farther afield in their search for better, fresher ingredients. Bouncing around in her beat-up Dodge Dart, Alice and Jeremiah were perpetually on the prowl, continuously combing the far reaches of San Francisco Bay. They bought olive oil and anchovies from the Italian delis of North Beach, fresh sardines from purveyors in Monterey Bay, tasty but non-commercial fish species from local fishermen they befriended. They dragooned friends into foraging for wild mushrooms from Mount Tamalpais, huckleberries from Point Reyes, "a better egg from some old farmer's lost race of chickens."

For a time there seemed to be no limits to their collaboration. "Sometimes I'd bring in something that I'd tasted, or some recipe I'd found, and Jeremiah and I would work through it, making it better, trying this and that accompaniment, till we were both satisfied. And we were both not easily satisfied. It was important for the sort of spiritual life of the restaurant that we be a little daring, that we not get too set in a path."

This meant continually expanding their repertoire, adding an extra layer of drama to the experience: more virtuoso cooking, more unusual menus. For one, they roasted a wild boar outside, on the sidewalk, in front of the restaurant. It weighed four hundred pounds and was turned on a spit. Jeremiah had some friends from Big Sur, Alice remembered, "and someone had hunted this huge object. (I think it was about four hundred pounds, and somehow we spitted this thing.) And then it started to rain, so we rigged up a tent. The spit was turned by hand. It did draw quite a crowd."

They embarked on a series of regional French dinners, paying homage to the specialties of Brittany, Alsace, Languedoc, Champagne. This was followed by a second cycle celebrating the great chefs of France—Escoffier, Urbain Dubois, Marie-Antoine Carême, Prosper Montagné, Henri-Paul Pellaprat. They added theme dinners: a hundredth birthday dinner in honor of Gertrude Stein, with text and recipes from *The Alice B. Toklas Cookbook*; a British dinner, to commemorate the Battle of Waterloo; a week of "Salvador Dali–inspired" offerings, the climax being the night they served "*l'entre-plat drogué et sodomisé*," a leg of lamb "drugged and sodomized" with a mixture of Madeira, brandy, and tangerine juice injected by syringe.

The festival dinners were beginning to attract press attention. Herb Caen, the respected *San Francisco Chronicle* columnist, wrote up the Stein-Toklas dinner. Someone brought in the celebrated James Beard, who described Chez Panisse in his syndicated column as a "fascinating" new place to watch. The clientele, in turn, was growing more chic. "They were all so glamorous," Ruth Reichl remembers of the scene. "There were all these really sexy-looking people at the restaurant, including the staff. In many ways, Jeremiah and Jerry [Budrick] were very similar—you know, bad boys."

Jeremiah was indeed every bit the bad boy. If there was an air of casual hedonism at Chez Panisse before his arrival, by the time he'd made the kitchen his own, the scene had grown considerably more sybaritic. But now there were new heights to scale. The write-ups had raised the bar on what the kitchen produced and everyone, including Alice, was feeling the pressure. Jeremiah was putting in long hours, often pushing himself to the brink of collapse, and so too was the staff. In response to the stress, the kitchen was growing more decadent by the day. "There were magnums of Sauternes and Champagne all the time," remembers one of the waiters, Bill Staggs. "[Jeremiah] used to keep these nitrous oxide canisters around for whipping cream. The waiters would take a hit of nitrous oxide before delivering the entrees to the dining room."

There was also cocaine. Jeremiah had long since grown accustomed to his morning hangovers. But the enervation he now felt was new: it was getting in the way of his cooking. And then one day, as he tells it, a friend sauntered into the kitchen with an antidote: "a big bag of blow" that he proceeded to cut into neat little lines. With a few quick snorts, it was a new world. Reinvigorated, Jeremiah could return to his station at the stoves and put in long, late hours, and afterward, still go out drinking until dawn. Others were soon indulging in the cocaine fest too. "I think the exuberance overtook everyone," recalls Willy Bishop, "and with coke you were able to do things you couldn't really do. Nothing was daunting."

Friends say Alice looked down on the drug scene. But she wasn't above the ongoing bacchanal that in those days was the kitchen: the magnums of Champagne, the late-night dancing, the merry-go-round of romantic couplings and uncouplings, the casual sex between friends. By now Alice was "semiofficially" with Jerry Budrick, though she wasn't living with him. As Jeremiah later wrote in his memoir, "Drugs were easier to organize than sex, unless it was casual, which usually meant with one another. Who else would put up with us?"

And then came the review that put Chez Panisse on the map: a rave that ran in the October 1975 issue of *Gourmet* by Caroline Bates, the magazine's West Coast correspondent. "Chez Panisse is joyously exploring *la vraie cuisine francaise* in all its vigor, freshness and variety and ignoring those French dishes that turn up elsewhere with such monotonous regularity," Bates wrote.

The review was a turning point, vaulting Chez Panisse to national fame almost overnight. Though Alice should have felt elated, in truth she felt "mostly dread," she remembered. "Everybody and his mother were going to want to come in and see what all the fuss was about . . . One of my dearest friends sent us a funeral wreath."

And Alice was correct. Everyone *did* want to come in: a new breed of "doubting, demanding" customers, many of them from out of town. Suddenly the phones were ringing off the hook. Bookings were coming in faster than they could take them. The kitchen was swamped. A portion of these bookings were now "no-shows," however, people too inconsiderate to cancel when they weren't coming. For the first time ever, tables at Chez Panisse were going empty. Despite what could only be described as a "dream" review, Chez Panisse was doing less well than before the glowing *Gourmet* write-up.

Alice was feeling rattled and worn thin. Her dear friends Claude and Martine had moved back to Vence by now. In the summer of 1975, in need of recharging, she decided to join them in a rented farmhouse in Provence. For four glorious weeks, they cooked and sat on pergolaed terraces, ate in local restaurants and dined with mutual friends, among them a French couple named Lulu and Lucien Peyraud, who owned the Domaine Tempier, a winery just outside the little port town of Bandol. "I felt like I'd walked straight into a Marcel Pagnol movie," Alice remembers. "That was when I really started to get it about French food. I realized how relentless and how careful that scrutiny was in the best places—and I don't mean necessarily the most expensive places. It was about caring."

The Peyrauds were producing some of the best wine in southern France and Lulu was also a superb cook. "Lucien became my surrogate father," Alice would later write, Lulu "my muse." Alice fell in love with Lulu's farmhouse kitchen and her robust, elemental style of cooking. Lulu did almost everything on a wood-fired hearth. The simple savory dishes she prepared embodied everything Alice admired about French country cooking: its earnest aromatic flavors, its earthy character, its emphasis on simple fresh ingredients. It was a style of cooking that Chez Panisse had moved away from, an integrity of taste built upon ingredients more than technique—a simplicity, Alice realized, to which she longed to return.

Alice was not alone in feeling that Chez Panisse had lost its way. Despite admiring words from reviewers, many of the old customers were falling away, put off by what Tom Luddy called the "overripe, overrich, decadent" Parisian food Jeremiah was producing. Chez Panisse was no longer the simple, convivial little neighborhood bistro Alice had originally envisioned. "I always felt that the food Jeremiah cooked went against the philosophy of Chez Panisse and Pagnol," says Luddy. "If you read Jeremiah's menus from that time, they're incredibly pretentious." Other old friends went even further. Greil Marcus, who had been a partner from its inception, claimed Chez Panisse had become "a very closed place, a private club within a public restaurant. I couldn't believe it." He and his wife no longer wanted to be there. "And when we *were* there, we felt that we didn't belong."

The staff was divided. Some, including the artist Patty Curtan, who worked at Chez Panisse throughout those years, bristled at Jeremiah's insistence that the restaurant's culinary excellence was entirely his creation. In *California Dish*, his chatty memoir, Jeremiah "took a story that was rich and nuanced and interesting, and he reduced it down to just this one version of himself at the center," Curtan said. "He portrayed himself as this incredibly hardworking genius who worked twelve, eighteen hours a day, and that wasn't true either. There were so many people working really hard."

Other staffers disagree, arguing it was Jeremiah who transformed Chez Panisse from a good restaurant into a truly great one culinarily. "Anybody who says that Jeremiah made it too fancy is whistling in the wind to me," says Budrick, who was headwaiter at the time. "All I knew is we sold out every dinner. The more elegant Jeremiah made those dinners, the more the public responded. And it wasn't just the snobby elite." The local Berkeley crowd came too.

Ruth Reichl is more circumspect. "Jeremiah had a much more sophisticated vision of what food should be," says Reichl. "And Alice was entranced by that. But she was influenced by other people too. He was never in the Chez Panisse aesthetic. He was about caviar and lobster, while she was about the perfect, freshly picked salad greens. It was a time when everybody was learning from everybody else. But I think that the uncompromising quality, the purity of the ingredients, is her. Jeremiah was extremely talented, but he didn't bring that to the table."

Truth, of course, is a scratched lens, rarely absolute or unassailable. What *is* certain is that by the winter of 1976, despite Alice's abiding admiration for Jeremiah's culinary genius, the two were no longer perfectly attuned. Their relationship was growing tortured and they were agreeing on less and less.

That winter, Jeremiah took a leave of absence, traveling to France. When he returned two months later, he was flush with new ideas. He wanted to start offering not one, but four set menus each night, he said. And he wanted the café upstairs to have its own separate menu, to serve dishes that were simpler and less expensive than the food downstairs, still classically French, but with more choices. Alice resisted. She worried that the restaurant's focus would be "diffused," that Chez Panisse's special identity would be lost. Finally, she turned down his proposals. Jeremiah announced that he would take his leave at the end of that year.

The culinary fireworks continued apace, even so. The knowledge of his imminent departure seemed to stoke Jeremiah's imagination, pushing him to greater heights. The restaurant was now consistently busy and the food more innovative than ever. But the mood in the kitchen had gone sour. Many of the patrons were now tourists, some of them extremely difficult, a portion of them downright surly. The easy familial spirit that had always characterized

the place, the warmth between the staff and the diners, was eroding. "These new customers, many were so rude," said Alice. And of
course those that didn't show, "that was costing us money." What
was hardest was "the wasted food. Seeing our beautiful food go
into the garbage! And empty tables—we weren't used to that at all.
There's a particular feeling when everybody's in there together, and
a big empty table absolutely spoils that."

Alice was feeling burned out: she was tired of the new clientele,
the petty squabbles, the low morale among the staff. Though Chez
Panisse was not yet five years old, she was ready to give it up. In
May 1976, an article in the *San Francisco Chronicle* reported that
Chez Panisse was for sale.

JUNE PASSED AND THEN JULY, AND THERE WERE STILL NO BUYOUT OFFERS.
But by then the mood had lightened. That July, it was a special
dinner in honor of Bastille Day that would lift Chez Panisse from
its funk, giving it a second life. The evening, July 14, 1976, was a
wild success. The garlic harvest in California always seemed to fall
close to Bastille Day, so a friend had proposed that every course
include garlic, an idea so popular that everyone present that night
insisted it be repeated. (Dessert was fresh figs, white cheese, and
garlic honey.) But more importantly, the dinner signaled a shift
in emphasis. Alice and Jeremiah had been trying for years to find
ingredients that approximated unobtainable French ones, sometimes ranging far afield to find them. For the first time, Jeremiah
wrote, it dawned on them that it would be just as interesting "to
use American ingredients for themselves," rather than as proxies
for foods they couldn't find. Though they didn't abandon their
French cookbooks, from that summer forward, they began looking for inspiration closer to home: drawing directly from the exquisite fresh ingredients of Northern California, inventing dishes
built expressly around them.

"We were doing some of the simplest food we'd ever done, and
I loved it," Alice says. "And the customers loved it . . . We had . . .

fresh pea soup. Steamed clams . . . avocado with walnut oil and lemon juice."

Alice and Jeremiah were also discovering better and better local sources. In just a few short years, the quality and variety of ingredients they could find had markedly improved: "We were really foraging and finding the most wonderful things," Alice remembered. Crayfish raised in California. Wild salmon that was caught just outside the Golden Gate. "People gathered wild mushrooms for us, and we just grilled them wrapped in grape leaves—Napa grape leaves. We were getting our oysters from Tomales Bay [just north of San Francisco]. More and more of what came out of that kitchen was about the fruits and vegetables that grew within just a few miles of the restaurant. Avocados, peaches, watercress, peppers, beans. We got rabbits from Amador County, squab from Sonoma, live trout from Big Sur. This was when the farmers' markets were springing up all over the place."

Beginning that summer of 1976, farmers' markets would become essential not only to the food at Chez Panisse, but also to its larger mission. "Forging a connection between cooks and farmers, between city and country, now it's a link that seems so logical," says Alice. "Now we take it for granted. But in the 1950s and '60s, with specialization, with so many middlemen, with the industrialization of agriculture, most people were cut off from the producers. I mean there were people who dropped out and had farms and lived on them, people who lived on communes. But it wasn't in the city, in Berkeley. It wasn't in most places."

Word, however, was getting around. Friends had long shared the bounty of their gardens with the chefs at Chez Panisse. But now it was small farmers too. Producers were beginning to show up at the kitchen door with boxes of lettuces, bulbs of fennel, perfect peaches. "People knew that we were looking for certain things," says Alice. "One of them happened to be radishes, another was haricots verts. A friend sent us a box of tiny, real string beans from the Chino ranch, where everything is done organically, and when I tasted them, I said, 'Oh my god, we have to have them every week.

We'd love to buy those.' They had ten varieties of beans alone, and they were all raised so beautifully."

Alice was beginning to see a second important connection: the farmers who were "obsessive" about growing the best-tasting produce were also concerned about the "health of the soil," the well-being of the beneficial insects, the purity of the water running off their land. They were interested in bringing back older varieties that were harder to cultivate, but much tastier, "and which brought a sense of continuity with the past to both their fields and our tables." What they all had in common, she saw, was they were doing things organically, meaning raised without agricultural chemicals. This revelation would change everything.

Alice hadn't been looking for organic per se. It was *taste* she was after, she insists. "I had associated organic with the health food stores of the sixties," she explains. "The vegetables always looked overgrown and bruised. And the variety was always limited. I wanted to have nothing to do with that, especially when I got back from France. Nothing to do with that kind of bread—so heavy, dry. The smell of those places always seemed a little medicinal. It put you off. The people who were making that bread, who were raising all that stuff, hadn't studied it. They hadn't refined it. And then we began to meet the local farmers. Friends of friends would say, 'I know someone who grows this. I know someone who grows that.' It was this search that ultimately led us to the doors of the organic farmers and producers. What they were raising was so beautiful, and had so much *more* flavor."

Alice had discovered this seemingly simple but essential connection not through ideology, but through direct, sensuous, first-hand experience. The integrity of a food's taste was inextricably bound to the care and nurturance of the soil in which it was grown. "It was a consistent pattern," she says. "When we looked for the freshest and best-tasting ingredients, we found that the people who produced them were always the most environmentally responsible." Alice, like Carson, Jacobs, and Goodall, was seeing her passions in the context of the earth's larger life systems; she was seeing the interconnectedness of the living world. The search for good ingredients was "pointless," as she would later write, "without a healthy agriculture and a healthy environment." It was the beginning of a new, more politically engaged chapter for Chez Panisse.

IN OCTOBER 1976, HAVING FEATURED NEARLY EVERY REGION IN FRANCE, Alice and Jeremiah did their first Northern California regional dinner. On the menu that evening were dishes focused exclusively on the bounty of Northern California: "cream of fresh corn soup, Mendocino style, with crayfish butter"; "Spenger's Tomales Bay bluepoint oysters on ice"; "Big Sur Garrapata Creek smoked trout steamed over California bay leaves"; "Monterey Bay prawns sautéed with garlic, parsley, and butter"; "Preserved California geese from Sebastopol." Not only were all the ingredients fresh and local, their provenance was noted on the menu, a practice common today but at the time completely novel.

That January 1977, Jeremiah hung up his apron and left Chez Panisse. Though Alice was now at the helm, her new sous-chef, Jean-Pierre Moullé, was a superlative cook, capable of producing anything she imagined. They cooked mostly as a team, with the addition of a third new hire, Mark Miller (who would later go on to fame with the Coyote Café in Santa Fe, New Mexico). Alice was full of new ideas. Though the restaurant continued to do special dinners—another Gertrude Stein birthday party, a Richard Olney menu, a suckling pig roast for Jerry Budrick's birthday—they were more occasional now. Alice had begun to refocus her energies. She was thinking about the restaurant in new ways. She had always been driven to find fresh local food. Now she wanted to find food grown without chemicals, in naturally enriched soil, using the core principles of sustainable agriculture—a much greater challenge. She wanted the food she served to be healthy for both people and the land.

Jerry Budrick, with whom Alice was now semiofficially living, had purchased 160 acres in the foothills of the Sierra Nevada. He and Alice envisioned a one-acre garden to grow the hot-weather crops that the farms of Marin and Sonoma Counties couldn't support. They wanted it to be completely organic. "My idea of organic was to grow everything so, so carefully," Alice says, "pulling off the bugs by hand if you had to, and bringing the food to the restaurant so fresh it was really still alive. That's what distinguishes a great salad from a good one—a great one is alive." But the garden was a bust. "We bought the seeds, but we didn't really know how to do it," Alice remembers. "And there were gophers and they ate everything."

That July, Chez Panisse did another Bastille dinner, which morphed into a weeklong festival of garlic-infused dinners. It was a big, delirious success, kicking off what would become an annual tradition. Afterward, Alice decided she needed a break, and so did everyone on the staff. At the end of August, a day before Chez Panisse's sixth birthday, Alice closed the restaurant for a month and bought a ticket to France. "I went to see Claude and Martine Labro in Vence," she remembers. "And then all of us descended on the Domaine Tempier," where Lulu and Lucien Peyraud were living. Later, Alice would remember it as "one endless meal, outside, under the arbor." "I was so happy to be with those friends. They seem to know so much, just instinctively, about how to live, how to be happy."

Alice returned in October and by January 1978, a massive wood-burning oven and grill had been added to the kitchen, an inspiration from Lulu Peyraud. Chez Panisse was still just getting by financially, but otherwise it was thriving. It was always full, always bright and bustling with people. Long-standing friends were returning, and so too was the old Chez Panisse spirit of generosity and community ties, the sense that it was a large and supportive family.

"Alice is a very loyal person," says Barbara Carlitz. "She doesn't leave people behind. She continues to gather people around, but not at the expense of old acquaintances or old friends . . . She has absolute conviction, mixed with timidity, which is a beguiling style." Though Alice and Tom Luddy were long separated, he continued to bring in friends from the film world—George Lucas, Roberto Rossellini, Satyajit Ray, Akira Kurosawa, Jean-Luc Godard—and would do so regularly, down through the years.

In 1972, to commemorate Chez Panisse's first birthday, David Goines had designed an elegant, art-deco-inspired poster, the first of what would soon be a restaurant tradition.

"In those days," remembers Goines, "there was a very strong sense of community and involvement. People would come by and shell peas and talk." When it was decided that the restaurant needed a wine cellar, "I dug it with my own two hands. When garlic week came around, people would come by and peel garlic and drink wine. They'd be there for three or four days, peeling garlic and talking."

Judy Rodgers, who would later achieve acclaim in her own right as the chef and co-owner of Zuni Café in San Francisco, remembers feeling this inclusive spirit during a dinner in celebration of her twenty-first birthday. Rodgers, at the time an art history major at Stanford, had spent her final year in high school living with Jean Troisgros in France, passing thrilling days in the Troisgros restaurant's kitchen, absorbing every culinary tip she could.

"I showed Alice my Troisgros recipes that night," Rodgers recalls. "We only talked a little, but I told her that I was bereft, ever since leaving Troisgros. I said, 'I'd love to just come and occasionally spend a day in your kitchen, just watching.' Alice was keen on that and said yes. We kind of negotiated—no, we didn't negotiate, it wasn't a negotiation. You don't negotiate with Alice. Nothing happens like that." It's always a kind of courtship.

Alice proposed that Rodgers come every Saturday morning to help with Chez Panisse's lunch. Rodgers was thrilled; and to her amazement, within weeks she was doing most of the lunches herself. "Alice trusted me to do it on my own," Rodgers remembers. "That's one of her characteristics, to trust people. She's less worried if you don't have the competence if your heart is in the right place and you're shooting for the right aesthetic.

"I don't think anybody would pretend that everything produced by that restaurant was stellar, three-star cuisine every day," Rodgers adds. "What people loved about Chez Panisse was the generosity, the spirit."

But Alice was still searching. She had decided she wanted to return to France, to go for four months this time. "I wanted to taste, travel, take it all in, and just be away," she remembered. In Alice's absence, Jeremiah returned to Chez Panisse. The two had remained cordial and Alice wanted someone she trusted to oversee the kitchen and the staff. For two weeks, Alice and several friends toured the great restaurants of Europe, sometimes going to two a day. And then in October, Jerry Budrick and two other friends joined her in Rome, where they rented a car and set off through Italy. It was there that Alice would find the special *something* she was searching for.

They had stopped in Turin, with plans for a late dinner at an upscale restaurant. But they were already starved and it was only midafternoon. They ducked into the first open place they spotted, a pizzeria. "We could see the fire burning inside, and it pulled us in," Alice remembered. "And there I had my first pizza out of a wood-burning oven. We all thought it was the best thing we had eaten on the whole trip."

Alice asked to see how it was done. She was led to a deep oven with a firebrick floor and a heap of glowing, "white-hot" coals piled in the corner. The temperature inside was 1,000 degrees Fahrenheit. The pizza went in on a long wooden paddle, where it got a quick blast of heat from the smoldering coals, "with flames swirling around under the ceiling of the oven." It was ready in less than two minutes, Alice remembered. That was the moment of her epiphany.

An open kitchen with a pizza oven in the upstairs café: this was what Chez Panisse needed. It would mean the return of neighborhood locals who could no longer pay Chez Panisse's prices. The ingredients would be just as fresh as those downstairs, the cooking just as creative and surprising, but there would be "no truffles, no foie gras, no classic French sauces." And no obligatory three-course meal. The menu, as always, would change daily, according to what they found in the market. But people could order as much or as little as they wanted. They could linger with friends, stay as long as they liked. It would be the quintessence of all she had originally envisioned—a version of Chez Panisse that had never really been tried.

Alice returned from her gastronomic tour of Europe more energized than ever. From Chez Panisse's kitchen now came a blitz of new dishes inspired by her travels. And once again the critics were in a swoon:

"For eight years now, Chez Panisse in Berkeley has been turning out a different menu each night," wrote Patricia Unterman in the *San Francisco Chronicle*. "That's in the neighborhood of 3,000 new dishes, and they keep doing it. What is so remarkable is that the food is always exciting, very rarely poor, occasionally merely good, and usually superb."

Lois Dwan of the *Los Angeles Times* was equally doting. "In some eight years, Alice Waters has made Chez Panisse a place of pilgrimage for all serious gastronomes," she pronounced.

"If you could eat in only one restaurant for the rest of your life, Chez Panisse would be the one to choose," advised Colman Andrews in *New West* magazine.

Mesclun lettuce was at the time unknown in America. Alice had brought back seeds from Provence and many of her friends' backyard gardens were now planted with frisée, mizuna, mâche, and fragrant baby arugula. Nearly every review made note of the delicate new lettuces, as well as the nuggets of warm goat cheese Alice was adding to the little plate of greens. No one had ever combined baked goat cheese with salad before, and it was creating a culinary stir.

Chez Panisse's initial joie de vivre had returned—in the kitchen, but also in the dining room. On one occasion, at the screening of Errol Morris's new film *Gates of Heaven*, a vow was made that would enter cinematic history and soon became Chez Panisse legend. For years, the director Werner Herzog had kidded Morris, claiming the film would never be finished. If it ever opened, Herzog told his friend, he'd eat one of the shoes he was wearing at that moment. Now Morris was holding Herzog to his word.

"I remember distinctly the cooking process," Alice recalled, "and the crazy idea I had—to put it in fat, cook it like confit, because I thought it would soften it up. I cooked it, and I cooked it, and I cooked it, probably eight hours, and it never happened. That shoe was something formidable. It wasn't some kind of little Italian loafer. It was a serious shoe. I remember watching him bite into it. He chewed on it for a long time."

THE REIMAGINED CAFÉ OPENED IN APRIL 1980 (THE UPSTAIRS HAVING BEEN closed for construction during the first half of that year). To pay for the renovation, the Chez Panisse corporation had secured its first bank loan of $50,000, and Alice's father had taken out an additional personal loan as well, which he lent to the restaurant. Chez Panisse was now $88,000 in debt. But Alice was convinced that the refurbished café would do well enough that they would soon make up the shortfall. And she was right. It was a roaring success, in both culinary and financial terms, drawing still more press attention. Alice was now being hailed as the mother of a new wave of cooking that was "a marriage of many and a mime of none," as Patricia Wells put it in a 1981 article in the *International Herald Tribune*, a style characterized by "a sense of experimentation, plus an uncompromising concern for good food and good dining that seems to have been lost in much of America, where fast food, fake food, and fern bar spinach salads are about as haute as many menus get."

And then one night calamity struck.

On March 7, 1982, fire broke out at Chez Panisse. Alice got the call at 4 A.M. and rushed from her house. When she reached Shattuck Avenue, she saw flames licking from the upstairs windows, fire trucks, arcs of water. Black smoke billowed from every aperture in the basement. She knew at once how serious it was. She felt, she remembered, as if she couldn't breathe.

"I remember very vividly going in the next day, when they had put out the fire, and the whole downstairs was dripping with water and completely charred. Everything."

The fire, it was theorized, had been started by a live coal that was put back into the wood box. "I was on the grill that night, so who knows. I might have done it myself," says Alice.

"Being in a fire is something . . . You can't get the smell out, you can't get the water out. You feel like you're never going to put it back together."

Alice was devastated. Some of the upstairs was saved, but the downstairs was gone. Everything would have to be rebuilt, even the floors. Ten more minutes, and the whole place would have burned. The fire was almost at the main beams, and when those go, the building collapses. They'd caught it just in time.

The fire was a turning point, Alice would say, "in that I kind of thought that restaurant was mine, you know. It wasn't mine. It was the people who came to eat there—it was theirs. They were part of it. I hadn't been paying enough attention, I think. I was in a little narrow, inner-circle Chez Panisse world, thinking we could just do it all on our own. We're going to grow our own, do our own, be our own." But after the fire she realized "our responsibility reached outside our doors."

Crowds of disbelieving, despairing onlookers thronged Shattuck Avenue the next morning, staring at the smoldering ruins. Alice, who at 7 A.M. had stood in tears at a friend's door, had already absorbed the shock and moved on. There was too much to do to look back. A lot of the food and wine untouched by the fire was still good, though it couldn't be legally sold. Alice gave it all to the staff. Full wages would also be paid during the restaurant's reconstruction, she insisted. Even to the dishwashers. And she wanted the place up and running again quickly.

Alice read *A Pattern Language* by Christopher Alexander soon after the fire. Alexander espoused a participatory approach to designing buildings and communities, in which the residents were active agents in their design, massing, and construction—surely a

Jane Jacobs idea if ever there was one. "I realized there was so much we could do," Alice remembered. "That book was my bible." Before the fire, a single narrow door had separated the kitchen from the dining room. Now both were gone and Alice saw that she liked it better that way. "I had always wanted to be connected, cooking and serving the customer in one room. So when that wall went down in the fire, I just said, 'Let it be.'"

Construction began at once. Friends appeared and lugged out rubble; strangers materialized from out of nowhere, joining the work crew. Lunch buffets popped up on the sidewalk to feed the volunteers. With construction in progress anyway, it made sense to cut in additional windows in the upstairs café. Alice and Jerry Budrick wanted to build a new wood-burning oven off the downstairs dining room. In the collectivist spirit that was Berkeley, a neighboring restaurant organized a benefit party to raise money for an employees' fund as well as the oven.

When the café reopened in March, "it felt completely different," Alice remembered. The kitchen was now white and gleaming, an integral part of the dining room, rather than a place that was separate and subordinate, sequestered. "To see outside, to see to the front of the restaurant. To see what was going on in the dining room and anticipate . . . It was fantastic." It was just the feeling of connectedness Alice had dreamed of, a physical expression of the inclusiveness at the heart of the Chez Panisse ethos.

ALICE WAS THIRTY-EIGHT NOW, BEGINNING TO FEEL HER AGE. FOR TEN years she had been wed to the restaurant, her sense of well-being rising and falling with its oscillating fortunes. Chez Panisse had become "her life, her family, her very identity." She wondered if it was enough, if there were other things she should be reaching for. Mindful that the years were passing, she knew there were certain decisions that needed to be settled. "I was in my late thirties, and I was very aware that I would either have to have a kid or I wouldn't have one. Really aware of that, and I think the tenth birthday of the restaurant and then the fire were the end of a certain chapter. That was a really difficult moment in my life."

Alice had yet to meet a man with whom she wanted to have a family. Nearly all of her lovers—and they'd been many—were in some way connected to the restaurant. Her latest had been one of the cooks, a smart, extremely sexy man fifteen years her junior. But he wasn't someone with whom she could imagine anything permanent.

One of Alice's friends, Patty Curtan, had mentioned there was someone that she wanted Alice to meet. He was an artist and he'd been renovating a loft to use as a studio. His name was Stephen and he was having a party to celebrate its completion. Alice couldn't recall ever having seen Stephen Singer. Apparently he'd come to the restaurant many times, but she didn't remember him. "And then," according to Alice, "we met that night, and we never parted."

It felt in some ways preordained, Alice would remember. There seemed to be no question. Stephen was slight, with dark thoughtful eyes and a lovely speaking voice. There was a playful quickness to his manner, an intellectual acuity, but also a gentleness, a softening around his eyes and smile. Stephen was twenty-seven and Alice had just turned thirty-eight, but in many ways he was the worldlier of the two. He was from a relatively well-to-do family and had traveled around Europe a lot. Like Alice, he loved and appreciated food, but it was wine that he really knew—art and wine. He was passionate about both.

Alice had always been wary of monogamy, but she sensed in Stephen someone who would make a wonderful father. "I had a lot of sort of desperate, difficult relationships right before Stephen," she told her biographer. "And they all ended in the same way. They began with this infatuation and this desire, and ended up not being satisfying in any other way. So I decided that desire wasn't at the top of the list, that I really wanted some other kind of relationship."

They moved in together immediately, first to the place where Alice was staying, then to Stephen's studio, finally to a house they bought together. Alice was pregnant six months later.

Fanny was born on the fifteenth of August 1983. She was early and small—colicky, which meant she couldn't sleep for more than an hour without waking in inconsolable pain. Neither Alice nor Stephen went to work. Neither slept. Neither cooked or ate especially well. For three months, they tended and fretted over Fanny, her comfort and happiness now their own. They had discovered that the only way to soothe her was through a kind of "sensory overload," Alice said. "We would turn on the shower in the bathroom, and I'd put her into her little Snugli, and I'd turn on the vacuum, and I'd sing to her all at the same time, and then she would go to sleep finally. All these things simultaneously. Which of course drove us over the edge too." Fanny cried for three straight months. "And then she stopped, magically." And became "the wonderful kid she is." But by then Alice had changed. She would never think about work in the same way.

Until the arrival of Fanny, McNamee writes, Alice's attention had been rooted mostly in the present. "[Her] focus had been largely on the moment, on pleasure for its own sake, on aesthetic refinement and sensual satisfaction." The search for better ingredients had led her to the doorsteps of the organic farmers, and to the recognition that good food was inextricably bound to the well-being of the land. Now, like Carson, Jacobs, and Goodall, she was seeing the world through a wider lens, using a longer time line. She was thinking "about the earth, about responsibility, about the future" and what she herself could do about it. She had begun to recognize that food had an ethical dimension. Alice was circling back to the political idealism of her first years in Berkeley. She wanted to reach out beyond Chez Panisse and engage the issues surrounding food and farming on a larger scale.

"My sense of the ethics and politics of food was coming to the surface," says Alice. "I mean, those values were instilled in me during the FSM, and in my early travels in France. But it didn't really start to come together till the early eighties . . . It had to do with becoming friends with the farmers and understanding deeply that the food at the restaurant was as good as it was because the produce and the ingredients were as good as they were. The farmers were the people who really got it, about the ethics of food."

"Getting it" meant seeing the connection between the health of the land and the health and well-being of the culture. It meant thinking in whole systems, rather than in parts, working toward a world in which Fanny and other children grew up understanding "what food meant to the survival of the planet."

Early in her pregnancy, Alice had hired a new chef, Paul Bertolli, to replace Jean-Pierre Moullé. Alice and Moullé had begun to quarrel, and Moullé, seeing the writing on the wall, had agreed to go.

Paul Bertolli was a native Northern Californian, though both his grandmothers were Italian born. He had studied music and composition at UC Berkeley but, after a trip to Italy, had fallen in love with food. Aware that opportunities as a concert pianist were few, in 1981 he'd decided to return to Italy, where he worked in three restaurants near Florence, and then as a private chef for Sir Harold Acton in his Villa La Pietra. The job would prove a dream. Beyond feeding Sir Harold and his guests, he had cooked for Acton's fifteen gardeners, the butler, the two full-time ironing women, and sundry others on the staff. His cooking skills now as fine-tuned as his piano, he had returned to Berkeley in the summer of 1982 and there wowed Alice with a meal of *vitello tonnato*, cooked lamb over fig branches, and a prune *semifreddo* with *nocino*. Alice hired him on the spot. Though his palate was more Italian than French, she recognized in Paul's cooking the simplicity and "clarity of flavor" she was after. "I loved that he had such an instinctual way of cooking. It was a beautiful thing to cross the border to Italy, into the land of olive oil and garlic and anchovies." And Paul got the importance of the farm-to-table connection.

IN AUGUST 1982, THREE MONTHS AFTER ALICE MET STEPHEN, RANDOM House published the *Chez Panisse Menu Cookbook*. Alice had been working on the project for two years and, like all her endeavors, it was a collaborative effort. David Goines had done the design; Patty Curtan had helped Alice assemble the recipes; others had tasted, tested, and refined them.

In her opening essay, Alice wrote charmingly about how she thought about cooking, beginning with her first cardinal rule: "It's a fundamental fact that no cook, however creative or capable, can produce a dish of a quality any higher than that of the raw ingredients," she said. She went on to describe her intuitive approach in the kitchen: "When I cook, I may pull a bunch of thyme from my pocket and lay it on the table; then I wander about the kitchen gathering up all the wonderfully fresh ingredients I can find. I look at each foodstuff carefully, examining it with a critical eye and concentrating in such a way that I begin to make associations . . . Sometimes I wander through the garden looking for something appealing, absorbing the bouquet of the earth and the scent of the fresh herbs."

Looking hard, trusting one's senses, letting the ingredients speak for themselves; paying attention to how elements combine and interact; allowing the meal in some senses to self-assemble—all notions strikingly similar to the way Carson, Jacobs, and Goodall approached their work. All were drawing upon deep reserves of empirical knowledge, intuition, and an unerring sense of the natural harmonies they uncovered through close observation. Gone were loyalties to received knowledge and preconceived attitudes. In their place was a fresh respect for seeing the world anew and entering a dance in which the physical and the firsthand were allowed to lead.

THE BOOK WAS A BIG SELLER, DRAWING LOTS OF GLOWING PRESS AT-tention. But its aftermath brought misunderstanding and bruised feelings. Early in the project, a casual friend, Linda Guenzel, had jumped in to help Alice with the writing, spending long hours transcribing their rambling conversations and then painstakingly rendering them into readable prose. But Alice hadn't been completely sure about the draft—she didn't feel it sounded like her—and without thinking, she had called in another friend to make revisions.

Linda, who felt she'd done all the heavy lifting, was wounded. Hardworking and hugely loyal, she expected to be named as the coauthor, which didn't happen. Though Alice gratefully acknowledged her tremendous input on the book's opening page, claiming, "This book is as much hers as mine," Linda felt betrayed. And she wasn't the only one. Jeremiah Tower was also angry. He felt Alice had taken credit for menus and innovations that were his. In Alice's defense, she did extend heartfelt thanks to both Linda and Jeremiah in her page of acknowledgments, crediting them for their contributions. But neither was able to forgive her. Going forward, Alice would never make that mistake again. She would name her collaborators as coauthors on all subsequent books.

WITH FANNY'S BIRTH, LOGISTICS HAD CHANGED, AS HAD ALICE'S FOCUS. Two friends, Bob Carrau and Sue Murphy, moved in with Alice and Stephen to help with the childcare. Alice, Patty Curtan, and another friend with kids turned Alice's garage into "an ad hoc day-care center," and in true Berkeley fashion, they all took turns minding the children, Bob and Sue a part of the mix. Both Alice and Stephen cut back their work to part-time.

Alice was spending less time at the restaurant now. Even so, she was still grappling almost daily with the challenge of finding untainted food. So much on the market was poisoned to some degree by pesticides or hormones or antibiotics; so much compromised by poor farming practices, or cruelty to animals, or neglect for the safety and well-being of agricultural workers. And so much was processed beyond recognition. To search the aisles of the average supermarket, reading the ingredients labels on any manufactured food, was to stumble into the laboratory of a mad scientist. The sheer number of chemical additives was alarming. Citric acid, lactic acid, glucose, fructose, maltodextrin, sorbitol, mannitol, xanthan gum, modified starches, dextrins, cyclodextrins, artificial colors, flavors, binders, emulsifiers, fillers, leavening agents: who knew what these compounds were, except that they might add a week, or a month, or a year to a product's shelf life. Even the fresh produce departments were compromised: the cucumbers waxed to retain their sheen, the tomatoes gassed to ensure they remained "firm and red" for weeks. It was increasingly hard to get beyond the long reach of industrial agriculture, and food as a factory product.

Alice was convinced that there were farmers engaged in sustainable practices out there, if only they could find them. All through the seventies, Chez Panisse had been building up a network of growers and producers on which they could rely, including a number of urban gardeners. Yet even with this deepening list, it was never enough to supply all of Chez Panisse's needs. And they were always dependent on the farmers' choice of crops, and on their planting schedules. That year, 1983, Alice decided to hire a friend, Sibella Kraus, as Chez Panisse's first official "forager." Sibella's formal, full-time position was now to go out in search of perfection: to find and cultivate relationships with local growers and suppliers invested in sustainable farming methods, and to encourage them to plant new varieties. It was an innovation that was completely unknown at the time, but that would eventually take on a life of its own.

Soon Sibella was meeting farmers who were growing "eight kinds of heirloom Japanese eggplant." These were products, she remembered, "that nobody had ever seen or heard of." Alice was, of course, elated. "Sibella would go out and just comb the hills looking for farmers who would plant these particular varieties I wanted," she remembered. "And everything was raised so beautifully and responsibly."

Sibella's efforts would prove to be groundbreaking. Driving the back roads of Northern California, it occurred to her one day that there had to be others like Alice who would be interested in forging direct relationships with these extraordinary farmers. So one afternoon she approached Alice and a couple of other restaurateurs and suggested they start a "Farm Restaurant Project." The idea was to bring local farmers and chefs together, to design their own little delivery system that could move produce straight from the farmer to the restaurant. "Nobody had ever done that," remembered Sibella. "For the first time, the farmers and the chefs were talking to one another about their particular needs." And for the first time, the conventional distribution system, which operates like a standard industrial supply chain, was being undercut. "It opened up a whole world."

"Sibella organized this very important tasting," remembers Alice. "It was called Tasting of Summer Produce. She was someone who knew the farmers, because she was one herself, and she knew the restaurateurs and she brought us all together. We figured out what they should be planting, and we had this meal together. That was the beginning of the collaboration."

Catherine Brandel, the restaurant's second "official" forager, soon joined Sibella. The three women were very much in sync. "Alice's philosophy, and mine, was always to seduce and educate," remembers Sibella. "Taste this peach. Now that the juice is dripping down your chin, let me tell you about the farmland where that comes from, and how endangered it is, and what you can do about it." It was the dawn of a new reality—that food is political: how it is grown, how it is distributed, how it impacts the health of the consumer as well as the land.

"When you're around Alice, you understand the sensibility of 'this bean might do for dinner today, but I know there's a better bean out there. Get me a better bean.'"

By 1985, Alice's father had jumped into the search too, making it his project to find a farm that Chez Panisse could work with year-round, growing the produce they wanted, in the quantities they needed. There were certain criteria: the farm had to be organic; its agricultural methods had to be virtuous, meaning environmentally sustainable; and it had to be no more than a ninety-minute drive from the restaurant. Pat Waters interviewed more than one hundred farmers before they settled on Bob Cannard, who to this day is Chez Panisse's chief supplier. Two times a day, someone drives the restaurant van to Bob's Sonoma County farm an hour away. There, after unloading vegetable waste from the restaurant into Bob's compost, he or she will help pick vegetables, load them into the van, and then make the return trip. "We buy everything Bob grows," explains Alice, "so he can count on his livelihood. It's a beautiful arrangement. It's a version of community-supported agriculture. It's restaurant-supported agriculture." Not only does it restore the link between those who grow our food and those who consume it, it's a way of taking responsibility for the "real-life consequences" of our "everyday acts."

CHEZ PANISSE WAS AT LAST SAILING ON AN EVEN KEEL. WITH THE HELP of Alice's father, it was fiscally stable. Pat Waters had convinced Alice that it was time to computerize the entire operation, including reservations, which Alice, who loathed all things high-tech, finally agreed to. For the first time, there were Christmas bonuses, and small raises all around. Alice was also receiving additional income from book royalties. A second cookbook, *Chez Panisse Pasta, Pizza & Calzone*, had come out in June 1984, this time with Alice's collaborators duly noted on the cover. Alice was beginning, says her biographer, "to dream again." Chez Panisse, she now said, had been "an unruly child" in the seventies, "just like a kid from age one to ten, difficult to take care of." By the eighties, it was becoming more professional, more consistently good. Its philosophy was also more talked about.

Newsweek was calling its food "a revolution in American cooking." *Vogue* described dining at Chez Panisse as "one of this country's most sensuously satisfying, highly personal eating experiences." And behind this awakening to the quality of its food was the growing recognition of the connections between taste and health.

Alice was now regularly identified as the mother of "California cuisine," as it was being called, a style of cooking based upon fresh, seasonal, local, and sustainable ingredients. "More than any other single figure," Marian Burros wrote in the *New York Times*, "Miss Waters had been instrumental in developing the exciting and imaginative style that has been labeled New American Cuisine."

And there was more to it than simply a name. It was an approach to cooking imbued with an ethical dimension, a sense of responsibility for the earth, as well as the wholesomeness of the food.

Alice wasn't wholly at ease as a public persona. A part of her liked the limelight, the respect, the veneration. But there were times when the attention was uncomfortable. For all her native charm, Alice was easily flustered, never completely relaxed speaking off the cuff, inarticulate sometimes. Charles Shere remembers flying to New York one time. He and his wife, Lindsey, were sitting three or four rows behind Alice, and somehow, the way the seats were aligned, he could see through the spaces to Alice. "I just happened to see her pick up the in-flight magazine," he remembered. "She opened it up, saw a full-page picture of herself, closed the magazine, and put it back in its pocket. And she was closing it in a real hurry lest one of the people next to her should see it. It was a very funny moment. She was embarrassed."

Alice was feeling uneasy for other reasons, her attention torn between Chez Panisse and her private life. Stephen wanted more time with Alice than she could give, which she found confusing. As much as she fought it, she wasn't feeling as close to him as before. Something between them had shifted, though it was still unspoken, felt but not fully discussed, a quiet tension in the air.

By late spring 1985, Alice was coming into Chez Panisse less and less. Some days she appeared briefly to taste the day's offerings, others to set up the downstairs dining room or tend to staffing issues. But the bulk of her time was now divided between Fanny, Stephen, and a big new project she was proposing: Alice wanted to make the restaurant at the Oakland Museum of California into "a living exhibition of her philosophy." Food would become an interactive part of the museum's program, she hoped. There would be a teaching garden, where children could learn how crops were grown and converted into food. In the museum dining room, the kids would be able to eat what they'd just studied. Fanny was now twenty-one months old and Alice was thinking hard about children's food, and the poor quality of American food in general. When the museum people pressed her about costs, her answer was forthright: "There's always money for good ideas," she said.

Spring turned to summer, and the days grew more carefree. While the strain between Alice and Stephen hadn't entirely lifted, they were determined to get beyond it, "if only for Fanny's sake." On September 23, 1985, on a lovely cloudless fall afternoon, Alice Waters and Stephen Singer were married in New York's Central Park, in the herb garden, with two-year-old Fanny in tow. "[Alice] had all the Pagnol fantasies going on at the same time," her sister Ellen remembered. "She wore a dress that could have been a double for the dress Fanny wore in the wedding in *César*." Her own daughter, Fanny, was the flower girl. They left for three weeks in Tuscany shortly after, Fanny still part of the festivities.

All that fall and winter, Alice was filled with high spirits. The restaurant was always busy, and the kind reviews continued. Happy to rely on Stephen's wine expertise, she hired him to lay out the wine list and to handle the ordering and tracking of what—and how much—was consumed. (Somehow, five hundred bottles a month were still going missing.) In anticipation of Chez Panisse's upcoming fifteenth birthday, she approached North Point Press, who agreed to publish a special edition of Pagnol's memoirs, for which she wrote a short foreword. In it she described the gentle dream she'd taken from Pagnol: "an ideal reality where life and work were inseparable and the daily pace left you time for the afternoon anisette or the restorative game of petanque, and where eating together nourished the spirit as well as the body—since the food was raised, harvested, hunted, fished, and gathered by people sustaining and sustained by each other and by the earth itself." It was the essence of everything Alice believed was important.

IN OCTOBER 1986, ON A BLUE-SKIED SUNDAY, THE ENTIRE CHEZ PANISSE staff gathered for a retreat in the Napa Valley. They swam and ate organic hamburgers; they teased and toasted one another. They talked about butchery and knife sharpening and pastry making. They aired their gripes. And then Alice stood up, her voice quavering slightly as she described what she saw ahead. She was feeling increasingly "our responsibility to the rest of the world," she said. Our goal, she stressed, is "to educate ourselves and the public."

Alice had always given generously to the causes closest to her heart: famine in Ethiopia, local hunger programs, aid for political refugees from El Salvador. In 1981, she and David Goines had helped raise funds to thwart Pacific Gas and Electric plans to build a nuclear power plant in Diablo Canyon. But she was beginning to feel that giving money wasn't sufficient. She needed to do more. She wanted to begin actively working against industrial farming and the diminishing quality of America's food. And then crisis struck even closer to home. Tom Guernsey was diagnosed with AIDS.

Kind, gentle, and "unpretentiously princely," in Greil Marcus's words, Tom Guernsey had long been the social glue of Chez Panisse. He "could and did do everything." He was peacemaker, sometimes handyman, genial host, and a dear and cherished friend. "Despair doesn't begin to describe what we felt," Alice remembers. "We had to do something." AIDS had been identified just four years before, but it was already ripping its way through the gay community, especially in the Bay Area, taking lives with rapacious speed. Tom's life couldn't be saved, but money could be raised, help dispensed. "As *famille Panisse*, we'd all take care of each other if we got sick," remembers a friend. Alice teamed up with the owner of Zuni Café and they organized a giant benefit to raise money to care for those with the disease. They called it "Aid and Comfort."

With loss now so near, Alice was feeling more than ever the importance of community and the life of food. "Those of us who work with food," she wrote in her 1989 essay "The Farm-Restaurant Connection," "suffer from an image of being involved in an elite, frivolous pastime that has little relation to anything important or meaningful. But in fact we are in a position to cause people to make important connections between what they are eating and a host of crucial environmental, social, and health issues."

Alice was addressing an issue that comes up from time to time.

Wasn't her vision elitist? Didn't it exclude the poor? The average Joe, said critics, could little afford a dinner at Chez Panisse, which can cost eighty-five dollars a person. (Or organic ingredients, which were more expensive than the industrial supermarket variety.) "That's silly," Alice insists. "That's like saying that a really good soup can't be as divine as blinis with caviar. You can make polenta for one hundred people for eight dollars. There are twenty different varieties of shell beans that are healthy and delicious and cost next to nothing. One of my favorite things is grilled bread with mashed fava beans. Wholesome good food should be an entitlement of all Americans, not just the rich."

Those close in refute the elitist claim in different terms. In truth, say insiders, Chez Panisse has never made much money. This is partly because Alice and the board have always felt it extremely important that everyone who works there, from cooks to dishwashers to waitstaff, receive a living wage and decent health insurance, and that all get generous vacation time. There are two chefs, each of whom takes off half the year, with pay, so they are able to have lives, so they don't burn out, so they can continue to give their best when they *are* at Chez Panisse. This is part of the lifeblood of the place, its philosophy and ethics. And Alice has always had the good fortune of working with people who shared her ideals, who cared more that a place like Chez Panisse exists in their community than that it makes a lot of money. This too is part of the restaurant's DNA: that the people involved in its governance have always thought of it less as a bottom-line business venture than as a shared mission, a model for a set of values, for a humane and sustainable way to live. Alice, as many charge, may be a purist with lofty aspirations, but she has never betrayed her politics. This includes never capitalizing on Chez Panisse's fame by mass merchandizing of any sort. There are no Chez Panisse frozen dishes, no Chez Panisse pasta, no Chez Panisse condiments. There are only cookbooks and "a line of granola."

"Alice is the least elitist person I know," offers Ruth Reichl. "Here's this woman who is the only person of her fame who hasn't cashed out. Look at her. She lives in the same house she has lived in forever. She drives a Prius, and drives *herself* wherever she's going. She's not rich. She hasn't opened a chain of restaurants. You look at Wolfgang Puck and Mario Batali and Daniel Boulud. She could be any one of them. And she hasn't."

"What you have to understand about Alice," says Greil Marcus, "is that she has never given a damn about money. She really hasn't. She has always been a person who *looks* for chances to take. She believes that unless you take chances, nothing will ever change. And it is Alice who has always proposed the radical changes at Chez Panisse that more conservative people, sometimes like me, have said, 'Oh no, that will never work, or we could blow the whole thing.' For Alice it's always been worth it to blow the whole thing, in order to step into a new situation, where new people can be involved, where ideas that you never would have thought of suddenly become obvious and necessary. And she's done this over and over again. She's never flagged."

In truth, Alice sees the issue of food costs in a larger context. The problem of affordable food is actually one of perspective, she argues. The point isn't that "organic" is too expensive, but that the highly chemicalized industrial food chain is too cheap. Large industrial growers are able to "externalize" their costs by depleting soils, polluting waterways, and compromising human health. Those costs are never factored into the price of cheap food.

IT WAS NOW 1986, FIFTEEN YEARS AFTER THE OPENING OF CHEZ PANISSE, and the health and environmental impacts of industrial farming had only grown more grave. Agribusiness was now the norm in America, as was its dependence on cheap, federally subsidized corn, synthetic fertilizers, and a flood of pesticides. Genetically modified corn (along with its hybrid cousins) had given rise to high-fructose corn syrup, which could be produced for a whole lot less money than other sugars. It was now routinely added not only to soda and desserts, but also to a host of other foodstuffs: canned soups, spaghetti sauces, frozen dinners, mustard and ketchup, hot dogs and ham. Consumers weren't always aware of the added sugar, but they developed a taste for it. And its low cost allowed food companies to supersize portions at little to no added expense, which also pleased shoppers. Meanwhile, "food technologists" in white lab coats were perfecting another sort of alchemy. Complex grains could now be reduced to their molecular parts and then reconstituted into new edible concoctions of any shape, texture, or taste, giving birth to such Frankensteinian wonders as nondairy creamers, fruitless fruit juices, "meat analogs" spun from "textured vegetable proteins," and the ubiquitous corn-and-fat-based chicken nugget (which contains twenty-eight ingredients, according to Michael Pollan, most of them created in the lab). Not only could such artificial foodstuffs be engineered to "resist decay," or remain "ever-gooey" or "perpetually crisp," they could also be sold for much less than the real thing they imitated. It was an insidious cycle: the cheaper industrial food got, the more of it Americans were eating. (And the more their health, in turn, was declining.) Schools that once served nutritious hot lunches cooked on-site were now offering ready-to-eat microwavable meals pulled from vending machines. It was easier on school budgets. And most students preferred to leave campus rather than to eat in the cafeteria anyway. It was no accident that fast-food joints always seemed to be close to schools. "These guys find where the exits are on the freeway before they've been built," said Alice, "and they buy the land next to the school."

Still more worrisome was the silent health toll these foods were taking. According to the *New York Times*, "high-fructose corn syrup can hinder the body's ability to process sugar, and can promote faster fat growth than sweeteners derived from cane sugar . . . Since the advent of the syrup, consumption of all sweeteners has soared; the average American's intake has increased 35 percent . . . Furthermore, the rise of type 2 diabetes since 1980 had closely paralleled the increased use of sweeteners, particularly corn syrup."

Americans were eating their way to obesity, getting fatter and sicker by the year. Fast food—long on calories and short on nutrition—sold briskly. It was by far the most robust sector of the food economy. It was cheap and it was convenient. "Food shouldn't be fast," Alice says. "And it shouldn't be cheap. It's only cheap if somebody is losing out. And it's always the farmer. The whole industrial food system works on the backs of undocumented migrant workers who are totally exploited. So we have to pay the real cost of food, we have to really decide we're going to do that."

But there was another dimension to the problem. Fast food was but one face of a hydra-headed monster, one arm of a much larger agro-industrial system that was depleting the natural environment to a degree few people grasped. The model for factory farms was to plant thousands of acres with a single crop (a monoculture) and then add vast quantities of synthetic chemicals, stripping the soil of nutrients, reducing biodiversity, poisoning the water, and wiping out wildlife habitat. Raising farm animals was just as industrialized, if not just as environmentally grim. Modern hogs now spent their lives in aluminum sheds suspended over vast, methane-emitting manure pits. Chickens were reared in crowded, fluorescent-lit indoor factories and cattle in feedlot cities, or, in agribusiness parlance, CAFOs (confined animal feeding operations). Fed and "finished" on corn, liquefied beef fat, chicken feces, and an array of pharmaceuticals—this to combat the diseases that came of living in such close quarters—the modern animal lived its unnatural modern life mostly indoors. As Michael Pollan notes in his wry and masterful natural history of eating, *The Omnivore's Dilemma,* "While America's human population found themselves leaving the cities for the suburbs, our farm animals found themselves traveling in the opposite direction," from green, widely dispersed farms to "densely populated new animal cities." Small family farms, unable to compete on the scale of Big Ag (or afford these chemical and technological shortcuts), were being driven out of business.

Alice was deeply worried by all this. But she wasn't naive. She knew restaurants like hers, which cared more for quality than for profit, and which were willing to run at a little more than a break-even basis, were few and far between, no match for the long reach of

Big Ag, which was steadily co-opting every facet of the nation's food system. It was hard to see how she or other like-minded souls could ever go toe-to-toe with such a multitentacled giant.

But someone, as she told her biographer, had to try. At stake were issues too critical not to. And time was not on their side. Alice was also coming to another hard truth. Advocating on real food's behalf meant accepting an increasingly public role; it meant stepping into the media glare, courting reporters and critics, enlisting celebrities—becoming the public figure she had never wanted to be. It was the only way to reach large numbers of people, the price of successful advocacy. But it also meant less time for a private life. These days, what little she had she devoted mostly to Fanny, which was painful for Stephen, lonely and dispiriting.

ALICE'S BIG DREAM FOR THE OAKLAND MUSEUM WAS VOTED DOWN. There would be no living classroom, no organic sack lunches for kids, no learning-by-doing lessons about how the stuff of nature was transformed into delicious things to eat. Alice had wanted children to see where their food came from, to trace the lines of connection between the earth and what they ate, to revel in the thrill of seeing things grow. But it was not to be. At least not yet.

She moved on, a second, equally ambitious project now in her sights. Her new idea was to organize a big food market in downtown Oakland. It would be modeled on Les Halles in Paris, the now-demolished central market, a place where farmers could sell directly to the public. The produce would be local and seasonal, mostly organic, cheaper than supermarket fare because of the absence of any middlemen. The purveyors would share in the market's ownership. It would be an expansion of Sibella's idea of connecting cooks to those who grew the food.

And then, quite suddenly, the financing fell through and overnight the project died. By then, however, Alice was quietly relieved. She had realized that the "insanely high" costs of construction were going to make it unaffordable to the locals she'd hoped would shop there. (It was a conundrum Jacobs had observed too: "Old ideas can sometimes use new buildings. New ideas must use old buildings.")

She shifted gears again, throwing her support behind a number of organizations working to get food to the poor: the Food First Institute (a think tank that explores the root causes of world hunger); Share Our Strength, an association of restaurants that donates leftovers and unused provisions to the hungry; the Daily Bread Project, a sister group doing the same.

"It was very distressing realizing poor people were buying these highly processed industrial foods. They were much more expensive than good organic produce! We had beautiful farmers' markets coming in all over the Bay Area, all over the country, in fact, but most of them were in affluent areas selling to affluent people."

Alice kept at it, but she was seeing how hard it was to get beyond the social stigma that made many poor people wary of the area's farmers' markets. The cushy foreign cars, the expensively dressed couples, the urban poor's worries over their own substandard English—all were inhibitions. The people without access to healthy good food just weren't coming.

But Alice was learning invaluable lessons. She was seeing that every endeavor, even those that were only marginally successful, helped win hearts and minds.

"You had to start somewhere," she says, "and show that it could work . . . People need to see it in practice. Poor people in this country, for the most part, eat badly because they are victims of an intense campaign to keep them eating industrial processed food. We've got to start countering that."

That moment was about to come.

ALICE WASN'T LOOKING FOR THE SCHOOL; THE MARTIN LUTHER KING JR. Middle School found her. She drove by it every day, not only on the way to work, but en route to everywhere else she went. It almost looked abandoned. Graffiti tagged the windows. The grass wasn't cut—what little grass there was, that is. Most of the yard had been asphalted over. It was cracked and riddled with weeds. During an interview, Alice happened to mention the school. She'd recently been thinking about Fanny and school: where she would go, what she would eat when she got there. Fanny was eleven now. It was one thing to have "the life of the restaurant," she told the reporter, "with all those very special suppliers, this little idealistic world." But she knew one couldn't really exist as an island. "How could we have allowed this to happen?" Alice asked. "How could public schools look like this in the city of Berkeley, where the great University of California lives?" A few days later, Alice got a call. It was the King school principal. He asked her, she remembers wryly, "to come to his office." And Alice did.

"He thought I would come and plant a flower garden out front," she remembers, "make the place look more pretty. That's really what he had in mind."

Alice and the principal walked the grounds. The King school had been built in 1921, on seventeen acres. It was meant to hold five hundred students, but now there were one thousand. Portable classrooms had been trucked in to accommodate the extra kids. The cafeteria was closed because they couldn't feed all the children in the allotted time. Instead, there was a little "snack shack" at the end of the yard. Its "kitchen" consisted of a microwave, where prepackaged pizzas and hamburgers could be heated.

As they crossed the playground, Alice started talking. "I have this great idea," she said. "We could plant a vegetable garden, so the kids could see the cycle from the garden to the table, and then back to the garden—the whole life cycle of plants, from seed to sprout to food to compost and soil. We could do the whole thing. We could build a new cafeteria and feed all the kids with *real* food that they'd grown here, on the school grounds, in their own garden." All the kids would eat together, she added. They would sit at tables and help do the serving; it would be an everyday experience that changed their behavior. Alice pulled "all the Montessori ideas into it," she recalls. Every child, she insisted, would get a free lunch.

"You can't have some kids eating potato chips," she told the principal. "Everyone eats the same food."

The principal thanked her very much. I'll give you a call, Alice remembers him saying. "And I didn't hear from him for six months," she says. "But then he called."

No one has ever accused Alice of thinking small. From the very beginning, she insisted that it was "either all or nothing" at the King school. "I want it all," she remembers the principal saying, "but let's keep the part about free school lunch for all kids our little secret, because it's going to frighten people for you to talk about this." For a school worried about buying enough chairs for the children, enough desks, she explains, "the idea that you would spend money on the cafeteria for the food—it was unthinkable then. This was twenty years ago."

So Alice started slowly, beginning with the garden piece, which she was calling the Edible Schoolyard. She asked three friends to donate seed money: $25,000 for three years. She invited the teachers to Chez Panisse for a meal. "I told the math department they could have their meetings there, every month, in the dining room," she remembers. "Then I invited the principals from *all* of Berkeley's lower schools. It's so unheard of to give things to teachers and principals. These people are doing God's work, and yet they are so underpaid. It was a surprise for them to be asked to come for a meal. But it worked like a charm. It's a powerful thing to bring people into an unexpected place, and then to really have a conversation about something they haven't thought about before."

With Alice, it's always a "seduction," says food writer Dorothy Kalins. She has a certain "breathless way of letting you in on a delicious secret." She never pushes, never strong-arms. It's always about gently winning people over to her views.

"I think real food has a way of impressing just about everyone," Alice offers. "You kind of begin with an incredible savory soup, or a hot, freshly baked baguette. But after that, it comes forth. I think anyone can be won over." If you're "seduced by something that's beautiful and nourishing, you want to repeat that experience again."

Ruth Reichl tells a story about Alice that perfectly captures this faith. She and Alice had taken a trip to the Chino ranch, which supplies much of Chez Panisse's produce. On their last morning, they went out and picked strawberries for Chez Panisse for dinner that night. "It's so hard to remember," Reichl says, "but this was when the only strawberries you could get in America were supermarket strawberries, which were huge and white in the middle, and had no flavor or aroma. We were picking these tiny little berries. And then we got on the plane. We were flying from San Diego to Oakland and it was a fairly small plane, and the aroma of these berries starts spiraling up and fills the plane. My recollection is that one by one everyone on the plane got up and came over and begged for a strawberry. We kept hearing, 'I forgot what strawberries taste like.' Alice was ecstatic. This was her dream. I am going to show you just how great food can be. And it's not even *mediated* in any way; it's just a strawberry. People literally lit up. I watched Alice essentially give away that night's dessert for the restaurant. And she was thrilled by it."

Alice, says Greil Marcus, "is the person who says, of course it's possible, when everyone else is saying it's impossible . . . She really does believe that she can change the world, she can change individual people one by one, she can improve individual lives. This isn't just rhetoric she trots out whenever she is given an award."

Alice approached the school board next. At the King school, she wrote, the Edible Schoolyard would be "a comprehensive solution to both the neglect and the underutilization of the physical plant and its surroundings." It would teach kids where their food comes from, how to feed themselves and tend the land, and how to pull together as a team, "because we all live here together." The "core of the intended learning experience," she added, would be "an understanding of the cycle of relationships that exists among all of our actions":

> *The tangerine peel that gets tossed into the compost pile*
> *becomes a feast for the organisms that will turn it into*
> *humus, which enriches the soil to help produce the fruit and*

*vegetables that the students will harvest, prepare, serve, and
eat. The health and well-being which they derive from the
garden is recycled back into their attitudes, relationships, and
viewpoints. Thus the discarded peel becomes the vehicle which
provides tomorrow's city planners, software engineers, artists
and master gardeners their first adult understanding of the
organic concept of interconnectivity.*

Alice continued to refine her plans for the Edible Schoolyard.
Though she still thought of herself primarily as a restaurateur, she
was often away now, speaking about sustainable agriculture, about
conservation, about the importance of "farm-to-table connectivity,"
preaching the gospel of local, fresh, organic ingredients wherever
she could. She had become, despite her initial wariness, an increasingly public personality, with all that this entailed. The invitations
to speak, to accept awards, to appear as a celebrity chef poured in,
more than she could possibly manage. And yet her sense of mission, her genuine belief in a life lived in consonance with the earth's
cycles and natural systems, was such that it was hard for her to say
no, even when she should have. To all appearances, she seemed to
be managing her crusade on behalf of "real food" and the evils of
industrial farming well. Yet it was coming at a price. She was finding she had less time for those she loved, and less time for the restaurant. Chez Panisse was approaching its twentieth birthday, and
Alice wondered at moments if perhaps it had run its course. Some
of this had to do with her general state of mind, a vague feeling of
estrangement from her own life. For some months now, she and
Stephen had been sleepwalking through their marriage. Stephen
had grown more distant, perhaps out of self-preservation. While he
still did all the wine purchasing for Chez Panisse, he had decided
to team up with one of the former chefs, to open a restaurant of his
own in Napa—perhaps to give himself more space, a second world
in which he felt more agency. While he and Alice remained deeply
connected to Fanny, their time together now revolved less around
each other than around her.

IN LATE MAY 1994, STEPHEN HAD A NEAR-FATAL ACCIDENT. AN AVID CY-
clist, he was riding in Berkeley, near their home, soaring down a
hill, when his wheel hit something that threw the bike. At least
that's what he later assumed. He had no actual memory of the crash.
He woke up in the hospital. His nose had been ripped from his face
and reattached; his lip split in two, almost to his nostrils. His front
teeth were gone and his eyes, mercifully spared, were oozing. Had
he not been wearing a helmet, said friends, "he would have been
dead."

Alice was away at the time. Bob Carrau reached her by phone.
He and Sue tried to explain what had happened, but also not to
scare her. Stephen was all right, they said, he was not going to die,
and they were there. Alice asked if she should come home immediately. Bob remembers saying yes; Sue is less sure. Perhaps, to shield
Alice, they hadn't been as clear as they should have been.

Alice returned not the next day, but the following one. As soon
as she saw Stephen, she knew she shouldn't have waited. She hadn't
understood how serious it was. "They should have said to me, 'You
need to get on a plane right now,'" she says. "They were trying to
protect me."

Stephen's recovery was slow and dispiriting. He endured months
of pain, multiple plastic surgeries, a spate of restorative dental work.
It was an agonizing interlude and he needed Alice to be there in
ways he didn't think she could be. "I used to have a notion that I
needed to be pretty self-reliant," he told Alice's biographer, "because
if I ever really needed Alice, I wasn't sure if she would be able to
even see that I needed her, let alone be up to the task, and I know
that she thinks that she wanted to help me."

Stephen continued to keep his own counsel. But having now
named the discontent, he could no longer wholly escape it. He saw
the distance between them for what it portended: he was beginning to let go. Alice, meanwhile, buried her head, retreating to those
parts of her life that fed and sustained her: Chez Panisse, her many
devoted friends, her growing public platform. In her own way, she
was trying to put things right with Stephen, confident, as always,

that she could. Several times a week, she gathered their closest friends together, to cook and eat at home with Fanny, often in the garden surrounded by tea lights. When she and Stephen dined at Chez Panisse, Fanny always came along, making even small moments feel festive and full. But the strain in the marriage persisted. Alice turned her attention to the Edible Schoolyard, assuming they would get beyond the rough patch, as they always had before.

EVEN IN THE EARLIEST DAYS, ALICE NEVER ENVISIONED THE EDIBLE Schoolyard as a gardening or cooking class, per se. "It was always connected to the curriculum," she says. "But it was learning by doing, using the senses: an outdoor science lab in the garden, a history lab in the kitchen. You were studying the foods of Egypt and preparing and eating them, multiplying fractions to work out measurements. So it was an interactive experience. And of course, the kids were cultivating the garden and eating the raspberries at the same time."

Beyond teaching sustainability and nourishment, it was Alice's hope to foster other fundamental values too. "This is the first generation of kids who haven't been asked to come to the table," she reflects twenty years into the project. And we're seeing the results. "Kids are disconnected. Many haven't been outside before. All children now are on their cell phones. They are watching TV. They are sitting in front of their computers. They are deprived of the natural world. Their parents are too busy; they are working two jobs. Everything for these kids is mediated, and they're perpetually being fed the values that places like McDonald's put out."

"Fast-food values," as Alice calls them, are a corrosive force in the culture: insidious, unhealthy, and destabilizing. They're the cause of a great national malaise, not only a loss of the social role food has played for thousands of years, but also our links to the land. "I've been giving this talk about the really systemic problem we're facing that we don't understand," she reflects. "We see all the problems, yet we don't see where it's coming from. It was Brillat-Savarin who said, 'We are what we eat; the destiny of nations depends on how we nourish ourselves.' I never got that until I really *got* it," she adds. "When you eat fast food, you're *digesting* the values of a culture that says there are no seasons, that everything should be available 24/7, that food isn't important, that advertising confers value, that waste is fine. Cooking and farming, don't bother with those. Someone else will take care of that. We don't need to do these hard jobs. Sitting at the table, that doesn't matter either. You can eat in your car. Time is money. More is better. Everything should be fast, cheap, and easy. That's what fast-food culture tells us, and we are eating these values. Fast food is literally addictive. And it's cheap, cheap, cheap. But it's not really factoring in the real costs."

Those real costs include what the surgeon general recently called an "epidemic" of obesity, costing the health care system an estimated $90 billion a year. Three out of every five Americans today are overweight; one of every five is obese. Diabetes is on the rise, as are cancers of the breast, colon, and prostate. According to the Centers for Disease Control and Prevention, there has been an "unimaginable 655 percent increase in the percentage of Americans with diabetes." Meanwhile, every single day, a third of all U.S. kids are chowing down on fast processed foods. They are regularly, and unwittingly, ingesting carcinogens, neurotoxins, and endocrine disruptors, all of which pose serious dangers. "What's going on with children is so *so* serious," Alice says. "With childhood hunger, obesity, general unhappiness. The low cost of food doesn't factor in human health. It doesn't factor in the damage to our environment. It doesn't factor in the fracturing of our communities."

For Alice, the tiny silver lining to all this—the subversive part, if you will—was having the chance to teach kids a whole *different* set of values: the pleasures of meaningful work, healthful eating, *slowness*, cooperation, respect, patience, personal responsibility, orderliness, gratitude, wonder, a love of beauty, in addition to the "socializing effect" of the rituals of the table. "I've always believed in public education," she says. "When Gloria Steinem said it is 'our last truly democratic institution' I heard that. It's the one place where we can reach every child."

Which is just what Alice intended to do. The Edible Schoolyard was soon up and running despite its ambitious scope. And it was soon modeling these ideals. Alice recruited builders, bakers, master gardeners, even curriculum developers to share their expertise. She approached the Berkeley Horticultural Nursery, which generously offered to donate plants and teach the kids how to cultivate them. She created the Chez Panisse Foundation, to help with funding.

And then one fine day, bulldozers arrived and clawed up the asphalt. The earth was dug and cover crops planted to restore the nutrient-starved soil. Sowing the first seeds soon followed. The kids swung tools, cleared brush, sifted soil, dug furrows, counted seeds. They learned about compost, about pollinators, about butterflies and bees. They built a chicken coop, examined grubs, "named and renamed chickens." They weeded and collected eggs. And finally the first crop was harvested—mâche, arugula, mustards, lettuces, kale, bok choy, carrots, turnips, beets, garlic, fava beans, tomatoes.

The garden continued to grow, and so too did the program. At first, only the math and science teachers were involved. But with time, the humanities teachers joined in too. New buildings were added, including a kitchen classroom, with long worktables and multiple sinks, nooks for aprons and knives, hot plates and chopping boards, backpacks and boots. The kids built a tool shed and an outdoor pizza oven, an open porch and a pergola with climbing vines for shade. They planted orange and lemon trees; raspberry bushes and apples; then plums, figs, kiwi, scarlet runner beans, ground cherries, and jasmine. And all the while they cooked and ate the vegetables they grew.

In the beginning, Alice remembers, it was hard to get the kids to want to go outside. "They were afraid to get their shoes dirty. So we got them boots. But it was about six weeks to kale. That's all! It's the tactile part that engages them—engages their senses. They become so *conscious*."

Getting a conversation going "at table" took time too. "We began by teaching them to set the table, putting out linens, arranging
the silverware, adding flowers for beauty." And then they learned
how to serve. But once they sat down to eat, they didn't know how
to make conversation. "So we decided to create conversation cards
and put questions on them. They don't need them anymore," Alice
adds. "But in the beginning, we were teaching them how to ask each
other questions, how to talk to each other about real things."

The same went for cleaning up. In the early days, they were as-
signed cleaning tasks, so it seemed a chore. "So we designed a card
game where the kids chose cards; 'Oh, I got the sweep the floor
card!' someone would excitedly say." It changed everything. "It's
like the difference between exercise and dancing. Everyone wants
to dance." The point was, "it wasn't someone telling them to do
something, it was them *choosing*." That's the way the entire curriculum was designed: to make the experience elective. "We're empowering them to make decisions, we're not telling them what to
eat; we're just bringing them to this different relationship with food.
And then they make those choices, and once they make them, they
make them for life."

The one truth about the Edible Schoolyard, Alice reflects, is that
"if they grow it and they cook it, they *all* want to eat it. That's the
most compelling part of the program. Kids fall in love with nature
instantaneously. There's a kind of magic that occurs. And once they
connect with nature and understand that our food comes from that
place, they never go back."

"It's why these models are so valuable," she adds. "They allow
people to walk in. Once it's modeled, you see it's right as rain. It's
not a Berkeley thing. It belongs in every school."

The Edible Schoolyard was soon proving to be a powerful in-
cubator of community. Parents began to appear, asking for recipes,
many expressing their desire for a garden at home. The kids—
sixth-, seventh-, and eighth-graders—began to hang out in the
garden after school, "boys and girls together, in little packs," says

Alice. "They go there and just sit in the garden. They are eating mulberries under the tree. They know it's their place, and they love it. In the garden they have a purpose; they feel engaged. I think that's just it: We don't have anyone helping us define our purpose on this planet. We don't even have John Lennon singing 'give peace a chance.' But these kids, I feel as if any one of them could give a TED Talk by the time they graduate, at the end of three years. They express all that we wanted them to absorb. They feel connected to the garden, to the cycles of the earth, to the larger world. They have a sense of what they can give."

Alice, however, still wanted to do more: she wanted to shake up the entire food chain. It was her hope that eventually *every* ingredient used in the lunch program would be sustainably grown, and that every student, in every grade, at every Berkeley grammar school, would be fed a free organic lunch. Those foods that couldn't be raised in the school garden would be bought from organic farms in the Bay Area. Like Chez Panisse, the Edible Schoolyard would help strengthen and deepen the local food economy. It would create a bridge between city and country, culture and nature, if you will, "making visible the lines of connection," in Michael Pollan's words. "Schools buy so much," Alice points out. "And kids make up twenty percent of the population. One lunch at a medium-size school uses two hundred and fifty pounds of potatoes, and that's just potatoes. Think of the organic farmers that could sustain! Think how much cleaner the air and the water and the earth would be! If we could really redirect who and where schools bought from, it could change the world."

The idea isn't so radical, she insists. Most of us, she writes in her essay "The Farm-Restaurant Connection" have "become so inured to the dogmas and self-justifications of agribusiness that we forget that, until 1940, most produce was, for all intents and purposes, organic, and, until the invention of the refrigerated boxcar, it was also of necessity fresh, seasonal and local. There's nothing radical about organic produce: It's a return to traditional values of the most fundamental kind." The aberration wasn't the advent of organic, but the invasion of industrial agriculture that prevailed under the pretense of progress. What better place to start that return than with schoolchildren, our best hope for the future?

Alice's ambitions for the Edible Schoolyard continued to deepen. She had always envisioned the project as a model that could be taken nationwide. Now she wanted to see that happen. In late 1995, she wrote to President Bill Clinton, describing the Edible Schoolyard and urging him to rethink the government's policy on school lunches, where the very worst processed foods still made up the bulk of the program. "Help us nourish our children by bringing them back around the table," she pressed. "Help us create a demand for sustainable agriculture, for it is at the core of sustaining everyone's life. Talk about it, promote it as part of the school curriculum; encourage the spread of farmers' markets; and demonstrate it with organic gardens on the grounds of the White House."

Both Bill and Hillary Clinton responded in separate, cordial letters. But neither made any firm commitments. Unwilling to give up, Alice wrote a second time.

> *Dear Mr. President,*
>
> The prospect of your second term fills me with hope— hope that you'll seize this opportunity . . . I continue to believe that the way to bring people together is by changing the role food plays in our national life . . . A program like the Edible Schoolyard ought to be in every school in the country.
>
> *Respectfully yours,*
> ALICE WATERS

But even presidents have only so much power. Despite Bill Clinton's sympathies, school lunches across America continued to be sold to the lowest bidder. "At best, we serve our children the surplus of government-subsidized industrial agriculture," Alice laments. "At worst we invite fast-food restaurants to open on school grounds . . . I think Clinton wanted to do something about it, but he didn't have the will to really put any money behind it."

CHEZ PANISSE TURNED TWENTY-FIVE IN THE SUMMER OF 1996, A YEAR after Alice's letters to the president. To mark the occasion, Alice decided she wanted not one party, but a week of celebrations that would include the extended Chez Panisse community—one night a gathering for fellow restaurateurs and chefs, another for family, friends, and ex-staff; a third for suppliers, growers, and wine people; and on the actual anniversary, a dinner with "old and dear friends" from the restaurant's first days. Alice was both jubilant and exhausted by the end. "It was right at the time when Stephen and I were having huge problems," she remembers. "So I was in a state, and I think the series of parties kind of pushed us over."

Stephen didn't need much pushing. He was feeling agitated for his own reasons. His restaurant in Napa had failed, but he was determined to try again. He wanted to open a tapas bar, which he was calling "Bar César." It was a nod to another of Marcel Pagnol's Marseille trilogy, but it seemed uncomfortably near to what Alice had done. Odder still, he was planning on taking the space next door to Chez Panisse, a choice that friends found confounding, at once needlessly competitive and awkwardly close. "He was so angry at [Alice]," Bob Carrau would recall in a conversation with Alice's biographer, yet he couldn't seem to get out of her "shadow." It all seemed so self-defeating.

Alice and Stephen held their life together for another year, despite the tension. And then, in the summer of 1997, Stephen dropped a bombshell. He had fallen in love with someone else, he told Alice. He wanted her to know, even though he claimed he still loved her too. Alice was shattered. She should have seen it coming, but she hadn't. Though she'd been aware for a long time that Stephen was unhappy, she had always believed they were fundamentally connected, that whatever the rupture between them, it could ultimately be worked through. It was the first time, say friends, that Alice the nurturer, the "problem solver," the fixer, couldn't make things right. Late that summer, she and Stephen agreed to divorce.

It was an "excruciating" time, Alice told her biographer. "Very painful . . . Fortunately I had a kid, and fortunately I had a family at Chez Panisse . . . I don't know how people manage. I was really lucky. I called all my friends. I was crying for help from everybody. Anybody."

Alice's friends became her lifeline. Sharon Jones, one of her closest friends, walked with her each morning, often for miles, talking as much as Alice needed. Others came to the house and made dinner. "Alice didn't even want to cook for a while," remembers Bob Carrau, which was completely out of character. Sad, shaken, uncharacteristically enervated, Alice struggled to regain her bearings. It was a long, slow trudge back from the despair she felt. But when at last her spirits began to lift, she was more certain than ever of her path going forward: she wanted to devote her best energies to advancing the cause of edible education, changing the way the world ate by teaching children the "healing magic" of the garden and the table.

MOVING HEARTS, AS ALICE WAS LEARNING, IS EASIER THAN MOVING government policy. If the dream for a 100 percent organic school lunch for every student, in every American school, still hasn't come to pass, Alice has had other important successes. Early on, she succeeded in convincing the Berkeley school district to stop using all foods made from milk injected with bovine growth hormone, and all foods derived from genetically modified crops. In 2003, after a citywide referendum, funds were approved to build a magnificent new cafeteria on the King school grounds, a soaring, light-filled space with sustainably harvested redwood trusses and windows on all sides. Designed to double as a learning center, where each year educators from across the globe convene for hands-on training in edible education, it's been so successful that the city of Berkeley now uses it to make food for the entire district. Eventually, however, every school will have its own cafeteria and a full-service kitchen, Alice explains. "It's only now," in 2016, she says, "that we have a superintendent of schools who's really excited and wants to buy everything locally." And that's in progressive Berkeley.

Even more encouraging is the speed with which the Edible Schoolyard has spawned sister programs—just as Alice hoped. At a little over twenty years old, the Edible Schoolyard is not only prospering, its core ideas continue to spread. If not yet national policy, as Alice continues to work for, it's a grassroots movement with legs. Requests from "start-up" school gardens pour in from across the country—enough so that a website has been created. "We haven't even begun to document all the programs," Alice says, "but already schools are getting ideas from each other." To date there are 5,500 known school garden programs, and these are only the schools that have chosen to post their curriculums online. "I know there are thousands more programs doing this off the grid."

Progress overseas has been swifter still. In 1999, the Edible Schoolyard officially went global after Carlo Petrini, founder of Italy's Slow Food Movement, visited the Edible Schoolyard and was smitten. From that moment forward, school gardens would be folded into Slow Food's mission of promoting sustainable agriculture, local food, the old-fashioned family meal, and public awareness of the ecological havoc wrought by industrial farming.

A similar enthusiasm took hold at Yale. When Fanny matriculated there in 2002, Alice helped organize the Yale Sustainable Food Program, which began with an organic garden large enough to provision one of the school's twelve dining halls. Students were soon forging meal passes to dine there, forcing the college to buy additional food from local organic purveyors to make up the shortfall. Since then, the garden has been greatly expanded, along with the program.

Determined to take these ideas a step further, one of Alice's current projects is now to expand the Edible Schoolyard to include high school. We know a lot about what can happen in K–8, she says. "And we have lots of examples of the curriculum. But the part that we really don't know too much about is what you do in nine to twelve? What happens to these kids in high school? In California, many of them drop out of school, and for every kid who drops out, that's $85,000 if they end up in jail. So the mayor of Sacramento agreed to take on this idea I had, which was to design a program in which the kids ran the whole cafeteria themselves. As a project, for their whole time in high school. The students would do everything: the outreach, the math, the statistics, the dining room training, the cooking, even the design of the room. I think it would be a wonderful experiment. It fits into the most radical and important pedagogy for success that's out there right now."

Alice has also been working for the last six years on a project in Rome at the American Academy, a storied institution where artists and American scholars have historically come together to enjoy an interdisciplinary conversation at the crossroads of the world. It's a place of magnificent architecture on one of the highest hills in Rome. But the food was always "steam-table" American cafeteria style, says Alice, so as not to distract anybody from the "important" work they were doing. People who went consistently wrote letters afterward describing what a beautiful time they'd had, except for the food, which was famously awful. Finally the director decided to do something about that; she called Alice.

When Alice arrived, she remembers, she eyeballed the kitchen, the miserable food, the almost-empty dining room, and said, "Well, do you want to change the food? Or do you want a revolution? Because I'm only interested if it's a revolution."

The director's answer was perfect. "I want a revolution . . . You can plant organic gardens. You can serve fresh fruits and vegetables to the scholars and residents. You can find local organic farmers and pay them the proper amount. You can bring in interns from around the world to work in the kitchen. You can write books about this, put them out into the world. You can have the moon." So Alice agreed.

"We worked hard on that project," Alice adds. "I was very lucky to find Mona Talbott, who's a great cook. But the most amazing, gratifying part is that within two days of the project's inauguration, everyone came back to the table."

ON A MILD JANUARY DAY IN 2016, ALICE WATERS SITS WITH A VISITOR. Chez Panisse happens to be closed today. A few updates are being made to the kitchen, so she's invited her guest home for tea. The house is simple but elegant, a modest 1908 Craftsman cottage with a lovely, old-fashioned kitchen. A grand piano graces the front room, surrounded by floor-to-ceiling bookshelves. The kitchen is painted a deep forest green. There are two wood-burning ovens built into one wall, as well as an open hearth for grilling. Dark stone counters, a copper sink, a long prep table, and a professional stove complete the work area. A farmhouse table stands off to one side; the chairs are charmingly mismatched.

Alice has lived in this house for more than thirty years, restoring it room by room. The attention shows. Small moments of beauty loom everywhere the eye alights: antique baskets on a shelf, an exquisite milk-glass window in the bathroom, a delicate heirloom textile on a daybed. The view from her library is of the back garden, which even in winter is planted with herbs. Two glass teapots sit on the coffee table, each filled with hot water and fresh mint—a generous fistful, which has just been plucked from the garden. It's a small detail, but a meaningful one. The mint flavors the water in a way no teabag could. The impulse is to savor each sip, to slow down and notice. Which is just as Alice intends.

Alice has a lot on her mind. She's a remarkable character by any measure. Not simply because she's a celebrated chef and the founder of Chez Panisse; or because she's the mother of a philosophy of cooking based on fresh, local, seasonal ingredients; or because she's America's most visible champion of organic farming, ethical agriculture, and healthily fed schoolchildren—all of which she is, but because she continues to crusade. Alice hasn't stopped fighting for our collective good health. At seventy-two, still faintly elfin, her honey-colored hair wispy and tomboy short, her face handsome and unlined, Alice remains a revolutionary, still the Berkeley activist of her FSM days. And yet, still also the sensuous Alice who fell in love with food and France so long ago. It's a potent combination.

"Eating is an agricultural act," she insists, quoting Wendell Berry, but it is also "a political act." We need to "eat with intention." Every decision we make has consequences. "You can support the people who are taking care of the land, or you can support the people who are destroying it: you make those decisions every time you buy food."

Alice believes we must help sustain the people who sustain us. We must treasure the farmers who are caring for the land, the water, our collective future. She wants to know where the salmon swam and what the baby calf chomped. She wants, if possible, to know the farmer who grew her food. And she wants that farmer to be paid a living wage.

"It's why I want to do this big event in Washington for the National Park Service," she says, her voice quickening.

As always, Alice had a big new project up her sleeve. She wanted to put on a birthday party for the one hundredth anniversary of our national parks. The idea was to celebrate America's local farm traditions, and to honor the National Park Service's legacy of land preservation and agricultural stewardship. There were hundreds and thousands of acres of agricultural land within our national parks, she explained, and most were sustainably farmed or ranched. The date she'd chosen for the festivities was August 25, 2016—Founder's Day—and she had elaborate plans. She envisioned an interactive garden running the length of the Mall, featuring edible plants from all fifty states; she wanted campfires and cooking, music and storytelling, art projects and films, a pop-up museum, and a stage with live performances. She'd already asked Patti Smith to be there, to sing "This Land Is Your Land," and Patti had said yes. Wendell Berry agreed to speak about community and the land, Robert Redford about a sustainable future for the planet, Michael Pollan about the national food consciousness. How could anyone say no? one wonders. Hearing the scope of Alice's plans, feeling the depth of her passion, her commitment, "you'd follow her anywhere," Dorothy Kalins once wrote. Millions already have.

Alice imagined a larger version of the edible garden she planted on the Mall in 2005, for the Smithsonian Folklife Festival. That project caught the attention of politicians with a say in national farm policies, who started dropping by for lunch. She hoped this celebration would advance the cause of sustainable farming too.

In the meantime, Alice has traveling to do, other irons in the fire. After tea, she will stop by Chez Panisse to check on today's progress. She wants to be sure everything's on track, as she's leaving for Italy soon. She won't be back in Berkeley for almost three weeks, which is not unusual. Alice's mission these days, like Goodall's, is global. The world is her stage, the table her platform. "I want to live the change I want to see," she says. "That's what Gandhi said. I want to live the values that are important to me." Alice has the ear of statesmen and senators, artists and presidents, and she continues to expand the Chez Panisse family. Mikhail Baryshnikov is her friend, and so is the Dalai Lama. She's cooked for the Clintons and corresponded with the Obamas. Yet if Alice is a star among stars, she uses her celebrity to promote the causes she supports. On this trip, she will stop in London to confer with her friend Prince Charles, England's most vocal champion of sustainable agriculture. She will speak at fund-raisers for the Chez Panisse Foundation, for the Edible Schoolyard, for Slow Food International, where she's now vice president. In Italy, she will join Carlo Petrini to talk through their plans for Terra Madre, this year's Slow Food International conference in Turin. Five thousand producers from across the globe are expected, including several from Berkeley. There's been good news there recently: the regional governments of Emilia-Romagna and Tuscany have adopted Slow Food's platform of biodiversity and sustainability, which will determine agricultural policies going forward. She still dreams that similar changes will be adopted in America's farm states.

Alice, of course, will eventually head home. On her way, she will check in at the American Academy in Rome, to see that all is well there. She'll visit Fanny, who lives in London now. If she gets the green light on the National Park Service's celebration, she'll stop in Washington for a day or two. If not, she'll begin cooking up another big idea designed to increase the nation's appetite for fresh, wholesome, sustainably grown food, hoping as always to advance America's food revolution. And once back in Berkeley, she'll do as she always does: she'll swing by Chez Panisse to tend and taste and flutter and perfect—to sample the night's soup, to feed a log onto the fire, to fuss over a table or deliver a prosecco to a friend.

Alice isn't finished with her "delicious revolution," she insists. "My daughter did a little cartoon of me one time that's so poignant," she says. "I told her, I'm not going to die quietly over there," she jokes, gesturing drolly in the direction of nowhere. "I'm going to go out fighting. Fanny called the cartoon 'Going Out with a Bang.' She had me as an old lady in a supermarket, saying, 'If you don't go organic, I'm going to let it blow.'

"I really do feel that way," she adds. "I can't *not* help to make this change happen. It's their lives, their future."

If Alice's dream for a little restaurant where friends could gather and share good food has grown into a revolution, she herself is not so changed. Four and a half decades later, Alice is still devoted to the same ideals: beauty, excellence, generosity, and the "healing magic" of the table. We need to have everyone speaking as loudly and as creatively as they can, she says. "To bring the world back to its senses."

chapter FIVE

....

HOPE IN THE SHADOWS

Today, the life and vitality of American cities seem anything but endangered. Work and performance spaces occupy once-abandoned warehouses; artists' studios colonize former factories; showrooms and tech start-ups fill once-tenantless industrial lofts. Bars, shops, bistros, galleries, and other grassroots ventures light up blocks that once felt dangerous and deserted. These days, such vital signs of dynamic urban life are if anything imperiled by their own enormous success, increasingly threatened by the sort of developer-driven hypergentrification they helped to inspire. Indeed, in cities from New York to San Francisco to Boston, it can seem today as if the pendulum has swung in the opposite direction. Though American cities comprise just 3.5 percent of the nation's landmass, they are home to 62.7 percent of the population, driving up real estate values and putting tremendous pressure on urban environments. Unlike in Jacobs's era, where "white flight" coupled with broad and comprehensive disinvestment left every urban center in America in danger of being abandoned, cities today are considered attractive places to live—so attractive, in fact, that they run the risk of being unaffordable to all but the affluent. Given all this, it can be hard to envision the plight of cities at the time that Jacobs began her brave crusade. But it's key to remember that by the middle of the last century, every city in America was literally dying. That they've made such an extraordinary comeback is a tribute to the power of Jacobs's thinking. Cities now are, if anything, victims of their own desirability. But we have to bear in mind how much better they are for it, even as we look for solutions to address a different kind of livability issue—that of affordability. If overheated gentrification remains the new challenge, it is arguably preferable to the bad old days when city governments engaged in massive social engineering, razing whole districts in the name of "slum clearance" and then warehousing the displaced in dismal public housing towers, which inevitably meant racial and economic segregation. The complicated problem of more demand than supply of livable urban neighborhoods with human-scale streets, affordable buildings, and a proximity to ordinary amenities is ongoing, of course—and largely market driven (which puts social agendas on the back burner). Theoretically, the conundrum could better be solved, as Jacobs often argued, by giving local communities more say in the changes that impact their own neighborhoods; by sensible zoning that isn't easily overridden by a developer's financial clout; by creative use of land trusts, nonprofit co-ops, and other ingenuities that retain neighborhood diversity. Which is to say, by gradual, "internally driven" improvements, rather than rapid upscaling by outside forces in the name of profits and tax revenues, which nearly always sells out the poor and the middle class. As with all change, it's two steps forward and one step back. The problems that bedevil cities these days are often the consequence of their own monumental appeal.

The state of America's food is getting better too—despite the nation's ongoing appetite for fast and processed foods. Beyond Whole Foods and other such upscale markets, where sustainably raised fish and meat, artisanal cheeses, and heirloom fruit and vegetable varieties are standard fare; beyond the flowering of gourmet food trucks and start-up organic wineries, specialty mushroom farmers and self-taught sausage makers, craft beer producers and small-batch pickle stands, the quality and variety of ingredients everywhere have truly and markedly improved. Today, wandering the aisles of even the most pedestrian supermarket, from Food Lion, to Kroger, to Safeway, one is likely to find an organic produce section, grass-fed beef, a sushi bar, a place to sample gelato, an in-store bakery offering "artisan breads." But these changes are not solely the province of yuppie foodies, and their importance can get lost in the veneer of easily satirized hipsterism. The availability of healthy and sustainably produced food has trickled down to the broadest common denominator. Fast-food chains like McDonald's and Burger King, responding to changing tastes, now offer a few healthy alternatives to the standard Whopper or Big Mac: apple slices in lieu of french fries in a more wholesome version of a Happy Meal, a salad featuring "spring mix lettuces" that includes arugula, radicchio, and frisée. Even the great demon Walmart, the nation's largest mass-market retailer, has an aisle dedicated to pesticide-free products, minimally processed meats, baked goods with no artificial ingredients. If these developments have lowered the bar for what qualifies as "organic," as some would contend, they nonetheless represent a very real shift in consciousness and, as such, are part of the complicated push and pull of progress. They stand as a vital first step in the right direction.

Indeed, tucked away in rural pockets of New York's revitalized Hudson Valley, or the back roads of Vermont or Virginia or California or Maine, it's no longer unusual to find enterprising young farmers growing new vegetable varieties from heirloom seeds without using chemicals, raising chickens and beef in sustainable and humane ways, making cheeses and charcuterie as delicious and interesting as anything in Tuscany or Provence. There is a sense of idealism and craftsmanship in these burgeoning endeavors, a pride of place, yes, but also, more tellingly, an understanding that the quality of food is inextricably linked to how we care for the land.

Even conservative Kansas farmers who profess not to believe in climate change have turned to environmentally friendly farming methods of late, from "no-till" techniques that decrease soil erosion, to planting a rotation of grasses that anchor the soil and conserve dwindling water. They understand the need to "mimic" the diversity of what was once the natural prairie ecosystem, and to capitalize on sustainable sources of available energy. As of 2014, Rebecca Solnit reports, conservative Iowa gets "28% of its electricity from wind alone." Thirty percent more solar power was installed in 2014 than the year before in the U.S. It's no longer just left-leaning tree huggers who speak about protecting soil fertility with "mob grazing," or leaving corn stubble standing after harvest (which keeps precious topsoil from blowing or washing away), or who worry about shrinking aquifers and "weather" they warily admit has grown more extreme. Faced with severe drought followed by freakish storms over the last decade, these heartland farmers understand the logic, if they don't always agree with the liberal politics they say swirl around the subject. They know their fate as farmers is inextricably tied to the fate of the land. Adopting restorative farming practices has become a question of survival.

Even in the poorest regions of Africa and Latin America, there is a rising awareness of the need for sustainable land-use policies. Brazil, home to the largest piece of the Amazon, has moved from clear-cutting forests to policies aimed at rehabilitating the damaged acreage. According to the United Nations, the country was able to reduce its deforestation rate by 74 percent between 2004 and 2009, albeit with the help of international donors. (This is especially important because deforestation, according to the Union of Concerned Scientists, "generates one-tenth of total global warming emissions.") Similarly, Indonesia, for years accustomed to handing out timber concessions and promoting the expansion of palm oil plantations, has recently embarked on efforts to curtail deforestation. Against still more formidable odds, progress has been reported in the Democratic Republic of Congo as well. Long considered a "fragile state," where corruption is rampant and the judicial system a sham, the country has initiated measures to protect the Congo forest and other natural resources. Globally, say UN researchers, "protected areas" grew by 38 percent in 2010 from their level in 1992. Though a fraction of what needs to be done, it's a beginning.

Given all this, it's easy to forget how imperiled these urban and rural landscapes once were, both physically and metaphysically, or the monumental role that Rachel Carson, Jane Jacobs, Jane Goodall, and Alice Waters played in igniting these transformative changes. Which is to say, it's easy to forget how far we've moved from the reckless plunder of the 1950s, even with recent political setbacks.

America *then* was a different place from America today. It was racing blindly and ruinously toward more automation, more standardization, more chemicals, more nuclear arms with no sense of consequence. Sleek and big-spending, it was a narrower, less worldly, more homogenous nation: a place of corporate strivers and stay-at-home moms; soaring birthrates and spreading suburbs; intense social conformity and blighted cities; Cold War paranoia and a swelling military-industrial complex. The nation was in the thrall of technology and big science, aided by an army of specialists. The economy was growing at a gallop, spreading prosperity at a pace never before seen. If social inequities were mounting, material ingenuity seemed boundless, economic expansion a promise without end.

"To a populace whose forebears had within living memory colonized the interior of a vast continent and whose country had never lost a war," writes E. O. Wilson, "arguments for limit and constraint" didn't figure into the mainstream political agenda. Conservation issues commanded little to no attention; ecology was an unknown concept. The scientific community was preoccupied instead with the stunning achievements of the "molecular revolution," which highlighted chemistry and physics, a direct consequence of the triumphs of World War II. "Researchers were learning to reduce living processes to their molecular components."

Science at that moment was deemed infallible. That it was capable of giving mankind "the whip-hand over nature," as Aldo Leopold ruefully observed, was seen as "progress," a mark of modernity and national vigor. Indeed, aided by ingenious new technologies, insects could now be eliminated with chemical poisons; farmland treated with synthetic fertilizers to enhance efficiency; animals fed an array of pharmaceuticals to speed up growth; industrial foods concocted to save time and money; cities reengineered to accommodate the highway and the car. The downside, of course, was that the same technological cunning also made total annihilation by nuclear war possible.

It was a schizophrenic age. If Americans were privately fearful of nuclear Armageddon, they were also distracted by the nearness of the American dream. In the words of one *New York Times* reporter writing at the time, the culture was "drowning in a sea of luxury and mesmerized by the trivialities of the TV screen." The typical male striver could expect a well-paid job and a split-level house in the suburbs with an emerald green lawn and a backyard barbecue, a family of kids and a garden shed full of chemicals, whether he was blue collar or scaling the corporate ladder. The prospects for his stay-at-home wife, of course, were more limited. Above all, she was expected to look pretty and care for the house, to be a discerning consumer and an attractive sidekick to her husband's career. She might volunteer to work in a political campaign, or tutor in a settlement house; she might start a cooperative nursery school or even work part-time toward a degree in social work or teaching, but it was understood that she'd be back in the kitchen before the kids got home from school. Her job definition was helpmate and support staff. "You may be hitched to one of these creatures we call 'Western man,'" Adlai Stevenson told the Smith class of 1955, "and I think part of your job is to keep him Western, to keep him truly purposeful, to keep him whole." A housewife's task, he added, was an important one, since "we will defeat totalitarian, authoritarian ideas only by better ideas"—the assumption being that it was her role to instill those ideas in her husband's head, rather than to act on them herself.

It's easy to forget how few options a woman had at that moment. Few aspired to law or medicine, and for good reason. Most medical and law schools banned female students or severely limited their numbers. If a woman was a journalist, she was unwelcome in the newsroom, shunted off instead to the women's pages. If she was a scientist, she could teach, but her work rarely if ever included lab or field research. She might become a secretary, of course; scores of women did (including Jacobs and Goodall). And she could be a stewardess. But if she married she had to quit. In 1961, the year Jane Goodall returned to Gombe for the second time and *Death and Life* was published, a year before Rachel Carson's *Silent Spring* and three before Alice Waters transferred to Berkeley, many states still had "head and master" laws, stating that a wife was subject to her husband's will. If a woman and her husband owned joint property, he could sell or mortgage it without seeking her permission. If they applied for a mortgage, her income (if she was fortunate enough to have an income) was only considered if she was "at least age 40 or could prove that she had been sterilized." If she owned property herself, once she married, any rental income was solely his. America in the 1950s was a nation with seemingly limitless horizons, but a woman's place was still in the home.

And then into this blustery, all-male world of patriarchs and company men, technocrats and cold warriors, walked four women who saw things differently and were unafraid to say so; four women joined not by friendship, or generation, or even their fields, but by the stunning convergence of their ideas, which, taken together, would not only shape and inform the ethos of the emerging counterculture, but also become the building blocks for today's progressivism, in effect all but changing our world.

Instead of seeing science and advanced technology as the "holy grail," Rachel Carson, Jane Jacobs, Jane Goodall, and Alice Waters saw perils, sounding the alarm on the culture's increasingly "reckless assault" on the environment, both the natural biota and the built world. Instead of strict hierarchies and separations, they saw unities and connections, the world as a holistic system. Instead of sweeping generalizations, they saw complexity and fine-grained detail. Instead of the world as an inert place, they saw movement and flow, evolution and process. Instead of mathematical abstractions, they saw living communities.

Each of these women was acutely observant, each in her own

right attuned to networks and systems, to how things work and how they work together, whether it was the dynamics of an urban community, or an ocean ecosystem, or a group of chimpanzees, or microbes in the soil. Each displayed a profound respect for intuition and the wisdom of direct engagement, what the writer Verlyn Klinkenborg has called "thinking with all your senses," keenly aware of the relationship between the body and the physical environment.

Close observers of process and patterns, all saw the world as a web of interactions and exchanges, rather than a strict hierarchy; a mesh of dynamic, interconnected communities, rather than a place of closed and separate compartments, which at the time was the dominant view. Instead of counting and categorizing phenomena— oyster beds in the sea, citizens in a blighted neighborhood, vervet monkeys in a forest, the net return on a carrot—each was focused on mapping relationships, on understanding the interplay between living things and the systems that sustained them. All intuitively grasped the overarching idea of "connection," which is the basis of what we now call "web" or "systems thinking." If these insights seem self-evident today, it is only because of how thoroughly we have internalized their essence. For these women's revolutionary ideas about the interconnected nature of the earth, so foreign and threatening at the time, not only turned out to be prescient, but culture changing—the catalyst to a radical shift in consciousness. "The change that counts in revolution takes place first in the imagination," observes Rebecca Solnit.

Wary of the institutional face of power, deeply alarmed by the culture's reckless war on nature, each of these women warned against the regimented, the standardized, the faceless and centrally controlled, seeing in these elements the authoritarian impulse. Unafraid to discuss the ways in which powerful constituencies were destroying the natural and urban environments, not only physically, but also metaphysically, each was a champion of the "little guy against the machine," whether it was the machine of Big Ag, or city planning policies, or government pesticide programs, or efforts to engineer animals, recognizing the hubris of trying to remake nature to serve mankind's ends. Instead of embracing the status quo, each looked to whose interests were actually being served, arguing that specialization, coupled with corporate greed and ruinous government policies, had left people in the dark about the implications of decisions being made on their behalf. All believed that ordinary citizens, armed with the facts, were capable of deciding for themselves.

The culture at that moment gave priority to ideology and technical expertise, to quantifiable data rather than subjective experience. Each of these women, by contrast, drew upon lived experience, resisting received information and the sovereignty of specialists. All worked "practically and additively," beginning with the individual and the particular—the qualities of a single city block, the expressions of kinship between a mother chimp and her young, the taste of a single bean—and then built from there, looking for like patterns. Intuitively, all understood that in order to grasp the "big picture" one had to understand the small, "essential linkages" first, that it was often the detailed view, the idiosyncratic and irregular, that was the most revealing. Instead of the false neutrality of the design theorist or the traffic engineer, the agricultural technician or the academic zoologist, each of these women used the felt and observed as the template upon which to build her ideas. Each viewed big philosophical issues—the poisoning of the environment, the industrialization of agriculture, the wholesale razing of cities, the commodification of animals—from the perspective of individuals, unafraid to put a face on complex problems.

Exuberant, uncommonly focused, all of these women loved what they did. Each was a crusader by virtue of her visceral connection to her work, not because she set out to change the world. Each was driven by a sense of passion and curiosity, not by egotism or a lust for fame. Eloquent communicators, all understood that people will only protect what they love, that changing minds inevitably meant touching hearts. At a time when the culture was preoccupied with what needed to be suppressed, if not eliminated, these women were focused on what warranted nurture and saving (local, small-scale economies; imperiled habitats and endangered species; the social rituals of the table). Instead of looking at what was dying or should be killed, each looked at what was working and why.

"Traditional conservation is based in part on negative emotions," observes Thomas McNamee—"guilt and shame for damage we've visited to our land, anger towards those who despoil it." All of these women, by contrast, opted for a different way in; they began their critiques from an affirmative stance, appealing to people's emotions and fascinations by drawing upon their own. Jane Jacobs's arguments were predicated on the premise that cities were wondrous entities, capable of enriching lives. Rachel Carson celebrated the beauty and mysteries of the natural world, even as she warned darkly of the threats to its viability. Jane Goodall's hypnotic storytelling brought the world of primates to life, providing a rationale for why these creatures merited consideration. Alice Waters's love of food and friends and the bounty of the earth inspired as it educated. In each case, these women gave readers a reason to care. They wrote books that inspired people, but also, more importantly, profoundly moved them.

We are divided as a culture between "exploitation and nurture," the farmer-poet Wendell Berry has written. "The standard of the exploiter is efficiency; the standard of the nurturer is care. The exploiter's goal is money, profit; the nurturer's goal is health—his land's health, his own, his family's, his community's, his country's. The exploiter typically serves an institution or organization; the nurturer serves land, household, community, place. The exploiter thinks in terms of numbers, quantities, 'hard facts,' the nurturer in terms of character, condition, quality, kind."

Rachel Carson, Jane Jacobs, Jane Goodall, and Alice Waters were nurturers. Instead of material expansion, each emphasized quality of life, the public good, what was sensible and ethical. Instead of measuring time by short-term markers, each was looking for what was sustainable in the long term. Instead of artificial order imposed from on high—centrally planned housing towers, carpet-bombing vast swaths of farmland with chemical poisons, an agro-industrial system in which every link in the food chain was under the same corporate umbrella—each argued for change that grew from the ground up, local, organic solutions rather than industrial manipulation. Instead of seeking dominion over nature, each advocated a respect for its processes. The hubris implicit in attempting to control nature, they warned, held within it the potential to still all life.

What all, in effect, grasped, and what connects each to the other in important and synergistic ways, were the unanticipated long-term implications of what had seemed ingenious new implements of progress—the chemical control of nature, the mass production of housing, the industrialization of food and farming, the commodification of animals. All were undermining the natural systems that sustained the living world and, by extension, the sensory and emotional quality of experience.

Put simply, they identified the moral crisis of modernity, and in so doing spoke to a rift that had divided the country since its first beginnings: the ongoing push and pull between reverence for untrammeled nature, and the rampant quest to exploit it. It was a clash that had played out in the culture before, but never with such stark implications. The fact of the bomb, and by extension the horrors of Hiroshima, had changed the calculus, fundamentally upending the scales, underscoring the link between "science-based technological progress" and the "long-term degradation of the environment." And yet, the existential threat from outside—the fear of the Communist menace, the accelerating nuclear arms race, the competitive push into space—had seemed to justify the peril. Until it didn't. In the age of nuclear weapons, Martin Luther King Jr. warned, the choice is either "nonviolence or nonexistence." It was the same message Carson had delivered.

As the deceptively placid world of the 1950s hurtled toward a rendezvous with the turbulent sixties, the culture—at least a part of it—began to recalibrate, driven by the first shock troops of a counterculture just beginning to coalesce, a cohort that was finding in these women's ideas—whether consciously or because they had entered the bloodstream of the culture—a platform upon which to build (the exception being Alice Waters, who, as *part* of the counterculture, was both channeling these ideas and adding to them). Unlike the previous generation, this gathering demographic was unconvinced by the ongoing posture of Cold War emergency, seeing it as a smoke screen to justify inaction at home; instead of the threat from *without* (which seemed overblown), it was focused on the problems *within*: racial injustice, inner-city poverty, DDT in milk, the growing quagmire that was Vietnam, the ongoing plunder of the environment. All were symptoms of a world knocked out of balance, the thinking went, a world imperiled by technological hubris. A world, in short, in need of "a total overhaul." To a generation of young, draftable males faced with either "quick death" in Vietnam or "slow strangulation" as a cog in some faceless corporate machine, the sense of crisis was acute; the idea of changing everything held obvious appeal. At that moment, the personal *was* political.

"A time comes when silence is betrayal," wrote Martin Luther King Jr., in a little-remembered speech delivered at New York City's Riverside Church in 1967. "In Vietnam," he added, "that time has come." To King and the insurgent young, Vietnam was a battle of technology against a small peasant nation we were bombing back to "the Stone Age," steadily and ruthlessly defoliating its forests and farmland, crushing its culture. Vietnam exemplified a terrible truth about the way "power and technology, once possessed," were inevitably loosed upon the modern world. "We have destroyed their two most cherished institutions," King went on, "the family and the village. We have destroyed their land and their crops . . . They watch as we poison their water, as we kill a million acres of their crops. They must weep as the bulldozers roar through their areas preparing to destroy the precious trees." It was a "madness," he added, that needed to stop. "I speak for the poor of America who are paying the double price of smashed hopes at home, and death and corruption in Vietnam . . . A nation that continues year after year to spend more money on military defense than on programs of social uplift is approaching spiritual death."

If King's challenge fell on deaf ears in the political mainstream, his assertion that injustice at home was inextricably bound to a policy of violence against nature and peoples abroad resonated with the dissident young, who already equated technological overreach with ecological disaster.

By 1969, seven years after *Silent Spring*, and a year after Jacobs moved to Toronto in protest of the war, the environment had become central to the politics of resistance. That year an oil spill off Santa Barbara had blackened beaches and buried birds alive. In Ohio, Cleveland's Cuyahoga River caught fire. Nearly every day it seemed, there were dire new environmental warnings: smog alerts in Los Angeles, news stories about DDT in breast milk, PCB poisoning in Japan (and later the Hudson River), mercury contamination in fish, beaches fouled with raw sewage, predictions of impending world hunger. "We must realize that unless we're extremely lucky, everyone will disappear in a cloud of blue steam in twenty years," wrote Paul Ehrlich in mid-1969.

That same year, in an act of peaceful civil disobedience, a group of self-proclaimed Robin Hoods seized a vacant lot owned by the University of California, Berkeley, and planted it with grass, vegetables, and trees, with the express purpose of feeding organic food to the poor and turning a wasteland into a garden. Unwilling to let a good deed go unpunished, Governor Ronald Reagan immediately called out the National Guard. Tear gas and bulldozers followed, beginning a political standoff freighted with meanings that would go far beyond the ownership of the garden.

To the counterculture, as Warren Belasco writes, the government's violent answer to what had seemed a gentle "bid for green space" symbolized a deep-seated American "disdain for nature," and, more immediately, the "mass defoliation in Vietnam." The confrontation represented everything that was wrong with America: its brutalization of nature, its accelerating industrialization, its compulsive need to conquer. It raised questions, as a later write-up put it, about "the quality of our lives," "the deterioration of our environment," and "the propriety" and "the uses to which we put our land."

People's Park, as it was dubbed, was both metaphor and model:

an effort to replace an attitude of dominion toward nature with one of respect for its systems, and a way of growing food without using agricultural chemicals. As such, it was also a repudiation of America's war machine, which deployed these same poisons in Southeast Asia.

People's Park "pointed away from violence, toward ecology," Belasco notes. ("Revolutionaries must begin to think in ecological terms," pronounced one underground newspaper.) It signaled the "greening" of the counterculture, the tilt toward nature that would lead to the back-to-the-land movement and, eventually, to the push for sustainable, chemical-free agriculture. Metaphorically, it was an embrace of the ecological idea that "everything is connected to everything else": planetary survival, the antiwar movement, civil rights. In this increasingly fragile, intertwined world, concern for the environment was a recognition not just of one imperiled habitat, but of all the problems that bedeviled the planet: starvation in India, chemical poisoning, crime in the inner city, smog in L.A., racial discrimination, the farms and forests of Vietnam. All were signs of breaks in the threads that bind life to life, as Carson had argued: a world thrown out of whack by unchecked technology.

It was this elemental truth that so aligned Rachel Carson, Jane Jacobs, Jane Goodall, and Alice Waters with their age: the dawning awareness of the interconnectedness of the living world. All life systems were interlaced, they argued, and all were threatened when this closely knit fabric was ripped apart. These four women, in their collective aversion to the industrialization of nature, their deep discontent with Western ideals of progress and power, their outcry over the exploitation of the commons, their embrace of nurture over exploitation, their willingness to marry action to idea, their push for decentralized economies and sustainable systems, not only anticipated the disaffections of the emergent counterculture (or, as in Alice's case, embodied them) but also affirmed many of its core ideals. What each in essence argued was that the way forward was not by reengineering nature, but by respecting, in the deepest sense, its cycles and systems, the fundamental interconnectedness that we now call "ecology." As such, their voices both seeded and illuminated their cultural moment, setting into motion ideas that would prove both transformative and enduring.

Paradigm shifts, especially revolutionary ones, rarely if ever occur without resistance. When they creep in from outside the academy, opposition is only that much keener. Each of these women faced down powerful and entrenched interests—in government, in science, in academia, in industry. Each suffered derision, dismissal, efforts to silence or discredit their arguments. And yet, to our good fortune, all tenaciously stood their ground, in this way changing the way we think about entire fields: architecture and the organic life of cities; conservation biology and the interconnectivity of the natural biota; zoology and the social behavior of animals; food and cooking and the links between sustainable farming and the health of the planet. These four visionary women—in many ways the unsung heroes of the counterculture—all but created the world we live in today. They dared to imagine a more humane and sustainable path forward, interrupting what had seemed like the inevitable march of history. Out of their arguments came social movements that would change the world.

The story, of course, isn't over. The battles these women ignited are still being fought. More than fifty years later, the partisan divide between "exploiters" and "nurturers" hasn't closed. Sadly, our putative leaders have "devolved" from what Carson called "little tranquilizing pills of half-truth" to "alternative facts." Well-financed fossil-fuel interests, coupled with their Republican allies, regularly and systematically undercut scientifically established research on everything from climate change to the perils of industrial chemicals. We are still victims of what Gro Harlem Brundtland, the former prime minister of Norway and the chairwoman of the World Commission on Environment and Development, calls "the tyranny of the immediate," putting short-term economic interests ahead of long-term "environmental necessity." Indeed, despite all we know, toxic chemicals continue to be dumped onto our land and into our water; industry still extracts the earth's resources for short-term gain; animals are still harnessed to serve mankind's ends. Foods are still genetically modified to gin up profits, historically significant buildings torn down by greedy developers, forests felled and wildlife species lost with chilling regularity—all with no thought to environmental consequences, or to our collective well-being or sense of place. And yet, we have also come a long way since these four remarkable women first raised their voices. The EPA was created in 1970 as a consequence of Rachel Carson's urgent call to arms, followed by the Endangered Species Act in 1973, galvanizing generations of future activists. Her brave and enduring work would ultimately be regarded as the "founding text" of the environmental movement.

Today Jane Jacobs's prescriptions for healthy cities—compact, densely settled neighborhoods; pedestrian-friendly streets; mass transit; reuse and adaptation of old structures—are guiding principles for the professional planning community, as well as the New Urbanism, an architectural movement dedicated to combating "placeless" suburban sprawl. Once-desolate city plazas now teem with people-friendly amenities, from farmers' markets, food trucks, and pop-up cafés, to seasonal skating rinks, film venues, and concert spaces. Similarly, ideas about new development have shifted too. Thanks to Jacobs's work, urban renewal and top-down redevelopment schemes are widely viewed with suspicion, and highway construction through city centers is no longer considered a viable option. Instead, the trend these days is to dismantle intrusive urban highways (as happened in Milwaukee, Denver, Baltimore, Buffalo, and Boston) or turn ill-conceived or outmoded arteries into pedestrian parks, such as New York's High Line, an elevated freight rail line on Manhattan's West Side that was transformed into a much-used, much-loved public space. "Eyes on the street" has not only been adopted by the design profession, but also as a successful strategy for law enforcement. And *Death and Life* remains required reading for anyone interested in urban issues. Jacobsian ideas about how decentralized, ground-up systems function have even shaped the platform designs of social media and online networks.

Animal studies have been similarly impacted by Jane Goodall's work. Five decades later, her methodology is still considered the gold standard for inquiries into the social systems of animals from ants to eagles. Zoo conditions, too often devoid of adequate space and sensory stimulation, have markedly improved in response to her outreach and educational efforts, as has the care and number of primates living in hopeless captivity in biomedical laboratories. Goodall's activism on behalf of endangered species and habitats has become a model to environmental leaders across the globe, from the directors of Save the Elephants to the Orangutan Project. Her global youth program, Roots & Shoots, continues to flourish, challenging young and old alike to a deeper appreciation of our collective responsibility to the earth.

Likewise, the sustainable food and farm-to-table movements show no signs of ebbing, thanks to Alice Waters's ongoing advocacy, reinforced by a fresh generation of chefs and educators who carry the torch, spreading the word and growing the community. Slow Food International now boasts 100,000 members, spread over 150 countries, from Kazakhstan to Mexico. School garden programs continue to multiply, many of them in low-income communities, teaching kids the cycles from seed to harvest to table. Community garden projects bloom in countless cities today, from Seattle, to Salt Lake City, to South Central L.A., cutting across class, race, and gender lines. New Roots, a program sponsored by the International Rescue Committee, helps refugees settle in the U.S. by literally growing "new roots" in American soil. Guerrilla Gardening, a London-based initiative, works to reclaim neglected public spaces and grow vegetables, flowers, and trees. Bronx Green-Up, the New York Botanical Garden's outreach program, has helped create community gardens and urban farms throughout the poorest sections of the South Bronx. A prime example of the movement is Green Bronx Machine. Started as an after-school program for high school students, today it provides some of New York's most economically disadvantaged students with an education that includes job training, urban farming, entrepreneurial skills, and community work.

Waters's influence on the nation's collective table is ubiquitous too. Chefs and home cooks these days seek out fresh, local, seasonal produce as a matter of course, and farmers' markets pepper the landscape of country and city alike, their numbers having increased by a whopping 180 percent between 2006 and 2014. Culinary excellence in America is no longer confined to a few select restaurants in a handful of cities. A good farm-to-table restaurant is just as likely to be found in a small town in West Virginia as it is in Sonoma, a decent market offering high-quality, sustainably grown produce in rural Kansas as in New York. No longer an obscure topic, the environment today looms front and center in our political consciousness. Even small gains move us forward.

Rachel Carson, Jane Jacobs, Jane Goodall, and Alice Waters appeared at a watershed moment in the culture, presenting us with a road map to the future that grows more relevant, if not more urgent, with time. Their work continues to move, to inspire, and to stir us to action, reminding us that the power of one voice can be transformative; that change can and does begin with the local, the particular, and the passionately observed; that the best ideas can and often do come from the bottom up. That one individual can make a difference.

ACKNOWLEDGMENTS

No book is ever a solo endeavor. I owe a great many people thanks for their support and contributions to this one. Thanks first to my writer friend Kurt Andersen, who gave me the seeds for this book; to Anne Kreamer, for her friendship and wise counsel along the way; to Aki Busch, for her humor and support and for passing on a book that was pivotal to this project early in its young life; and to Patty Cronin, for doing the same with an article that helped set my path.

Special thanks are due to Denise Oswald, my editor, for her smart and perceptive reading and wise editorial suggestions; she did what good editors do, she made this book better; to her assistant, Emma Janaskie, for her discerning comments and invaluable computer help; to Victoria Mathews and Trent Duffy, for their copyediting; and to my agent, Joy Harris, for her enthusiasm, her savvy, her kindness, and her unwavering support and professionalism.

Thanks also to the many friends who lent me books and steered me to authors and sources I might not have found. Special mention goes to Rhiannon Leo-Jameson, the wonderful librarian at Northeast-Millerton Library; my sister Kim Springer, who gave me access to her library in Wyoming; Heidi Cunnick, for her enormous generosity in sharing her access to the Columbia University Libraries; Peter Wheelwright; and Michael Pollan. Additional thanks to Caroline Stewart, for her insights, endless generosity, and steadfast friendship; Elisabeth Cunnick, for good company and conversation; Eliza Hicks, for photographs and friendship; Jill Choder and Michael Goldman, for lent books, levity, and laughter over numerous dinners when I was a work orphan; Brooke Allen for her extraordinary close and helpful reading; Lili Francklyn, who sent me articles throughout this journey; Susan Horton, for invaluable help with Skype; Eric Rayman, for wise counsel; and to Mary and Walter Chatham, for their friendship and support. Grateful thanks also to Jim Jacobs, for his warmth, time, and memories of his mother, Jane Jacobs; Alice Waters, for the hours she spent talking to me; and to Jane Goodall for the same, for taking precious time from her family Christmas for an interview. I am grateful to Greil Marcus, for sharing his memories of Chez Panisse; Ruth Reichl, for her descriptions of Berkeley in the seventies and her insights about her friend Alice Waters; and to Samantha Wood for her in-depth tour of the Edible Schoolyard. Thanks are also due to Barbara Wright, for her close and generous reading of the manuscript at a critical juncture, and to my daughter, Pippa White, for her smart and thoughtful suggestions on my proposal. I would also like to acknowledge the many generous library curators who shared their time and expertise during my search for photographs: Martha McClintock at Getty Images; Molly Tighe at Chatham University's JKM Library; Mary Paris, from the Jane Goodall Institute; Hannah Love, Alice Waters's winningly generous and warm aide-de-camp; Hannah Silverman, assistant to Cervin Robinson; Stephen Goldsmith, at the Center for the Living City; Erik Huber and Mary Grace DeSagun at the Queens Library archive; Paul Civitelli and Dolores Colon at the Yale Beinecke Library archives; Benjamin Panceira at the Linda Lear Center for Special Collections & Archives at Connecticut College. Thank you also to Sarah Yake of the Frances Collin Literary Agency, trustee of the Rachel Carson estate, for her help with permissions. Finally, I remain eternally grateful to my extraordinary husband, Kit White, for his unflagging generosity and support, for patience in the face of long work nights, gourmet dinners that never stopped, close readings when I most needed them, and humor I could always count on. This would have been a lonely project without him.

SELECTED BIBLIOGRAPHY

Ackerman, Diane. *A Natural History of the Senses*. New York: Random House, 1990.

Alexander, Christopher. *A Pattern Language*. New York: Oxford University Press, 1977.

Alexiou, Alice Sparberg. *Jane Jacobs: Urban Visionary*. New Brunswick, N.J.: Rutgers University Press, 2006.

Allen, Max, ed. *Ideas That Matter: The Worlds of Jane Jacobs*. Toronto: The Ginger Press, 1997.

Anderson, Sara F. "The View from the Outside: How Three Women Contributed to Changes Toward Equity and Human Rights: A Study of the Work of Rachel Carson, Jane Jacobs and Betty Friedan." *New England Journal of History* 52, no. 1 (Spring 1995).

Angier, Natalie. "A Society Led by Strong Females." *New York Times*, September 13, 2016.

Aubrey, Allison. "About a Third of U.S. Kids and Teens Ate Fast Food Today." *The Salt* (blog). NPR. September 17, 2015.

Baldwin, James. *Nobody Knows My Name*. New York: Vintage, 1961.

Barr, Luke. *Provence, 1970: M. F. K. Fisher, Julia Child, James Beard, and the Reinvention of American Taste*. New York: Clarkson Potter, 2013.

Belasco, Warren J. *Appetite for Change: How the Counterculture Took On the Food Industry*. Ithaca, N.Y.: Cornell University Press, 2007.

Berman, Marshall. *All That Is Solid Melts into Air: The Experience of Modernity*. New York: Penguin, 1988.

Berry, Wendell. *The Unsettling of America: Culture and Agriculture*. San Francisco: Sierra Club Books, 1977.

Bloom, Alexander, and Wini Breines, eds. *"Takin' It to the Streets": A Sixties Reader*. New York: Oxford University Press, 1995.

Bolois, Justin. "The 10 Dishes That Made My Career: Alice Waters." FirstWeFeast.com, April 20, 2015.

Broms-Jacobs, Caitlin, ed. *Jane at Home*. Toronto: Estate of Jane Jacobs, 2016.

Bromwich, David. "Martin Luther King's Speech Against the Vietnam War." Antiwar.com, March 16, 2008.

Brooks, Paul. *The House of Life: Rachel Carson at Work*. Boston: Houghton Mifflin, 1972.

Carson, Rachel. *Always, Rachel: The Letters of Rachel Carson and Dorothy Freeman, 1952–1964*. Edited by Martha Freeman. Boston: Beacon Press, 1995.

———. *The Edge of the Sea*. Boston: Houghton Mifflin, 1955.

———. *Lost Woods: The Discovered Writing of Rachel Carson*. Edited by Linda Lear. Boston: Beacon Press, 1998.

———. *The Sea Around Us*. New York: Oxford University Press, 1951.

———. *Silent Spring*. New York: Houghton Mifflin, 2002.

———. *Under the Sea-Wind*. 1941. Reprint, New York: Oxford University Press, 1952.

Collins, Gail. *When Everything Changed: The Amazing Journey of American Women from 1960 to the Present*. New York: Little, Brown, 2009.

Coontz, Stephanie. *A Strange Stirring: The Feminine Mystique and American Women at the Dawn of the 1960s*. New York: Basic Books, 2011.

Cronon, William, ed. *Uncommon Ground: Rethinking the Human Place in Nature*. New York: Norton, 1996.

Curtis, Olga. "Time-Saving Modern Kitchen Miracles." *Washington Post and Times Herald*, April 22, 1957, C2.

Dallek, Robert. "What Made Kennedy Great." *New York Times*, November 22, 2013.

Deitz, Paula. "Literature and the Environment." *The Hudson Review* 66, no. 1 (Spring 2013).

Dickstein, Morris. *Gates of Eden: American Culture in the Sixties.* New York: Basic Books, 1977.

Dreifus, Claudia. "In 'Half Earth,' E. O. Wilson Calls for a Grand Retreat." *New York Times*, February 29, 2016.

Eisler, Benita. *Private Lives: Men and Women of the Fifties.* New York: Franklin Watts, 1986.

Ellis, W. S. "A Way of Life Lost: Bikini." *National Geographic*, June 1986, 813–34.

Epstein, Jason. "Way Uptown." *New York Times Magazine*, April 11, 2004.

Evans, Sara. *Personal Politics: The Roots of Women's Liberation in the Civil Rights Movement & the New Left.* New York: Knopf, 1979.

Fisher, M. F. K. *The Art of Eating.* New York: Collier, 1990.

———. *With Bold Knife and Fork.* Berkeley, Calif.: Counterpoint Press, 1969.

Flint, Anthony. *Wrestling with Moses.* New York: Random House, 2009.

Foundation for Economic Education. "Jane Jacobs." Sandy Ikeda, September 2006. https://fee.org/articles/Jane-Jacobs/.

Fradkin, Philip L. *Fallout: An American Nuclear Tragedy.* Tucson: University of Arizona Press, 1989.

Friedan, Betty. *The Feminine Mystique.* New York: Norton, 1963.

Friedman, Thomas. "Stampeding Black Elephants." *New York Times*, November 23, 2014.

Gaddis, John Lewis. *The Cold War: A New History.* New York: Penguin, 2005.

Galanes, Philip. "The Fights of Their Lives: Ruth Bader Ginsburg and Gloria Steinem Recall the Long Campaign for Women's Rights." *New York Times*, November 15, 2015.

Gallagher, Winifred. *Rapt: Attention and the Focused Life.* New York: Penguin, 2009.

Goines, David Lance. *The Free Speech Movement: Coming of Age in the 1960s*. Berkeley, Calif.: Ten Speed Press, 1993.

Goldsmith, Stephen A., and Lynne Elizabeth. *What We See: Advancing the Observations of Jane Jacobs*. Oakland, Calif.: New Village Press, 2010.

Goodall, Jane. *Africa in My Blood: An Autobiography in Letters; The Early Years*. Edited by Dale Peterson. Boston: Houghton Mifflin, 2000.

———. *Beyond Innocence: An Autobiography in Letters; The Later Years*. Edited by Dale Peterson. New York: Houghton Mifflin, 2001.

———. *In the Shadow of Man*. Boston: Houghton Mifflin, 1971.

———. *Reason for Hope*. New York: Grand Central, 2000.

———. *Through a Window: My Thirty Years with the Chimpanzees of Gombe*. Boston: Houghton Mifflin, 1990.

Goodall, Jane, with Gail Hudson. *Seeds of Hope*. New York: Grand Central, 2014.

Goodall, Jane, with Thane Maynard and Gail Hudson. *Hope for Animals and Their World*. New York: Grand Central, 2011.

Goodall, Jane, with Gary McAvoy and Gail Hudson. *Harvest for Hope: A Guide to Mindful Eating*. New York: Grand Central, 2006.

Gopnik, Adam. "Annals of Gastronomy: The Millennial Restaurant." *The New Yorker*, October 26, 1998.

———. "Street Cred: What Jane Jacobs Got So Right About Our Cities—and What She Got Wrong." *The New Yorker*, September 26, 2016.

Gould, Kira, and Lance Hosey. *Women in Green*. Washington, D.C.: Ecotone Press, 2007.

Gratz, Roberta Brandes. *The Battle for Gotham*. New York: Nation Books, 2010.

———. *Cities Back from the Edge*. New York: John Wiley and Sons, 1998.

Grimes, William. "Bill Berkson, 76, Poet and Art Critic of '60s In-Crowd." *New York Times*, June 20, 2016.

Griswold, Eliza. "How 'Silent Spring' Ignited the Environmental Movement." *New York Times*, September 21, 2012.

Gross, Terry. "Forty Years of Sustainable Food" (interview with Alice Waters). *Fresh Air*, NPR, August 22, 2011.

Halberstam, David. *The Fifties*. New York: Ballantine, 1993.

Halliwell, Martin. *American Culture in the 1950s*. Edinburgh: Edinburgh University Press, 2007.

Hannah-Jones, Nikole. "Choosing a School for My Daughter in a Segregated City." *New York Times Magazine*, June 12, 2016.

Harris, Gardiner. "The Public's Quiet Savior from Harmful Medicines." *New York Times*, September 14, 2010.

Helgesen, Sally. *The Female Advantage: Women's Ways of Leadership*. New York: Doubleday, 1990.

Helgesen, Sally, and Julie Johnson. *The Female Vision: Women's Real Power at Work*. San Francisco: Berrett-Koehler, 2010.

Horowitz, Daniel. *The Anxieties of Affluence: Critiques of American Consumer Culture, 1939–1979*. Amherst: University of Massachusetts Press, 2004.

Huxtable, Ada Louise. "Noted Buildings in Path of Road." *New York Times*, July 25, 1965.

Isserman, Maurice, and Michael Kazin. *America Divided: The Civil War of the 1960s*. New York: Oxford University Press, 2000.

Jacobs, Jane. *Cities and the Wealth of Nations*. New York: Vintage, 1984.

———. *The Death and Life of Great American Cities*. 1961. Reprint, New York: Modern Library, 2011.

———. *The Economy of Cities*. New York: Vintage, 1970.

———. *The Last Interview and Other Conversations*. Brooklyn: Melville House, 2016.

———. *Systems of Survival*. New York: Random House, 1992.

———. "Violence in the City Streets." *Harper's Magazine*, September 1961.

———. *Vital Little Plans: The Short Works of Jane Jacobs*. Edited by Samuel Zipp and Nathan Storring. New York: Random House, 2016.

Johnson, Steven. *Emergence: The Connected Lives of Ants, Brains, Cities, and Software*. New York: Scribner, 2001.

Jones, Abigail. "Jane Goodall's Jungles." *Newsweek*, October 23, 2014.

Kahn, Howie. "Alice Waters Makes the World a More Edible Place." *Wall Street Journal*, November 6, 2013.

Kalins, Dorothy. "Alice Waters: True Believer." *Town and Country*, January 2005.

Kamp, David. "Cooking Up a Storm." *Vanity Fair*, October 2006.

———. *The United States of Arugula*. New York: Broadway Books, 2006.

Kapur, Akash. "Couldn't Be Better: The Return of the Utopians." *The New Yorker*, October 3, 2016.

Kent, Leticia. "Persecution of the City Performed by Its Inmates." *Village Voice*, April 18, 1968.

Kisseloff, Jeff. *Generation on Fire: Voices of Protest from the 1960s*. Lexington: University Press of Kentucky, 2007.

———. *You Must Remember This: An Oral History of Manhattan from the 1890s to World War II*. San Diego: Harcourt Brace Jovanovich, 1989.

Klinkenborg, Verlyn. *Several Short Sentences About Writing*. New York: Vintage Books, 2013.

Kramer, Jane. *Off Washington Square*. New York: Duell, Sloan and Pearce, 1963.

Kristof, Nicholas. "Are You a Toxic Waste Disposal Site?" *New York Times*, February 13, 2016.

Kuhn, Thomas S. *The Structure of Scientific Revolutions*. Chicago: University of Chicago Press, 1962.

Kunstler, James Howard. *The Geography of Nowhere*. New York: Simon and Schuster, 1993.

———. "An Interview with Jane Jacobs, Godmother of the American City." *Metropolis*, March 2001.

Lapp, Ralph E. *The Voyage of the Lucky Dragon*. New York: Harper and Bros., 1957.

Lear, Linda. *Rachel Carson: Witness for Nature*. New York: Henry Holt, 1997.

Leopold, Aldo. *A Sand County Almanac, and Sketches Here and There*. 1949. Reprint, New York: Oxford University Press, 1987.

Marcus, Robert D., and David Burner, eds. *America Since 1945*. New York: St. Martin's, 1981.

Martin, Douglas. "Jane Jacobs, Urban Activist, Is Dead at 89." *New York Times*, April 25, 2006.

———. "Manny Roth, 94, Impresario of Café Wha?, Is Dead." *New York Times*, August 3, 2014.

Marx, Leo. *The Machine in the Garden: Technology and the Pastoral Ideal in America*. 1964. Reprint, New York: Oxford University Press, 2000.

Matthiessen, Peter, ed. *Courage for the Earth: Writers, Scientists, and Activists Celebrate the Life and Writing of Rachel Carson*. Boston: Houghton Mifflin, 2007.

McKibben, Bill. "The Pope and the Planet." *The New York Review of Books*, August 13, 2015.

McNamee, Thomas. *Alice Waters and Chez Panisse*. New York: Penguin, 2007.

Menand, Louis. "Books as Bombs." *The New Yorker*, January 24, 2011.

———. "It Took a Village." *The New Yorker*, January 5, 2009.

———. "What the Beats Were About." *The New Yorker*, October 1, 2007.

Miller, Donald. *Lewis Mumford*. New York: Weidenfeld and Nicolson, 1989.

Miller, Richard L. *Under the Cloud: The Decades of Nuclear Testing*. New York: The Free Press, 1986.

Muir, John. *Nature Writings*. Edited by William Cronon. New York: Library of America, 1997.

Nash, Roderick. *Wilderness and the American Mind*, rev. ed. New Haven: Yale University Press, 1973.

Olney, Richard. *Simple French Food*. 1974. Reprint, Hoboken, N.J.: Wiley, 1992.

O'Neill, Molly, ed. *American Food Writing: An Anthology with Classic Recipes*. New York: Library of America, 2007.

Pareles, Jon. "Pete Seeger, Champion of Folk Music and Social Change, Dies at 94." *New York Times*, January 28, 2014.

Paumgarten, Nick. "The Mannahatta Project." *The New Yorker*, October 1, 2007.

Peterson, Dale. *Jane Goodall: The Woman Who Redefined Man*. Boston: Houghton Mifflin, 2006.

Petrini, Carlo, ed., with Ben Watson and Slow Food Editore. *Slow Food: Collected Thoughts on Taste, Tradition, and the Honest Pleasures of Food*. White River Junction, Vt.: Chelsea Green, 2001.

Pollan, Michael. *Cooked: A Natural History of Transformation*. New York: Penguin, 2013.

———. *Food Rules: An Eater's Manual*. New York: Penguin, 2011.

———. *The Omnivore's Dilemma*. New York: Penguin, 2006.

Popova, Maria. "The Writing of *Silent Spring*: Rachel Carson and the Culture-Shifting Courage to Speak Inconvenient Truth to Power." *Brain Pickings* (blog). January 2017.

Radio Bikini. New York: First Run/Icarus Films, 1987. Video.

Radosh, Ronald. "It Took the Village." *New York Times Book Review*, August 9, 2015.

Rauber, Paul. "Conservation a la Carte." *Sierra*, November 1994.

Reich, Charles A. *The Greening of America*. New York: Random House, 1970.

Rich, Nathaniel. "Poisoned Ground." *New York Times Magazine*, January 10, 2016.

Roszak, Theodore. *The Making of a Counter Culture*. New York: Doubleday, 1968.

Ruse, Michael. *The Gaia Hypothesis: Science on a Pagan Planet*. Chicago: University of Chicago Press, 2013.

Schlosser, Eric. *Fast Food Nation*. New York: Houghton Mifflin, 2001.

———. "What Goes In, What Comes Out." *New York Times*, November 23, 2014.

Schubert, Dirk, ed. *Contemporary Perspectives on Jane Jacobs: Reassessing the Impacts of an Urban Visionary*. Farnum, Eng.: Ashgate, 2014.

Shabecoff, Philip. *A Fierce Green Fire: The American Environmental Movement*. New York: Hill and Wang, 1993.

Sharman, Russell Leigh. *The Tenants of East Harlem*. Berkeley: University of California Press, 2006.

Showalter, Elaine. *A Jury of Her Peers: Celebrating American Women Writers from Anne Bradstreet to Annie Proulx*. New York: Knopf, 2009.

Slotnik, Daniel E. "Jerry Berrigan, 95, Leader of Protests Against Wars." *New York Times*, August 2, 2015.

Solnit, Rebecca. *Hope in the Dark: Untold Histories, Wild Possibilities*. Chicago: Haymarket Books, 2016.

Souder, William. *On a Farther Shore*. New York: Crown, 2012.

Stahl, Lesley. "Interview with Alice Waters." *60 Minutes*, CBS, March 2009.

Stone, Robert. *Prime Green: Remembering the Sixties*. New York: Ecco/Harper Perennial, 2008.

Strausbaugh, John. *The Village: 400 Years of Beats and Bohemians, Radicals and Rogues*. New York: Ecco, 2013.

Strom, Stephanie. "A Clamor for Organic." *New York Times*, July 15, 2016.

———. "Foodies Know: Boulder Has Become a Hub for New Producers." *New York Times*, February 4, 2017.

Tabuchi, Hiroko. "The Banks Putting Rain Forests in Peril." *New York Times*, December 4, 2016.

———. "Talk About the Weather." *New York Times*, January 29, 2017.

Tanenhaus, Sam. "In Kennedy's Death, a Turning Point for a Nation Already Torn." *New York Times*, November 22, 2013.

Taubes, Gary. "Big Sugar's Secret Ally? Nutritionists." *New York Times*, January 13, 2017.

Thomas, Evan. *The War Lovers: Roosevelt, Lodge, Hearst, and the Rush to Empire, 1898*. New York: Little, Brown, 2010.

Thomas, Robert Mcg., Jr. "Raymond S. Rubinow, 91, Master of Civic Causes." *New York Times*, April 7, 1996.

Thurman, Judith. *Isak Dinesen: The Life of a Storyteller*. New York: St. Martin's, 1982.

Tullis, Paul. "Wild at Heart." *New York Times Magazine*, March 15, 2015.

Tytell, John. *Naked Angels: The Lives and Literature of the Beat Generation*. New York: McGraw-Hill, 1976.

Waters, Alice. *The Art of Simple Food*. New York: Clarkson Potter, 2007.

———. *Chez Panisse Menu Cookbook*. New York: Random House, 1982.

———. *Chez Panisse Vegetables*. New York: HarperCollins, 1996.

———. "A Delicious Revolution." Ecoliteracy.org. September 23, 2013.

———. "The Farm-Restaurant Connection." *The Journal of Gastronomy* 5, no. 2 (Summer/Autumn 1989): 113–22.

———. "Tea and Cheese in Turkey." In *The Kindness of Strangers*, edited by Don George. Melbourne: Lonely Planet, 2003.

Waters, Alice, and Friends. *Forty Years of Chez Panisse: The Power of Gathering*. New York: Clarkson Potter, 2011.

Wetzsteon, Ross. *Republic of Dreams: Greenwich Village; The American Bohemia, 1910–1960*. New York: Simon and Schuster, 2002.

White, E. B. *Here Is New York*. 1949. Reprint, New York: The Little Bookroom, 1999.

The WPA Guide to New York City. Introduction by William H. Whyte Jr. 1939. Reprint, New York: Random House, 1982.

NOTES

PREFACE

xi "in the shells of birds": Shabecoff, Philip, *A Fierce Green Fire: The American Environmental Movement* (New York: Hill and Wang, 1993), 79.

xii "feedbacks" and "narrow purposes": Ibid., 98.

INTRODUCTION: THE AGE OF WRECKERS AND EXTERMINATORS

2 "will not start a chain": "Case 3 Atomic Testing at Bikini Island," Georgetown University: Kennedy Institute of Ethics, High School Bioethics Curriculum Project, p. 1.

2 "1000 times more powerful": Ibid., 2.

3 "diamonds and emeralds": Carson, Rachel, *Always, Rachel: The Letters of Rachel Carson and Dorothy Freeman, 1952–1964*, ed. Martha Freeman (Boston: Beacon Press, 1995), 187.

3 "glowing in the sand": Ibid.

3 "no future peace": Lear, Linda, *Rachel Carson: Witness for Nature* (New York: Henry Holt, 1997), 328.

4 "the courage you showed": Ibid., 420–21.

4 "assumed that someone was looking": Ibid., 423 (from National Park Association speech).

4 "jeopardize the nation's food": Ibid., 437.

4 "why a spinster with no": Ibid., 429.

5 "Miss Carson's book": Ibid., 419.

5 "no housewife would reach": Ibid., 435.

5 "food additives, thalidomide, radioactive fallout": Ibid.

5 "gospel of technological progress": Ibid., 429.

5 "He was flying so low": Carson, *Always, Rachel*, 187.

6 "crowned with foam": Ibid., 186.

7 "the terror of every politico": Alexiou, Alice Sparberg, *Jane Jacobs: Urban Visionary* (New Brunswick, N.J.: Rutgers University Press, 2006), 113.

7 "Queen Jane": Ibid., 112.

7 "spaghetti dish": Jacobs, Jane, *The Death and Life of Great American Cities* (1961; repr., New York: Modern Library, 2011), 367.

8 "colonies of prairie dogs": Ibid., 444.

8 "sidewalk ballet": Ibid., 50.

8 "tangibly and physically" and "metaphysical fancies": Ibid., 96.

9 "What kind of administration": Flint, Anthony, *Wrestling with Moses* (New York: Random House, 2009), xiii.

10 "The expressway would": Alexiou, *Jane Jacobs*, 109.

12 "proper goal": Menand, Louis. "Books as Bombs." *New Yorker*. January 24, 2011.

12 "rise to his capacity": Isserman, Maurice, and Michael Kazin, *America Divided: The Civil War of the 1960s* (New York: Oxford University Press, 2000), 12.

14 "spectacle" and "an experiment in political": Dickstein, Morris, *Gates of Eden: American Culture in the Sixties* (New York: Basic Books, 1977), 23.

CHAPTER 1: RACHEL CARSON

18 "nerve poison": Souder, William, *On a Farther Shore* (New York: Crown, 2012), 244.

18 "for months": Ibid., 245.

20 "problem assignment": Lear, Linda, *Rachel Carson: Witness for Nature* (New York: Henry Holt, 1997), 78.

23 "thistledown": Carson, Rachel, *Under the Sea-Wind* (1941; repr., New York: Oxford University Press, 1952), 26.

23 "little transparent worms": Carson, Rachel, *Lost Woods: The Discovered Writing of Rachel Carson.*, ed. Linda Lear (Boston: Beacon Press, 1998), 59.

24 "the quick sharp sibilance": Ibid., 132.

24 "soft tinkling": Carson, *Under the Sea-Wind*, 9.

24 "dark silhouettes": Carson, *Lost Woods*, 130.

24 "the breath of a mist": Ibid., 62.

24 "countless thousands of years" and "ageless as sun": Ibid.

25 "great antiquity": Ibid., 78.

25 "a little better perspective on human problems": Ibid., 62.

25 "larger diverse community": Souder, *On a Farther Shore*, 44.

25 "interplay": Ibid., 89.

25 "is connected": Ruse, Michael, *The Gaia Hypothesis: Science on a Pagan Planet* (Chicago: The University of Chicago Press, 2013), 134.

26 "all technology was progress": Gould, Kira, and Lance Hosey, *Women in Green* (Washington, D.C.: Ecotone Press, 2007), 10.

27 "civic responsibility" and "Christian motherhood": Lear, *Rachel Carson*, 10.

28 "city boy" and "developer": Ibid., 12.

29 "intricate design of the creator": Ibid., 14.

29 "the only man I knew who would steal": Ibid., 23.

30 "stern-looking Mrs. Carson": Ibid., 21.

33 "slyly observant": Souder, *On a Farther Shore*, 29.

34 "large bosomy woman": Lear, *Rachel Carson*, 28.

34 "intellect" and "stamina": Ibid., 43.

34 "all life was interconnected" and "holistic": Souder, *On a Farther Shore*, 34.

35 "I have gone dead": Lear, *Rachel Carson*, 39.

35 "silver slippers": Souder, *On a Farther Shore*, 36.

36 "glorious time": Ibid., 36.

36 "Miss Skinker was a perfect": Ibid.

38 "clung to Rachel" and "looked ill": Lear, *Rachel Carson*, 71.

38 "seated at the table" and "I don't have time": Souder, *On a Farther Shore*, 48.

38 "I've never seen a word of yours": Brooks, Paul, *The House of Life: Rachel Carson at Work* (Boston: Houghton Mifflin, 1972), 20.

38 "extremely shy": Ibid., 71.

39 "more dirt than had been dug out": Souder, *On a Farther Shore*, 98.

39 "swirling murk": Carlson, Avis D., *The New Republic*, from Wikipedia, http://en.wikipedia.org/wiki/Black_Sunday_(storm).

39 "A howling wilderness": Nash, Roderick, *Wilderness and the American Mind*, rev. ed. (New Haven: Yale University Press, 1973), 32.

40 "permanent contributions" and "harnessed, controlled": Souder, *On a Farther Shore*, 53.

40 "control"; "rebalance"; and "more friendly to modern man": Ibid., 99.

40 "For wolves" and "only a matter of time": Ibid., 101.

41 "The greedy mills": Nash, *Wilderness and the American Mind*, 97.

41 "promote manly sport": Thomas, Evan, *The War Lovers: Roosevelt, Lodge, Hearst, and the Rush to Empire, 1898* (New York: Little, Brown, 2010), 53.

41 "vigorous manliness": Nash, *Wilderness and the American Mind*, 150.

41 "rational laws and exploitable": Belasco, Warren J., *Appetite for Change: How the Counterculture Took On the Food Industry* (Ithaca, N.Y.: Cornell University Press, 2007), 39.

41 "of a general sort" and "the material rather took": Souder, *On a Farther Shore*, 67.

42 "I don't think it will do": Lear, *Rachel Carson*, 81.

42 "uncommonly eloquent" and "fire the imagination": Ibid., 86–87.

42 "everything else followed": Brooks, *The House of Life*, 30.

42 "If the underwater traveler": Carson, *Lost Woods*, 8.

43 "Individual elements are lost": Souder, *On a Farther Shore*, 72.

43 "continuum": Ibid., 68.

44 "There is poetry here": Lear, *Rachel Carson*, 104.

44 "so skillfully written": Ibid.

44 "lyrical beauty" and "faultless science": Ibid., 105.

44 "the world received the event": Ibid.

44 "that was home to twenty": Souder, *On a Farther Shore*, 7.

45 "tempered by grave": Ibid.

45 "excessive nervousness, loss of appetite": Ibid., 8.

45 "white or tinted": Carson, Rachel, *Silent Spring* (New York: Houghton Mifflin, 2002), 174.

46 "impregnated" and "pocket-sized": Ibid., 175.

46 "electric vaporizing device" and "bomb": Souder, *On a Farther Shore*, 9.

46 "infallibility of material ingenuity": Wilson, E. O., afterword to Carson, *Silent Spring*, 358.

47 "acres of tract houses": Isserman, Maurice, and Michael Kazin, *America Divided: The Civil War of the 1960s* (New York: Oxford University Press, 2000), 10.

47 "team player": Halberstam, David, *The Fifties* (New York: Ballantine, 1993), 488.

47 "place Adolf Hitler's head": Steingraber, Sandra, in Matthiessen, Peter, ed., *Courage for the Earth: Writers, Scientists, and Activists Celebrate the Life and Writing of Rachel Carson* (Boston: Houghton Mifflin, 2007), 54.

47 "war on insects": Lear, *Rachel Carson*, 119.

47 "previously unthinkable": Coontz, Stephanie, *A Strange Stirring: The Feminine Mystique and American Women at the Dawn of the 1960s* (New York: Basic Books, 2011), 47.

47 *Ladies' Home Journal* even went so far: This and other details about women's employment during the war from Halberstam, *The Fifties*, 588.

47 "A mother already": Coontz, *A Strange Stirring*, 47.

48 "rebuild her husband's self-esteem": Ibid., 49.

48 "To a populace whose forebears": Wilson, afterword to Carson, *Silent Spring*, 358.

48 "I'm definitely in": Brooks, *The House of Life*, 76.

48 "well ordered" and "know where she was going": Lear, *Rachel Carson*, 130.

49 "were often more pungent": Brooks, *The House of Life*, 78.

49 "Nothing could pass" and "Intransigent official ways": Ibid.

49 "her qualities of zest": Ibid., 77.

49 "serious consequences": Souder, *On a Farther Shore*, 113.

50 "could conceivably do more damage": Ibid., 114.

50 "first hand": Lear, *Rachel Carson*, 119.

50 "Practically at my backdoor": Ibid., 118–19.

51 "serial tragedy" and "repeatedly squandered": Souder, *On a Farther Shore*, 123.

51 "in isolation": Ibid.

51 "Wildlife, water, forests": Brooks, *The House of Life*, 101.
52 "conqueror of the land community": Nash, *Wilderness and the American Mind*, 197.
52 "a biotic community": Ibid., 196.
52 "A thing is right": Ibid., 197.
52 "was in charge of humanity": Souder, *On a Farther Shore*, 125.
52 "We fancy that industry supports": Nash, *Wilderness and the American Mind*, 198–99.
52 "We are only fellow-voyageurs": Ibid., 195–96.
52 "entirely synthetic": Ibid., 196.
54 "something about her": Brooks, *The House of Life*, 97.
54 "put herself in the role": Lear, *Rachel Carson*, 141.
54 "always attentive, always": Brooks, *The House of Life*, 98.
54 "got away from being": Ibid.
55 "ecological consciousness": Lear, *Rachel Carson*, 150.
55 "that if you jumped": Ibid., 134.
55 "sharp, staccato cries": Carson, *Lost Woods*, 36.
55 "The gulls go so high": Lear, *Rachel Carson*, 135.
55 "Suddenly the silken sheet": Carson, *Lost Woods*, 36–37.
56 "whooshing sounds" and "diving experiences": Souder, *On a Farther Shore*, 136.
56 "Despite everything": Brooks, *The House of Life*, 119.
56 "None of the present": Ibid.
56 "Not a single walk" and "Then we shall": Ibid., 119–20.
57 "I don't like Miss Carson's": Souder, *On a Farther Shore*, 140.
58 "If I'm not solidly": Ibid., 146.
58 "small cyst or tumor": Ibid., 143.
58 "The operation will probably" and "get it over": Ibid.
59 "the most marvelous": Lear, *Rachel Carson*, 200.
59 "hypnotic" and "biblical sweep": Ibid., 204.
59 "one of the most beautiful": Souder, *On a Farther Shore*, 153.
59 "write what is a first-rate": Lear, *Rachel Carson*, 206.
59 "half way between" and "a superb book": Souder, *On a Farther Shore*, 154.
59 "he had always been": Brooks, *The House of Life*, 132.
59 "both bold and feminine": Souder, *On a Farther Shore*, 156.
59 "Apparently there are few": Ibid.
59 "pity that the book's": Ibid., 153.
60 "backbone"; "just plain hard slogging"; and "exceedingly technical": Ibid., 150.
60 "almost unimaginable presence": Tytell, John, *Naked Angels: The Lives and Literature of the Beat Generation* (New York: McGraw-Hill, 1976), 8.
61 "I'm pleased to have": Brooks, *The House of Life*, 126.
62 "I admit I felt hardly": Ibid., 131.

62 "We live in a scientific": Carson, *Lost Woods*, 91.

62 "Perhaps if we reversed": Ibid., 92.

62 "Mankind has gone very": Ibid., 161–64.

63 "the more clearly we can focus": Ibid., 163.

63 "charming and thoughtful": Carson, Rachel, *Always, Rachel: The Letters of Rachel Carson and Dorothy Freeman, 1952–1964*, ed. Martha Freeman (Boston: Beacon Press, 1995), 3.

64 "tiny" and "wistful expression": Souder, *On a Farther Shore*, 173.

64 "Dear Dorothy": Carson, *Always, Rachel*, 10.

64 "disappointment": Souder, *On a Farther Shore*, 177.

64 "stepping off the train" and "going mad": Carson, *Always, Rachel*, 12.

65 "We didn't plan it this": Souder, *On a Farther Shore*, 179.

65 "the sweet tenderness": Carson, *Always, Rachel*, 15.

65 "Dorothy had sensed the same": Souder, *On a Farther Shore*, 179.

65 "the thirteen hours": Ibid.

65 "little oasis of peace" and "truly perfect": Carson, *Always, Rachel*, 15.

65 "not a single thing": Ibid.

65 "craziness": Ibid., 18.

65 "general and newsy": Souder, *On a Farther Shore*, 180.

65 "apples": Freeman, Martha, editor's preface, in ibid., xvii.

65 "put in the strongbox": Souder, *On a Farther Shore*, 180.

66 "white hyacinth for his soul" and "the Hyacinth Letter": Ibid., 181.

66 "over-concentrating": Ibid., 185.

66 "overarching narrative": Ibid., 186.

67 "As I write of it": Brooks, *The House of Life*, 158.

67 "having taken so long": Souder, *On a Farther Shore*, 190.

67 "How blind I was not": Carson, *Always, Rachel*, 26.

67 "each progressive stage": Souder, *On a Farther Shore*, 191.

67 "But, oh darling, I want": Carson, *Always, Rachel*, 26.

67 "stardust": Souder, *On a Farther Shore*, 191.

67 "symphony": Carson, *Always, Rachel*, 41.

67 "the Hundred Hours" and "Maytime": Souder, *On a Farther Shore*, 192.

68 "Darling, you and I": Ibid., 196.

68 "one great love": Ibid., 201.

68 "out of range": Ibid., 198.

69 "no regrets " and "a lovely interlude": Carson, *Always, Rachel*, 76.

69 "Sex seems not to have been": Souder, *On a Farther Shore*, 199.

69 "existed mainly on paper": Ibid.

69 "lucid yet poetic" and "direct crystal clear prose": Lear, *Rachel Carson*, 275.

70 "toppled over like": Ibid., 293.

70 "openly hostile": Ibid., 301.

70 "harmless shower": Ibid., 314.

70 "gaping" and "splayed claws": Brooks, *The House of Life*, 232.

70 "The testers must have": Ibid.

71 "undemocratic and probably": Lear, *Rachel Carson*, 314.

71 "biodynamic gardening": Ruse, *The Gaia Hypothesis*, 126.

71 "a digestive invalid": Lear, *Rachel Carson*, 319.

72 "utmost concern": Souder, *On a Farther Shore*, 276.

72 "lively as 17 crickets" and "hold him down": Lear, *Rachel Carson*, 300.

72 "harder still": Souder, *On a Farther Shore*, 278.

72 "But I have been mentally blocked": Carson, *Always, Rachel*, 248–49.

73 "experts" and "organics": Ruse, *The Gaia Hypothesis*, 136.

73 "You are my chief clipping": Ibid., 134.

73 "surprise witness" and "convinced": Souder, *On a Farther Shore*, 284.

74 "testes and ovaries": Carson, *Silent Spring*, 109.

74 "a heavy body burden" and "poison used": Souder, *On a Farther Shore*, 287.

74 "rain of death" and "what of other lives": Matthiessen, Peter, "Introduction," *Courage for the Earth*, 12.

75 "Poor little fellow": Lear, *Rachel Carson*, 337.

75 "It is an amusing fact": Brooks, *The House of Life*, 244.

76 "a quarter of a billion dollars": Carson, *Silent Spring*, 17.

76 "harmless" and "people shopping or": Ibid., 90.

76 "harmless to humans and will not": Ibid.

76 "severe diarrhea, vomiting": Ibid.

76 "nausea, vomiting, chills" and "to something else": Ibid., 91.

76 "soaked from head to toe": Souder, *On a Farther Shore*, 258.

77 "slow, cumulative and hard-to-identify": Brooks, *The House of Life*, 244.

77 "every child born today": Ibid.

77 "showed some content of DDT": Ibid.

77 "also scattered evidence": Ibid., 245.

77 "psychological angle to all": Ibid., 241.

78 "strange future" and "made her feel ill": Souder, *On a Farther Shore*, 268.

78 "man-made ugliness" and "trend toward a perilously": Horowitz, Daniel, *The Anxieties of Affluence: Critiques of American Consumer Culture, 1939–1979* (Amherst: University of Massachusetts Press, 2004), 153–54.

78 "commercial schemes": Ibid., 154.

78 "Beauty—and all the values": Ibid.

79 "It is one of the ironies": Ibid., 153.

79 "farther and farther into experiments": Carson, *Lost Woods*, 94.

79 "deeply disturbing": Souder, *On a Farther Shore*, 268.

79 "minimally radioactive fallout" and "as little as two years": Ibid., 293.

79 "over a more concentrated area": Ibid.

80 "subtle genetic mutations" and "was far below": Ibid., 294.

80 "a strange and chilling thing": Ibid., 11.

81 "Castle Bravo": Ibid., 229.

81 "The sun is rising in the west!": Halberstam, *The Fifties*, 346.

81 "It's an atomic bomb": Ibid.

82 "bleeding gums, falling white blood cell": Souder, *On a Farther Shore*, 231.

82 "hepatitis": Halberstam, *The Fifties*, 348.

82 "sufficiently radioactive": Souder, *On a Farther Shore*, 232.

82 "American government and people": Ibid.

83 "secondary contamination": Ibid., 234.

83 "upper-level wind currents": Ibid., 235.

84 "unbearably frustrating": Ibid., 229.

84 "unshakable foundation" and "violent controversies": Brooks, *The House of Life*, 257–58.

84 "weight of evidence": Souder, *On a Farther Shore*, 299.

84 "some sort of thyroid": Ibid., 300.

84 "Sometimes I wonder whether": Lear, *Rachel Carson*, 365.

85 "quite fascinating": Brooks, *The House of Life*, 262.

85 "by far the most difficult": Lear, *Witness for Nature*, 365.

85 "too complicated": Ibid.

85 "otherwise at the end": Souder, *On a Farther Shore*, 308.

85 "suspicious enough" and "a precautionary measure": Lear, *Rachel Carson*, 367.

85 "talked her way out": Souder, *On a Farther Shore*, 309.

85 "hospital adventure": Ibid., 308.

85 "There need be no": Ibid., 309.

86 "special friends" and "I suppose it's a futile": Lear, *Rachel Carson*, 367.

86 "In a sense, all this publicity": Ibid., 375.

86 "a curious, hard swelling": Ibid., 378.

86 "after being so sure": Souder, *On a Farther Shore*, 312.

86 "I know now that": Lear, *Rachel Carson*, 368.

87 "I have a great deal": Ibid., 380.

87 "You spoke of the moonlight": Carson, *Always, Rachel*, 333–36.

88 "never been sicker": Souder, *On a Farther Shore*, 314.

88 "devastated" and "slumped and sobbing": Ibid.

88 "I am working late at night": Brooks, *The House of Life*, 269–70.

88 "This is William Shawn": Carson, *Always, Rachel*, 394.

88 "a brilliant achievement" and "literature, full of beauty": Ibid.

89 "Every perfectly ordinary little": Ibid., 404.

89 "reluctantly restarted": Souder, *On a Farther Shore,* 11.

90 "doubled": Ibid., 326.

90 "in the heart of America": Carson, *Silent Spring*, 1.

90 "strange blight": Ibid., 2.

90 "as though swept by fire" and "fallen like snow": Ibid., 3.

90 "strange stillness": Ibid., 2.

91 "No witchcraft, no enemy action": Ibid., 3.
91 "Communist sympathies" and "peace-nut": Souder, *On a Farther Shore*, 331.
91 "live without birds" and "but not without business": Ibid.
92 "heedless overuse": Ibid., 332.
92 "wholly ignorant": Carson, *Silent Spring*, 12.
92 "leaped aboard the pesticide": Caro, Robert, "Pesticides: The Hidden Poisons," *Newsday*, August 20, 1962.
92 "famed biologist and author": Ibid.
93 "panic and hysteria" and "pesticides provided": Souder, *On a Farther Shore*, 343.
93 "nationwide surveillance": Ibid., 349.
93 "keep pace" and "blanket": Ibid.
93 "the most valuable" and "this formidable opponent": Lear, *Rachel Carson*, 420.
94 "the sort that will help turn": Ibid.
94 "all of a piece" and "Thalidomide and pesticides": Ibid., 412.
94 "a realist as well as a biologist": Souder, *On a Farther Shore*, 355.
95 "The basic fallacy": Ibid.
95 "timely": Ibid., 356.
95 "the most important chronicle": Lear, *Rachel Carson*, 419.
95 "dangerous, long-term" and "Miss Carson's book": Souder, *On a Farther Shore*, 4.
95 "sinister parties" and "east-curtain parity": Ibid., 346.
95 "America's food supply" and "obviously the rantings": Ibid., 357.
95 "unfair, one-sided and hysterically": Brooks, *The House of Life*, 297.
96 not "a professional scientist": Lear, *Rachel Carson*, 430.
96 "published in peer-reviewed journals" and "no academic": Ibid.
96 "bird and bunny lover" and "spinster": Lear, Linda, Introduction to Carson, *Silent Spring*, xvii.
96 "In short, [she] was a woman" and "overstepped the bounds": Ibid.
96 "I don't know of a housewife": Lear, *Rachel Carson*, 413.
97 "illegal quantities of penicillin": Souder, *On a Farther Shore*, 292.
97 "shared biology of all living things": Ibid., 350.
97 "easily and irrevocably": Griswold, Eliza, "How 'Silent Spring' Ignited the Environmental Movement," *New York Times*, September 21, 2012.
97 "delicate and destructible" and "capable of striking back": Carson, *Silent Spring*, 297.
97 "from the moment of conception until": Ibid., 15.
97 "exact and inescapable": Ibid., 208.
97 "We are rightly appalled": Ibid., 37.
98 "our genetic heritage, our link": Ibid., 208.
98 "The control of nature": Ibid., 297.
98 "Can anyone believe it is possible": Ibid., 7–8.

98 "The question is whether any civilization": Ibid., 99.
98 "nothing must get in the way": Ibid., 85.
98 "Lulled by the soft sell": Ibid., 174.
98 "When the public tried to hold": Ibid., 13.
98 "Man has lost the capacity": Horowitz, *The Anxieties of Affluence*, 155.
99 "The head and neck were outstretched": Carson, *Silent Spring*, 100.
99 "And what of human beings?" and "one turns away": Ibid., 126–27.
99 "of not speaking out": Gould and Hosey, *Women in Green*, 21.
100 "working with or working against": Horowitz, *The Anxieties of Affluence*, 158.
100 "Every once in a while": Griswold, "How 'Silent Spring' Ignited the Environmental Movement."
100 "No man who owns his house": Isserman and Kazin, *America Divided*, 11.
101 "legislative monument": Dickstein, Morris, *Gates of Eden: American Culture in the Sixties* (New York: Basic Books, 1977), 26.
101 "These have been years" and "when one could hardly": Mailer, Norman, quoted in ibid., 52.
101 "long-term health": Lear, Introduction, in Carson, *Silent Spring*, xiv.
102 "vain chase for satisfaction": Horowitz, *The Anxieties of Affluence*, 160.
103 "in the interest of the export": Souder, *On a Farther Shore*, 363.
103 "a menacing shadow": Carson, *Always, Rachel*, 414.
103 "all the well-known" and "said things it did not": Souder, *On a Farther Shore*, 365.
104 "serve the gods of profit": Lear, *Rachel Carson*, 426.
104 "When the scientific organization speaks": Ibid.
104 "It has been such a mixed year": Carson, *Always, Rachel*, 420.
104 "crowded": Souder, *On a Farther Shore*, 368.
104 "Because of you": Ibid.
105 "light-hearted": Carson, *Always, Rachel*, 435.
105 "an invalid's life": Ibid.
105 "The main things I want to say": Ibid., 437.
105 "too stern" and "relax" and "gentle nature": Souder, *On a Farther Shore*, 369.
105 "their friends in the chemical": Ibid., 369.
106 "a little easier for you": Lear, *Rachel Carson*, 442.
106 "I have had a rich life": Ibid.
107 "If man were to faithfully": Souder, *On a Farther Shore*, 374.
107 "We still talk in terms of": Brooks, *The House of Life*, 319.
108 "important": Lear, *Rachel Carson*, 450.
108 "the most controversial" and "a national quarrel": Souder, *On a Farther Shore*, 380.
108 "go down in history": Carson, *Always, Rachel*, 461.
108 "things I need to say to you": Ibid., 456.

109 "You are the lady who": Lear, *Rachel Carson*, 3.

109 "Our heedless and destructive acts": Ibid., 454.

109 "could have questioned her integrity": Ibid.

109 "rare person who was passionately": Ibid., 4.

110 "Would you help me search": Carson, *Always, Rachel*, 467.

110 "drawn by some invisible force": Ibid.

110 "It occurred to me": Ibid., 468.

110 "For ourselves, the measure": Ibid.

111 "I've been x-rayed": Lear, *Rachel Carson*, 459.

111 "middle-aged, arthritis-crippled": Ibid., 464.

111 "as sick as she had ever been": Ibid., 465.

111 "shock, dismay, and revulsion": Carson, *Always, Rachel*, 497.

112 "out of bed again" and "lovely": Souder, *On a Farther Shore*, 387.

112 "I had not, until recently": Carson, *Always, Rachel*, 506.

112 "something of a miracle": Lear, *Rachel Carson*, 475.

113 "a brilliant white light": Ibid., 479.

CHAPTER 2: JANE JACOBS

115 "a bloodletting": Flint, Anthony, *Wrestling with Moses* (New York: Random House, 2009), 23.

116 "were only getting worse": Alexiou, Alice Sparberg, *Jane Jacobs: Urban Visionary* (New Brunswick, N.J.: Rutgers University Press, 2006), 45.

117 "Make no little plans": Ibid., 38.

117 "a whole city in the free air": Ibid., 37.

118 "We must kill the street!": Berman, Marshall, *All That Is Solid Melts into Air: The Experience of Modernity* (New York: Penguin, 1988), 168.

118 "Cafés and places of recreation": Ibid., 167.

118 "moving spirit of modernity": Ibid., 294.

118 "dissonances": Ibid., 169.

118 "new type of street" and "machine for traffic": Ibid., 167.

118 "First he took me to a street": Goldberger, Paul, "Tribute to Jane Jacobs," speech at the Greenwich Village Society for Historic Preservation, October 3, 2006, available at www.paulgoldberger.com.

118 "Where are the people?": Flint, *Wrestling with Moses*, 20.

118 "They don't appreciate these things": Alexiou, *Jane Jacobs: Urban Visionary*, 40.

119 "Garden City": Jacobs, Jane, *The Death and Life of Great American Cities* (1961; repr., New York: Modern Library, 2011), 18–25.

119 "curtain wall": Flint, *Wrestling with Moses*, 21.

120 "the energy and enthusiasm": Berman, *All That Is Solid Melts into Air*, 302.

120 "one public figure" and "qualified to build": Ibid.

120 "quasi-mythological" status: Ibid., 294.

120 "moving spirit of modernity": Ibid.

121 "bloodletting": Flint, *Wrestling with Moses*, 23.

121 "I can remember the people": Jacobs, *The Death and Life of Great American Cities*, 15.

122 "hot, rich reality": Allen, Max, ed., *Ideas That Matter: The Worlds of Jane Jacobs* (Toronto: The Ginger Press, 1997), 62.

122 "Each machine" and "Nobody [was] concerned": Jacobs, *The Death and Life of Great American Cities*, 95.

123 "showed me a way of seeing": Ibid., 16.

123 "mountainous open garbage dump": Baker, Kevin, "City of Water," *New York Times*, October 12, 2013.

124 "Women of the Moose": Allen, *Ideas That Matter*, 3.

124 a station called Christopher Street: Kunstler, James Howard, "An Interview with Jane Jacobs, Godmother of the American City," *Metropolis*, March 2001, 4.

125 "I think that's the hardest time": Allen, *Ideas That Matter*, 12.

126 the "cool, sweet-smelling shops": Jacobs, Jane, "Flowers Come to Town," *Vogue*, February 15, 1937, as excerpted in Allen, *Ideas That Matter*, 35.

126 "twenty thousand dozen": Ibid., 36.

126 "conclude their dickering": Ibid., 35.

127 "All the ingredients": Ibid.

128 "Not a sound is heard" and "the proceedings are baffling": Jacobs, Jane, "Diamonds in the Rough," *Vogue*, 1937, as excerpted in Allen, *Ideas That Matter*, 36.

128 "a cross between hocus-pocus": Ibid.

128 "Upstairs, in small light rooms over the stores": Ibid., 36–37.

128 "Outside on the Bowery" and "raucous chaos": Ibid., 37.

128 "I would think": Flint, *Wrestling with Moses*, 8.

130 "Look at that oak tree": Allen, *Ideas That Matter*, 14.

131 "She was a free spirit": Ibid., 17.

131 "very good questions" and "too theoretical": Ibid., 14.

132 "I learned a great deal": Ibid., 3.

132 "never to promise to do anything": Ibid., 16.

132 "proselytizing" and "outlaw": Ibid.

132 "It gave me the feeling": Ibid.

132 "I was very suspicious": Ibid., 26.

133 "I am proud that my grandfather": Ibid., 170.

133 "outlandish" and "respectable law and opinion": Ibid.

133 "Perhaps it is partly because" and "I was brought up": Ibid.

133 "women's rights and women's brains": Ibid.

134 "Fortunately, my [high school] grades": Ibid., 4.

135 "Ex-Scranton Girl Helps": Flint, *Wrestling with Moses*, 10.

136 "entertainment to which he": Alexiou, *Jane Jacobs: Urban Visionary*, 24.

136 "loose and untrue allegations": Ibid.

137 "Cupid really shot that arrow": Flint, *Wrestling with Moses*, 12.

137 "could spell molybdenum": Allen, *Ideas That Matter*, 12.

140 "The click-clack": Broms-Jacobs, Caitlin, ed., *Jane at Home* (Toronto: Estate of Jane Jacobs, 2016), 6.

140 "In my generation women": Allen, *Ideas That Matter*, 23.

141 "the father of grassroots": Flint, *Wrestling with Moses*, 16.

141 "piss in": Wikipedia: https://en.wikipedia.org/wiki/Saul_Alinsky.

141 "power to the people" and "anti-government rhetoric": Flint, *Wrestling with Moses*, 16.

141 "troublemaker" and "chauvinist": Ibid.

142 "Upon first reading the questions": Allen, *Ideas That Matter*, 169.

142 "I was brought up to believe" and "I was taught": Ibid., 169–70.

142 "I abhor"; "I believe in control"; and "current fear of radical ideas": Ibid., 178–79.

143 "mania for internal security": Dickstein, Morris, *Gates of Eden: American Culture in the Sixties* (New York: Basic Books, 1977), 27.

143 "Remarkable collection of angels": Isserman, Maurice, and Michael Kazin, *America Divided: The Civil War of the 1960s* (New York: Oxford University Press, 2000), 138.

143 "poetic drifter": Ibid.

144 "a declaration of independence": Ibid., 140.

144 "Moloch whose eyes": Berman, *All That Is Solid Melts into Air*, 310.

144 "authentic" and "alternate routes": Isserman and Kazin, *America Divided*, 140–41.

144 "To dance beneath the diamond sky": Ibid., 141.

146 "You'll be our schools and hospitals expert": Alexiou, *Jane Jacobs: Urban Visionary*, 34.

146 "I was utterly baffled at first": Allen, *Ideas That Matter*, 4.

147 "Sometimes you learn more": Jacobs, Jane, "The Missing Link in City Redevelopment," talk before the conference on Urban Design at Harvard College, as excerpted in Allen, *Ideas That Matter*, 39–40.

147 "In New York's East Harlem": Ibid., 39.

147 "Planners and architects": Ibid.

148 "hand-to-mouth"; "institutions"; and "vestigial": Ibid.

148 "Do you see"; "in the new scheme"; and "This is a ludicrous": Ibid.

148 "lively old parts" and "Notice the stores": Ibid., 40.

148 "at least as vital"; "to respect—in the deepest"; and "We are greatly misled": Ibid.

149 "tower-in-the-park": Alexiou, *Jane Jacobs: Urban Visionary*, 34.

149 "Your worst opponents" and "keep hammering": Allen, *Ideas That Matter*, 95.

150 "a most inappropriate choice": Alexiou, *Jane Jacobs: Urban Visionary*, 61.

150 "These projects will not revitalize": Jacobs, Jane, "Downtown Is for

People," *Fortune Classic*, 1958, Fortune.com; available at http://features.blogs.fortune.cnn.com.

151 "Look what your girl did": Alexiou, *Jane Jacobs: Urban Visionary*, 63.

151 "one standard solution": Jacobs, "Downtown Is for People."

151 "the gaiety, the wonder": Ibid.

151 "Urban ecosystems" and "fragile": Jacobs, *The Death and Life of Great American Cities*, xxvi.

151 "a world-class": Flint, *Wrestling with Moses*, 27.

152 "This cultural superblock": Jacobs, "Downtown Is for People."

152 "My God, who is this crazy dame?" and "piece of built-in rigor mortis": Flint, *Wrestling with Moses*, 28.

152 "hubris": Fulford, Robert, "Abattoir for Sacred Cows," in Allen, *Ideas That Matter*, 7.

152 "elitist, top down policies" and "central control": Gratz, Roberta Brandes, *The Battle for Gotham* (New York: Nation Books, 2010), xxii.

153 "regenerative potential; "static form"; and "process": Ibid., xxiii.

153 "top-down command economy" and "the industrial model": Ibid., xxvi.

153 "interconnectedness and fragility of": Ibid., xxiv.

154 "Marshall Plan": Flint, *Wrestling with Moses*, 63.

154 "modern construction and wider streets": Ibid., 62.

155 "Five minutes from Wall Street": Ibid., 55–56.

155 "theater in the round" and "who [were] the spectators": Jacobs, *The Death and Life of Great American Cities*, 105.

155 "a bewildering sprinkling": Ibid.

156 "the beats, the hips": Kramer, Jane, *Off Washington Square* (New York: Duell, Sloan and Pearce, 1963), 11.

157 "in perpetuity for the public": Flint, *Wrestling with Moses*, 68.

158 "bathmat": Ibid., 73.

160 "if necessary": Ibid., 78.

160 "We weren't trying to embrace": Ibid., 79.

160 "People knew what they were getting": Ibid.

161 "there should be no negotiation": Ibid., 80.

161 "absurd" and "a process of mere sausage": Ibid.

161 "The attack on Washington Square": Ibid.

162 "I consider it would be far": Ibid., 81.

162 "an island of quietness": Ibid.

162 "Rebellion is brewing in America": Ibid.

164 "borrowed ways of thinking": Menand, Louis, "It Took a Village," *The New Yorker*, January 5, 2009.

164 "New Journalism": Ibid.

164 "It is our view": Wolf, Dan, "The Park," *Village Voice*, November 9, 1955.

164 "awful bunch of artists": Flint, *Wrestling with Moses*, 84.

165   "little soldier" and "little elves": Ibid.

165   "It's for picketing": Ibid., 85.

165   "She would bring the three children": Ibid.

166   "Fit To Be Tied": Allen, *Ideas That Matter*, 70.

166   "There is something to be said": Flint, *Wrestling with Moses*, 85.

166   "some rare public testimony" and "Save the Square": Ibid., 86.

166   "Washington Square Park": Ibid.

167   "There is nobody against this": Kunstler, "An Interview with Jane Jacobs," 8.

167   "Square Warriors": Flint, *Wrestling with Moses*, 87.

169   "There seem to be two dominant": Allen, *Ideas That Matter*, 47.

169   "Where it works at all well": Ibid., 48.

169   "You sort of fell in love with": Flint, *Wrestling with Moses*, 91.

170   "rebuilding of cities": Jacobs, *The Death and Life of Great American Cities*, 4.

170   "a little man in a plain brown overcoat" and "a crowd that evolved": Ibid., 53.

171   "blind-eyed" and "Eyes on the street": Ibid., 57.

171   "business lunchers": Ibid., 52.

172   "was the scene of an intricate": Ibid., 50.

172   "never repeats itself": Ibid.

173   "Nobody was going to allow": Jacobs, Jane, "Violence in the City Streets," *Harper's Magazine*, September 1961.

173   "different people": Jacobs, *The Death and Life of Great American Cities*, 53.

173   "Impersonal city streets": Ibid., 57.

173   "Lowly, unpurposeful and random": Ibid., 72.

173   "marvels of dullness": Ibid., 4.

174   "Project prairies": Ibid., 186.

174   "to make the mistake of attempting": Ibid., 373.

174   "trial and error": Ibid., 6.

174   "Incurious about the reasons": Ibid.

174   If you moved people: Gratz, *The Battle for Gotham*, xxviii.

175   "dangerous nonsense": Allen, *Ideas That Matter*, 6.

175   "Intricate minglings of": Jacobs, *The Death and Life of Great American Cities*, 222.

175   "have common fundamental principles": Ibid., xxvi.

175   "Ask a houser": Ibid., 90.

175   "In defective city neighborhoods": Ibid., 87–88.

175   "run a gauntlet of bullies" and "pure daydreaming": Ibid., 77.

175   "I know Greenwich Village": Ibid., 85.

176   "matriarchy" and "occasional playground appearance": Ibid., 83–84.

176   "Well-established, high-turnover" and "necessary to the safety": Ibid., 188.

176 "there is no leeway for such chancy": Ibid.

176 "Old ideas can sometimes": Ibid.

177 "Time makes the high building": Ibid., 189.

177 "ingenious adaptations" and "town-house parlor that became a craftsman": Ibid., 194.

177 "things that are inherently organic": Gould, Kira, and Lance Hosey, *Women in Green* (Washington, D.C.: Ecotone Press, 2007), 28.

177 "Two Blighted Downtown Areas": Flint, *Wrestling with Moses*, 99.

178 "inane, anti-city forces": Allen, *Ideas That Matter*, 4.

178 "neat a case study" and "intellectual idiocies": Ibid.

178 "just what you would expect": Alexiou, *Jane Jacobs: Urban Visionary*, 99.

179 "community renewal": Flint, *Wrestling with Moses*, 102.

179 "single function": Ibid., 101.

180 "the new era for urban renewal": Ibid., 102.

181 "routine request" and "a project": Ibid., 104.

181 "was the opening fraud": Ibid., 105.

182 "The aim of the committee" and "we will look for": Ibid.

183 "participating citizens": Alexiou, *Jane Jacobs: Urban Visionary*, 101.

183 "The city wants our houses": Flint, *Wrestling with Moses*, 109.

184 "got information almost sooner": Ibid., 108.

184 "a public character": Alexiou, *Jane Jacobs: Urban Visionary*, 102.

184 "But we were, everybody": Ibid.

184 "a deficiency" and "indicators of blight": Flint, *Wrestling with Moses*, 111.

185 "They brought this proposal out": Breyer, Sam Pope, "'Villagers' Seek to Halt Renewal," *New York Times*, March 4, 1961, 11.

185 "We couldn't go two": Chapman, Priscilla, "City Critic in Favor of Old Neighborhoods," *New York Herald Tribune*, March 4, 1961.

185 "precisely because of what was valuable": Flint, *Wrestling with Moses*, 110.

185 "had gotten all their ducks": Alexiou, *Jane Jacobs: Urban Visionary*, 99.

186 "we decided to support": Breyer, Sam Pope, "Project Foe Hits 'Village' Group," *New York Times*, March 14, 1961, 26.

186 "a puppet" and "invented and nurtured": Ibid.

186 "intend to level the area": Flint, *Wrestling with Moses*, 112.

186 "moles": Alexiou, *Jane Jacobs: Urban Visionary*, 106.

186 "only clearance and redevelopment": Flint, *Wrestling with Moses*, 112.

187 "remove the slum designation" and "complicit in the city's plans": Ibid., 109.

187 "comply with the requirement": Ibid., 113.

187 "An irresponsible boondoggle": Alexiou, *Jane Jacobs: Urban Visionary*, 103.

188 "Years before the body counts": Flint, *Wrestling with Moses*, 114.

188 "The backbone of renewal": Alexiou, *Jane Jacobs: Urban Visionary*, 104.

188 "renewed commitment"; "discredited their offices"; and "They don't want": Flint, *Wrestling with Moses*, 114.

189 "I want to say for the record": Ibid., 115.

189 "must conform to Village tradition": Alexiou, *Jane Jacobs: Urban Visionary*, 104.

189 "pious platitudes": Flint, *Wrestling with Moses*, 107.

189 "If the mayor cares": Alexiou, *Jane Jacobs: Urban Visionary*, 105.

189 "deeply concerned and sympathetic"; "shelve"; "earnest consideration"; and "independence": Flint, *Wrestling with Moses*, 115.

190 "Down with Felt!": "Villagers' Near-Riot Jars City Planning Commission," *New York Herald Tribune*, October 19, 1961. Reprinted in Allen, *Ideas That Matter,* 68.

190 "double-crossed": Flint, *Wrestling with Moses*, 116.

190 "By this reprehensible": "Villagers' Near-Riot Jars City Planning Commission."

190 "You are not an elected": Flint, *Wrestling with Moses*, 117.

190 "Your name will be remembered" and "You belong with Khrushchev": Ibid.

191 "This is the most disgraceful demonstration": Allen, *Ideas That Matter*, 68.

191 "You will be obliged to arrest": Flint, *Wrestling with Moses*, 117.

191 "no more hearings": Ibid.

191 "near riot"; "ignorant, neurotic"; "riot tactics"; and "an attack on the democratic": Alexiou, *Jane Jacobs: Urban Visionary*, 105.

192 "it was not the protesters": Flint, *Wrestling with Moses*, 118.

192 "We had been ladies and gentlemen": Asbury, Edith Evans, "Deceit Charged in Village Plan," *New York Times*, October 20, 1961, 68.

192 "creepers": Flint, *Wrestling with Moses*, 119.

193 "It's the same old story": Ibid., 120.

194 "redevelopment": Allen, *Ideas That Matter*, 49.

194 "a pioneering act": Flint, *Wrestling with Moses*, 121.

194 "Joan of Arc": Ibid., 128.

194 "a sort of Madame Defarge": Allen, *Ideas That Matter*, 50.

194 "*madonna misericordia*": Kramer, *Off Washington Square*, 58–59.

195 "There seems to be a notion": Allen, *Ideas That Matter*, 50.

195 "explosive": Alexiou, *Jane Jacobs: Urban Visionary*, 70.

195 "The City Planners Are Ravaging": Flint, *Wrestling with Moses*, 121.

195 "This book is an attack": Jacobs, *The Death and Life of Great American Cities*, 3.

195 "The economic rationale": Ibid., 5.

195 "monotonous, unnourishing gruel": Ibid., 7.

196 "hammered away at the bad old city": Ibid., 20.

196 "thoroughly physical places": Ibid., 95.

196 "engineering models for traffic": Gratz, *The Battle for Gotham*, xxvii.

196 "innate functioning order": Jacobs, *The Death and Life of Great American Cities*, 14.

196 "There is a quality even meaner": Ibid., 15.

196 "It is hard for muddled": Kent, Leticia, "More Babies Needed, Not Fewer," *Vogue*, August 15, 1970.

196 "statistical city": Gould and Hosey, *Women in Green*, 25.

197 "with its obvious elements of regimentation": Allen, *Ideas That Matter*, 84.

197 "repression of all plans": Jacobs, *The Death and Life of Great American Cities*, 25.

197 "paternalistic" and "great informal experts": Gould and Hosey, *Women in Green*, 25–26.

197 "People who get marked": Jacobs, *The Death and Life of Great American Cities*, 5.

197 "In the form of statistics": Gould and Hosey, *Women in Green*, 25.

197 "same universal principles": Ibid., 27.

197 "have marvelous innate": Jacobs, *The Death and Life of Great American Cities*, 447.

197 "contain the seeds of their own": Ibid., 448.

197 "is not a status quo" and "it is fluid, ever-shifting": Carson, Rachel, *Silent Spring* (New York: Houghton Mifflin, 2002), 246.

198 "a biological view of the built": Gratz, *The Battle for Gotham*, xxiv.

198 "The alternative isn't to develop": Gould and Hosey, *Women in Green*, 27.

198 "brashly impressive tour de force": Rodwin, Lloyd, "Neighbors Are Needed," *New York Times Book Review*, November 5, 1961, 10.

198 "Seminal"; "a major work"; and "rare books that make": Allen, *Ideas That Matter*, 5.

198 "a revolutionary and revelatory": Ibid., 6.

198 "will have as much impact . . . as the Armory show had on art": Ibid. (The Armory show of 1913 was the first show of Modernist art in America.)

198 "It won't matter that what this author": Rodwin, "Neighbors Are Needed," 10

198 "flawed" and "dangerous": Flint, *Wrestling with Moses*, 124.

199 "Robert Moses . . . has made an art": Jacobs, *The Death and Life of Great American Cities*, 131.

199 "an old sad story" and "The art of negating the power": Ibid.

199 "Libelous, intemperate, and inaccurate": Allen, *Ideas That Matter,* 97.

199 "demagogue": Ibid., 9.

199 "crank" and "a gadfly": Ibid., 11.

199 "wild bohemian" and "anarchist": Kramer, *Off Washington Square*, 63.

199 "misinformed"; "transparent gaps"; and "blind spots": Alexiou, *Jane Jacobs: Urban Visionary*, 83.

199 "blasé misunderstandings of theory" and "congenial": Ibid., 86.

199 "Mrs. Jacobs has presented the world": Allen, *Ideas That Matter*, 10.
199 "Unfortunately, this cannot be": Ibid., 52.
199 "housewife": Alexiou, *Jane Jacobs: Urban Visionary*, 86.
199 "an angry young woman": Allen, *Ideas That Matter*, 96.
199 "seemingly limited experience": Ibid., 52.
199 "mother from Scranton" and "sentimental Hausfrau": Kramer, *Off Washington Square*, 63.
200 "Mistakes in City Planning": Alexiou, *Jane Jacobs: Urban Visionary*, 86.
200 "How have the planners reacted": Ibid., 85.
201 "I held my fire for": Allen, *Ideas That Matter*, 6.
201 "sloppy novice" and "schoolgirl howlers": Flint, *Wrestling with Moses*, 126.
201 "She describes her folksy urban place": Allen, *Ideas That Matter*, 53.
201 "Mrs. Jacobs had visited Pompeii": Flint, *Wrestling with Moses*, 126.
201 "where factories nestle beside homes": Allen, *Ideas That Matter*, 54.
202 "sometimes we would drive past": Alexiou, *Jane Jacobs: Urban Visionary*, 91.
202 "Her book had no recipe": Allen, *Ideas That Matter*, 17.
202 "I was interested in process": Ibid.
202 "one of the great afflictions": Kramer, *Off Washington Square*, 60.
202 "big with fresh insights": Alexiou, *Jane Jacobs: Urban Visionary*, 94.
202 "I believe now that he felt hurt and betrayed": Ibid.
203 "Dear Jane, What have you": Allen, *Ideas That Matter*, 58.
203 "prophet and leader of a great": Kramer, *Off Washington Square*, 58–59.
203 "Forget the big parking garages": Flint, *Wrestling with Moses*, 128.
203 "bleak, miserable, and mean": Allen, *Ideas That Matter*, 52.
203 "is being rebuilt by city haters": Flint, *Wrestling with Moses*, 128.
203 "Planners always want to make a big": Ibid.
203 "They are hated almost": Baldwin, James, *Nobody Knows My Name* (New York: Vintage, 1961), 63.
204 "Their administration is insanely": Ibid., 69.
204 "This culture with its huge housing": Kramer, *Off Washington Square*, 62.
204 "White residents used Federal Housing": Hannah-Jones, Nikole, "Choosing a School for My Daughter in a Segregated City," *New York Times Magazine*, June 12, 2016.
204 "often because Negroes": Flint, *Wrestling with Moses*, 130.
205 "deductive"; "imaginative leaps"; and "pursued because": Allen, *Ideas That Matter*, 8.
205 "connection": Gould and Hosey, *Women in Green*, 28.
205 "strong enough to strip bark": Isserman and Kazin, *America Divided*, 83.
205 "We will wear you down by our": Ibid., 20.

206 "machine" and "less tangible, sensory": Marx, Leo, *The Machine in the Garden: Technology and the Pastoral Ideal in America* (1964; repr., New York: Oxford University Press, 2000), 383.

206 "put your bodies against": Ibid., 384.

207 "unshakable commitment to genocide": Roszak, Theodore, *The Making of a Counter Culture* (New York: Doubleday, 1968), 47.

207 "I like attention paid to my books": Allen, *Ideas That Matter*, 26.

208 "laces" and "spaghetti dish": Jacobs, *The Death and Life of Great American Cities*, 367.

209 "great hardship and suffering": Flint, *Wrestling with Moses*, 149.

210 "hitting us below the belt": Ibid., 150.

210 "naysayers taking pot shots": Ibid., 151.

210 "Don't approve this road" and "kill the mad visionaries'": Ibid.

212 "the Palaces of Trade": Ibid., 152.

212 E. V. Haughwout Building: Huxtable, Ada Louise, "Noted Buildings in Path of Road," *New York Times*, July 25, 1965.

212 "Tiffany of its day": Flint, *Wrestling with Moses*, 153.

214 "cancel land acquisitions": Ibid., 155.

215 "The expressway would": *Village Voice*, August 30, 1962.

215 "Every delay gives added hope": Decision on Expressway Urged; Autoclub Scores Delay by City," *New York Times*, July 9, 1962, 33.

215 "if no feasible relocation": Flint, *Wrestling with Moses*, 156.

215 "outraged at this cat-and-mouse": Ibid.

216 "engineering and economic data": Ibid.

216 "This isn't about *The New Yorker*": Alexiou, *Jane Jacobs: Urban Visionary*, 109.

216 "wonderfully effective letter": Ibid., 110.

216 "would be the first serious": Flint, *Wrestling with Moses*, 157.

216 "Except for one old man": Ibid., 158.

217 "monstrous and useless folly": Ibid.

217 "a piece at a time" and "we'll be fighting the tentacles": *Village Voice*, August 30, 1962.

217 "The most spectacular": Alexiou, *Jane Jacobs: Urban Visionary*, 110.

217 "We won! Isn't it" and "You can well imagine": Flint, *Wrestling with Moses*, 158.

218 "The rule of thumb": Alexiou, *Jane Jacobs: Urban Visionary*, 111.

218 "Well, here I have been arrested": Allen, *Ideas That Matter*, 72.

218 "We want Jane": Kent, Leticia, "Prosecution of the City Performed by Its Inmates," *Village Voice*, April 18, 1968.

218 "It's interesting the way": Ibid.

219 "Thank you, sir": Ibid.

219 "errand boys": Allen, *Ideas That Matter*, 74.

219 "Listen to this": Ibid.

219 "I'm the prisoner": Ibid., 15.

220 "bear no relation": Flint, *Wrestling with Moses*, 175.

220 "The inference seems to be": Seveso, Richard, "Mrs. Jacobs's Protest Results in Riot Charge," *New York Times*, April 18, 1968.

220 "I resent, to tell you": Allen, *Ideas That Matter*, 22.

220 "They would have preferred": Kunstler, "An Interview with Jane Jacobs, Godmother of the American City,"10.

221 "I don't make up my mind": Allen, *Ideas That Matter*, 159.

221 "Ideology is narrowing": Ibid., 23.

221 "I thought how ironic": Ibid., 29.

222 "Take a look at the world": Kent, "More Babies Needed."

222 "see the evolution of the piano": Allen, *Ideas That Matter*, 88.

222 "Everyone talks about the canals of Venice": Ibid., 89.

222 "jobs versus environmental": Ibid., 655.

223 "vital nearby neighborhood": Ibid., 200.

223 "she began by talking": Ibid., 207.

224 "Internally, I'm not any": Ibid., 13.

224 "It's a fact of life and also": Ibid., 27.

224 "Cities on the whole": Flint, *Wrestling with Moses*, 193.

225 "From this house, in 1961": Ibid., 195.

CHAPTER 3: JANE GOODALL

228 "sex, face color, etc.": Goodall, Jane, *Africa in My Blood: An Autobiography in Letters; The Early Years*, ed. Dale Peterson (Boston: Houghton Mifflin, 2000), 171.

228 "It does seem a long time ago": Ibid., 186.

228 "high powered tropical": Peterson, Dale, *Jane Goodall: The Woman Who Redefined Man* (Boston: Houghton Mifflin, 2006), 236.

229 "life's mentor": Goodall, *Africa in My Blood*, 82.

229 "Then you should meet": Goodall, Jane, *Reason for Hope* (New York: Grand Central, 2000), 44.

229 "a true giant": Ibid.

230 "Admittedly, we don't meet": Goodall, *Africa in My Blood*, 109.

230 "I remember wondering": Ibid., 53.

231 "less likely to arouse": Peterson, *Jane Goodall*, 120.

232 "laughed a lot" and "whirly kind of life": Ibid., 7.

233 "touched me only once": Ibid., 11.

234 "They need the earth": Goodall, *Reason for Hope*, 5.

234 "melancholic" and "all gray, crumbling stone": Peterson, *Jane Goodall*, 16.

235 "irreducibles": Thurman, Judith, *Isak Dinesen: The Life of a Storyteller* (New York: St. Martin's, 1982), 24.

235 "original qualities" and "define a mysterious ground": Ibid., 25.

236 "a small, disheveled": Peterson, *Jane Goodall*, 22.

236 "a round white object": Ibid., 23.

237 "somewhere in France": Ibid., 25.

237 "We had tea and still Daddy": Ibid., 26.

237 "bawdy": Ibid., 31.

237 "advantages" and "I was never, ever told": Ibid., 29.

237 "strong, self-disciplined, iron-willed": Goodall, *Reason for Hope*, 8.

238 "a beloved uncle": Peterson, *Jane Goodall*, 36.

238 "separate peace": Ibid., 29.

238 "lovely big slow-worm": Ibid., 39.

238 "racers": Goodall, *Reason for Hope*, 10.

239 "gently atavistic fascination": Goodall, *Africa in My Blood*, 4.

239 "I read it all": Peterson, *Jane Goodall*, 45.

239 "ecstatically": Ibid.

240 "tall, very strict-looking": Ibid., 53.

240 "Woke gloomily up to the dreary": Ibid., 54.

240 "I suppose that everyone": Ibid., 55.

241 "Foul things—stockings" and "my first Champagne cocktail": Ibid., 59.

241 "her first girdle, a red": Ibid.

241 "rather interesting": Ibid.

242 "monotonous": Goodall, *Africa in My Blood*, 41.

242 "a clever girl": Peterson, *Jane Goodall*, 71.

242 "I haven't given up the journalism": Goodall, *Africa in My Blood*, 46.

242 "just very boring filing": Peterson, *Jane Goodall*, 73.

242 "boredom of this foul job" and "Do you not think": Ibid., 76.

242 "a dingy basement room": Ibid.

242 "Oh, Sally I am having such a wonderful time": Ibid., 76–77.

243 "I have decided, even more": Goodall, *Africa in My Blood*, 71–72.

243 "too settled, too fond": Peterson, *Jane Goodall*, 80.

243 "I have decided that I must": Goodall, *Africa in My Blood*, 71–72.

245 "Right from the moment": Peterson, *Jane Goodall*, 92.

245 "I really do simply adore": Goodall, *Africa in My Blood*, 88.

245 "They are even taller": Ibid., 91.

245 "slightly degrading in its effect": Ibid., 88.

245 "surprising endurance, [an] iron stomach": Ibid., 4.

246 "constitution" and "sensibility": Ibid.

246 "spit": Goodall, *Reason for Hope*, 45.

247 "appallingly childish" and "absurd": Peterson, *Jane Goodall*, 99.

247 "a white African": Goodall, *Africa in My Blood*, 83.

248 "missing buttons" and "overloaded": Peterson, *Jane Goodall*, 100.

248 "to stink": Ibid., 101.

248 "the whole morning": Ibid., 102.

248 "naturally very interested": Goodall, *Africa in My Blood*, 93.

248 "Women came to him": Peterson, *Jane Goodall*, 103.

249 "the third Mrs. Leakey": Goodall, *Africa in My Blood*, 85.

250 "dreadfully familiar": Ibid., 97.

250 "a small, lean woman" and "a little distant": Ibid., 98.

250 "I hear you might like to come with us to Olduvai": Peterson, *Jane Goodall*, 106.

250 "The great aim": Goodall, *Africa in My Blood*, 108.

250 "rise & fall and are covered": Ibid., 107.

250 "grazing herds": Peterson, *Jane Goodall*, 109.

251 "the most beautiful": Goodall, *Africa in My Blood*, 105.

251 "dirty, greasy hair": Ibid., 102.

251 "mystery of evolution": Goodall, *Reason for Hope*, 49.

251 "intense excitement": Ibid., 48.

251 "cat-like yowls": Ibid., 46.

251 "the whole immense vastness of Africa": Goodall, *Africa in My Blood*, 114.

251 "utterly adorable" and "friendly & joking": Ibid., 109–10.

252 "blotto" and "One is liable to get": Ibid.

252 "full camp"; "such fun"; and "rather like a dormitory": Peterson, *Jane Goodall*, 115–16.

253 "charming & great fun": Ibid., 122.

253 "watching ostrich courtship" and "charming Portuguese professor": Ibid., 123.

253 "I begin to see why Mary": Goodall, *Africa in My Blood*, 119.

253 "Old Louis really is infantile" and "monkey business": Ibid., 118.

254 "I had to come back" and "quite ill": Peterson, *Jane Goodall*, 124–25.

254 "very young": Goodall, *Africa in My Blood*, 120.

254 "an external layer" and "the character underneath": Ibid.

255 "utterly remote": Goodall, *Reason for Hope*, 54.

255 "the first person I've": Peterson, *Jane Goodall*, 127.

255 "quite utterly and completely mad": Ibid.

255 "daring and game to try anything" and "doing the ton": Ibid.

255 "quite heavenly": Ibid., 128.

256 "But can you imagine": Goodall, *Africa in My Blood*, 138.

256 "Brian, in a lot": Peterson, *Jane Goodall*, 129–30.

256 "Oh, yes. He *is* in love": Ibid., 130.

256 "He has been sweet & kind": Ibid.

256 "We've had a little talk": Ibid.

257 "merely a father to me" and "trust him with everything": Ibid.

257 "an extraordinarily promising": Goodall, *Africa in My Blood*, 85.

257 "with a mind unbiased" and "someone with an open": Goodall, *Reason for Hope*, 55.

257 "Miss Jane Morris-Goodall": Peterson, *Jane Goodall*, 120.

258 "the Brian problem": Ibid., 141.

258 "a boy of great charm": Ibid., 142.

258 "a wonderful flow of conversation" and "loathe to tear": Ibid., 143–44.

258 "through miles and miles"; "uninhabited monkey-filled"; "sailed on"; and "a weeklong float": Ibid., 144.

258 "far exceeded in strangeness" and "cheetahs slumbering": Ibid., 145.
259 "no European woman": Ibid., 119.
260 "smiling London": Goodall, *Africa in My Blood*, 139.
260 "there are some animals which": Peterson, *Jane Goodall*, 152.
260 "an extremely interesting": Ibid., 161.
260 "charming": Ibid., 162.
261 "who has worked in Kenya": Ibid., 154.
262 "I saw Leakey" and "all fixed": Ibid., 159.
263 "servants of science": Ibid., 163.
263 "guided" and "probably capable": Ibid.
263 "alongside their own infant" and "remained mute": Ibid., 163–64.
264 "one-man expedition" and "blinds": Ibid., 164.
264 "surround a group of chimps": Ibid.
264 "the chimp is nomadic": Ibid., 165.
264 "shocked" and "age, sex, reproductive": Ibid., 166.
265 "More slaughter of": Goodall, *Reason for Hope*, 56.
265 "But there were only two bored": Ibid.
265 "The more he reads": Peterson, *Jane Goodall*, 166.
265 "first setback": Goodall, Jane, *In the Shadow of Man* (Boston: Houghton Mifflin, 1971), 7.
265 "a short trial study": Ibid.
266 "serious precedent, established": Peterson, *Jane Goodall*, 170.
266 "hunter's technique": Ibid.
266 "taught me a great deal": Goodall, *In the Shadow of Man*, 8.
266 "At 2:15 an adolescent": Peterson, *Jane Goodall*, 171.
266 "species" and "individual members": Ibid.
267 "mature female with very pink": Ibid.
267 "the huge female" and "very obvious pregnant": Ibid.
267 "I had no idea that this": Goodall, *Reason for Hope*, 74.
267 "typically consisted of": Goodall, *Africa in My Blood*, 156.
268 "In the form of statistics": Gould, Kira, and Lance Hosey, *Women in Green* (Washington, D.C.: Ecotone Press, 2007), 25.
269 "living in a dream": Goodall, *In the Shadow of Man*, 13.
269 "horrified" and "impenetrable appearance": Ibid., 14.
269 "packed up" and "terribly young": Peterson, *Jane Goodall*, 176.
270 "small gurgling stream": Goodall, *In the Shadow of Man*, 15.
270 "so stiff with dust": Peterson, *Jane Goodall*, 177.
270 "hired assistants"; "good-humored"; and "tall and lean": Ibid., 183.
271 "depressed and miserable": Goodall, *In the Shadow of Man*, 17.
271 "I do hope it's all": Peterson, *Jane Goodall*, 183.
271 "I wish you could be here": Ibid., 179.
271 "reclusive leopards": Ibid., 141.
272 "Fortunately the wind was": Goodall, *Reason for Hope*, 62.
272 "strange sawing call": Ibid.

272 "an ingrained illogical fear": Goodall, *In the Shadow of Man*, 31.

272 "hammering" and "almost nonexistent": Goodall, *Reason for Hope*, 63.

273 "a reasonable understanding": Ibid.

273 "learned to hate man" and "no more dangerous": Ibid.

273 "Men tend to react": Gould and Hosey, *Women in Green*, 41.

273 "There is a way": Author interview with Jane Goodall.

274 "the fast running stream"; "thick vegetation"; and "low, resonant": Goodall, *In the Shadow of Man*, 18–19.

274 "This was a mistake": Peterson, *Jane Goodall*, 185.

274 "We heard them": Ibid.

275 "How can I ever see" and "mood, a depression": Ibid.

276 "This is the trouble": Ibid., 187.

276 "the most ghastly, livid": Ibid., 189.

277 "Suffocating by 9 AM": Ibid.

277 "a measured tread" and "palish face": Ibid., 194.

278 "collapsed and unconscious": Ibid., 191.

278 "frantic" and "I felt I had": Goodall, *In the Shadow of Man*, 25.

278 "official displeasure": Ibid.

278 "Earth kept vanishing": Peterson, *Jane Goodall*, 193.

279 "I was very conspicuous": Goodall, *In the Shadow of Man*, 26.

279 "perched like jockeys": Ibid.

279 "square-faced": Peterson, *Jane Goodall*, 196.

280 "always looked immaculate": Goodall, *In the Shadow of Man*, 23.

280 "a rhythm" and "piece together": Goodall, *Reason for Hope*, 65.

281 "unnatural": Peterson, *Jane Goodall*, 197.

281 "The first chimp": Ibid.

281 "I thought we could get": Ibid., 198.

281 "in fairly thick forest": Goodall, *In the Shadow of Man*, 32.

281 "The first old boy": Peterson, *Jane Goodall*, 198.

281 "old & grizzled male" and "Absolutely no fear": Ibid.

282 "This is the first opportunity" and "The penis": Ibid.

282 "Each occasion": Ibid.

282 "They all went rather fast": Ibid., 199.

282 "acorn-like nut" and "the most unpleasant, bitter pungent": Ibid.

282 "Dry, dark brown, and fibrous": Ibid.

282 "It squatted in a leafy" and "Very comfortable": Ibid.

283 "We interact with the environment": Gould and Hosey, *Women in Green*, 87.

283 "There is no way to understand": Ackerman, Diane, *A Natural History of the Senses* (New York: Random House, 1990), xv.

283 "saw things differently" and "a bottom line": Helgesen, Sally, *The Female Advantage: Women's Ways of Leadership* (New York: Doubleday, 1990), 35.

283 "guns and grassfires": Peterson, *Jane Goodall*, 201.

283 "biological monolith" and "simple and definite": Ibid., 202.

283 "I saw one female": Goodall, *In the Shadow of Man*, 28.

284 "two adult males" and "youngsters having wild": Ibid., 29.

284 "small infants dangling": Ibid.

284 "After eating two fruits": Peterson, *Jane Goodall*, 206.

284 "two or more": Goodall, *In the Shadow of Man*, 29.

284 "Dark mark on left": Peterson, *Jane Goodall*, 201.

284 "as a collective": Ibid., 203.

284 "know some of them": Goodall, *Africa in My Blood*, 160.

285 "ate fruits and berries": Peterson, *Jane Goodall*, 202.

285 "usual haunts": Ibid., 204.

285 "sentient creatures with humanlike": Ibid., 203.

285 "really nice to talk": Goodall, *Africa in My Blood*, 163.

286 "George said he thought": Ibid., 164.

286 "angry little screams" and "something which looked pink": Peterson, *Jane Goodall*, 205.

286 "Suspected meat": Ibid., 206.

286 "no hair or fur"; "But *impossible*"; and "he lifted it": Ibid.

286 "No response" and "presented her bottom": Ibid.

287 "unidentified victim": Ibid., 207.

287 "a black object": Ibid., 208.

287 "picking up things"; "Very deliberately"; and "Then he got down": Ibid.

287 "plump"; "soldiers"; and "to create a long": Ibid., 209.

288 "termite-fishing": Goodall, *In the Shadow of Man*, 36.

288 "After a few minutes": Peterson, *Jane Goodall*, 209.

288 "Grey beard, fingers looked": Ibid., 209–10.

289 "examine the scene" and "So—they had been watching": Ibid., 210.

289 "I was able to observe" and "He then spent": Ibid.

289 "He had his back to me" and "a large male baboon": Ibid., 211.

289 "the first chimpanzee I saw": Ibid.

289 "quiet, almost thoughtful" and "alien, ghostly, ponytailed": Ibid.

290 "Because he lost his fear": Goodall, *Reason for Hope*, 75.

290 "Now we must redefine": Peterson, *Jane Goodall*, 212.

291 "unrequited love letters"; "everything was rather hopeless"; and "living with Dr. Leakey": Ibid., 232.

291 "unlike any sound" and "tail above the grass": Ibid., 231.

292 "Inanimate objects developed": Goodall, *Reason for Hope*, 73.

292 "to hear the pattering": Ibid.

292 "living, breathing entity" and "intensely aware of the beingness": Ibid.

292 "rough, sun-warmed bark": Ibid.

292 "a powerful, almost mystical knowledge": Ibid., 72.

293 "The longer I spent on my own": Ibid., 73.

293 "It was most organized": Goodall, *Africa In My Blood*, 172.

293 "in a pouring rain with thunder" and "primitive hairy men": Ibid.

293 "I felt all the time": Peterson, *Jane Goodall*, 235.

293 "brow" and "Silhouetted on the skyline": Goodall, *Africa in My Blood*, 172.

293 "He stood up, holding": Peterson, *Jane Goodall*, 236.

294 "revolutionary" and "sterile old masculine": Ibid.

294 "shot up": Goodall, *In the Shadow of Man*, 53.

295 "white & fungus-y" and "all day long": Peterson, *Jane Goodall*, 236–37.

295 "100 yards": Ibid., 237.

296 "who rode everywhere on her": Goodall, *In the Shadow of Man*, 32.

296 "Can it be true?": Goodall, *Africa in My Blood*, 180.

296 "the strangest awakening": Peterson, *Jane Goodall*, 241.

297 "demolished—skin and all": Ibid.

297 "There were times" and "full of foreboding": Ibid., 242–43.

297 "for ages": Goodall, *Africa in My Blood*, 187.

298 "manipulator and voyeur": Peterson, *Jane Goodall*, 228.

298 "realistic" and "to test the apes'": Ibid., 227.

298 "egg-laden nests" and "made to look": Ibid.

298 "observer and observed in the same field": Ibid., 228.

298 "on foot, unarmed, lightly": Ibid.

298 "in a parallel universe": Goodall, *Africa in My Blood*, 231.

299 "This young, scientifically naive": Ibid., 156.

299 "only if Miss Goodall": Ibid., 192.

299 "a big glassy eye": Ibid.

300 "I want to do my own photos": Peterson, *Jane Goodall*, 247.

300 "since he is no stranger": Goodall, *Africa in My Blood*, 193.

300 "knows the animals"; "a wide-angle"; and "top quality": Ibid.

300 "far too heavy and clumsy": Peterson, *Jane Goodall*, 249.

300 "a camera round her neck": Ibid.

301 "the shortest correspondence": Ibid., 250.

301 "a raging headache" and "If only they realized": Goodall, *Africa in My Blood*, 198.

301 "not having previous camera": Peterson, *Jane Goodall*, 251.

301 "as I feel that a determined": Goodall, *Africa in My Blood*, 199.

301 "Of 37 exposures": Ibid., 194.

302 "some experience of color": Ibid.

302 "The production of satisfactory": Ibid.

302 "it has got to be done": Peterson, *Jane Goodall*, 253.

302 "skeletal" and "crisped termites": Ibid., 255–56.

302 "I can't remember when I last wrote": Ibid., 258.

303 "I hate to write such depressing": Ibid., 259.

303 "not exciting" and "a lack of good pictures": Ibid., 260.

303 "this shortcoming is so serious": Goodall, *Africa in My Blood*, 195.

304 "appropriate format": Peterson, *Jane Goodall*, 262.

304 "Incredibly handsome": Peterson, *Jane Goodall*, 273.

304 "He was not aggressive": Ibid.

304 "terribly in awe" and "pointed out": Ibid., 274.

304 "I'd better go and do" and "filled with frustration": Ibid.

305 "ascribing personalities" and "Only humans": Ibid., 277.

305 "graphs & statistics": Ibid., 275.

305 "I had no undergraduate degree": Ibid., 276.

307 "This is an era of specialists": Carson, Rachel, *Silent Spring* (New York: Houghton Mifflin, 2002), 13.

307 "a sense of occasion": Peterson, *Jane Goodall*, 286.

308 "red face" and "perfectly parted": Ibid., 290.

308 "Sixty-one different vegetable": Ibid., 287.

309 "very hostile" and "amateurs": Ibid., 290.

309 "characteristic of all primate": Ibid., 291.

309 "there are those who": Ibid.

310 "dominance relations"; "primate sex ratios"; and "over-riding importance": Ibid.

310 "a non-carnivorous animal": Ibid.

310 "a useful point to remember": Ibid., 292.

310 "assure"; "I realize you were only"; and "should continue": Ibid.

310 "that her legs were too nice, her hair": Ibid., 294.

311 "great simplicity": Ibid., 293.

311 "monogamous pairs" and "sex and dominance": Ibid.

311 "choice, kinship, learning": Ibid.

311 "It's—it's—it's for": Tullis, Paul, "Wild at Heart," *New York Times Magazine*, March 15, 2015, 57.

312 "must make every effort": Goodall, *Africa in My Blood*, 195.

313 "familiar with wild animal": Peterson, *Jane Goodall*, 299.

313 "a face burnt red-brown by the African sun": Ibid.

314 "Hugo is charming and": Ibid., 308.

314 "crack of dawn" and "her blessings": Ibid.

315 "He IS a devil!": Goodall, *Africa in My Blood*, 221.

315 "the happiest: the proudest, of": Ibid., 220.

315 "have accepted the presence": Ibid., 226.

315 "Hugo might be too embarrassed": Ibid., 196.

315 "You will by now have stills": Ibid., 224.

316 "just the right person for": Ibid., 223.

316 "so far and remote" and "and our conversation": Peterson, *Jane Goodall*, 327.

316 "to make a fine layout" and "a first person": Ibid., 311.

317 "provisioning": Goodall, *Africa in My Blood*, 233.

317 "the most hideous old bag": Peterson, *Jane Goodall*, 325.

317 "millions of males": Ibid., 265.

317 "in the most fabulous way": Ibid., 260.

317 "rushed about with all his hair": Ibid., 250.

318 "Goliath has borne Hugo": Ibid., 259.

318 "I know you have complete": Peterson, *Jane Goodall*, 320–21.

318 "some interfering officer": Ibid., 321.

318 "Louis, can you really": Goodall, *Africa in My Blood*, 263.

318 "We now have 21 regular": Peterson, *Jane Goodall*, 324.

318 "To be able to follow": Goodall, *Africa in My Blood*, 265.

319 "more aggressive" and "manipulative traditions": Peterson, *Jane Goodall*, 323.

320 "been conducting cancer research": Ibid., 329.

320 "to question the state" and "I thought to myself, gee here": Ibid., 328–29.

321 "Whenever I think of Africa": Ibid., 329.

321 "Your fascinating account": Ibid., 330.

322 "WILL YOU MARRY ME": Goodall, *Africa in My Blood*, 270.

322 "You can't imagine": Ibid., 273.

322 "just a snob value waste": Peterson, *Jane Goodall*, 342.

322 "there was no way": Goodall, *Africa in My Blood*, 335.

323 "flat out on the final stages": Peterson, *Jane Goodall*, 385.

323 "Nescafe and the occasional apple": Ibid., 379.

323 "a delightful month": Goodall, *Africa in My Blood*, 337.

324 "hollow-eyed, gaunt": Goodall, Jane, *Beyond Innocence: An Autobiography in Letters; The Later Years*, ed. Dale Peterson (New York: Houghton Mifflin, 2001), 145.

325 "I was in a dream world": The account on the next pages is based on my interview with Jane Goodall, unless otherwise noted.

325 "That's when the shock hit me": Ibid.

329 "And I am not aggressive": Tullis, "Wild at Heart," 58.

329 "It never ceases to amaze": Ibid.

329 "I have to raise money": Author interview with Jane Goodall for the following conversation about Roots & Shoots and Goodall's good work there.

332 "but the windows are chicken wire": Tullis, "Wild at Heart," 58.

332 "People often ask me": Author interview with Jane Goodall.

CHAPTER 4: ALICE WATERS

334 "Elsewhere, even when I found": McNamee, Thomas, *Alice Waters and Chez Panisse* (New York: Penguin, 2007), 18.

334 "downstairs part": Ibid., 17–18.

335 "Everything in Paris": Ibid., 16.

335 "some experience with French men": Ibid., 14.

335 "my mother didn't really": Author interview with Alice Waters.

336 "The skirt was kind of itchy": Ibid.

336 "We played outside": Ibid.

336 "stand her shivering": McNamee, *Alice Waters and Chez Panisse*, 9.

336 "We climbed the trees": Author interview with Alice Waters.

336 "I could name all the trees": Ibid.

336 "The mulberries seemed": Bolois, Justin, "The 10 Dishes That Made My Career: Alice Waters," FirstWeFeast.com, April 20, 2015.

337 "Our whole concept": Halberstam, David, *The Fifties* (New York: Ballantine, 1993), 159.

338 "ones from the other side": McNamee, *Alice Waters and Chez Panisse*, 9.

338 "That was a dark period": Ibid., 10.

339 "That changed everything": Author interview with Alice Waters.

340 "There is a time": Bloom, Alexander, and Wini Breines, eds., *Takin' It to the Streets: A Sixties Reader* (New York: Oxford University Press, 1995), 111–12.

342 "it was just finely chopped up": McNamee, *Alice Waters and Chez Panisse*, 12.

342 "That was the time": Ibid., 15.

342 "They had pretty decent pâté": Ibid., 14.

342 "We would just sit there": Ibid.

343 "obsession": Ibid., 16.

343 "It was a wonderful time": Author interview with Alice Waters.

343 "I was still very intimidated": McNamee, *Alice Waters and Chez Panisse*, 16.

343 "What the fruit bowl": Ibid.

343 "a complete seduction" and "a whole different": Author interview with Alice Waters.

344 "I had a couple of French friends": Ibid.

344 "I'd hardly even heard": McNamee, *Alice Waters and Chez Panisse*, 17.

344 "That was our real introduction": Ibid.

344 "I'll never forget those": Ibid.

345 "the old gods": Barr, Luke, *Provence, 1970: M. F. K. Fisher, Julia Child, James Beard, and the Reinvention of American Taste* (New York: Clarkson Potter, 2013), 25.

345 "opened a door to pleasure": Ibid., 13.

345 "The atmospherics of desire": Ibid.

345 "When I got back": McNamee, *Alice Waters and Chez Panisse*, 19.

346 "God, it was a wild time": Ibid., 21.

346 "horrors of Babylon": Kisseloff, Jeff, *Generation on Fire: Voices of Protest from the 1960s* (Lexington: University Press of Kentucky, 2007), 158.

346 Trip Without a Ticket: Ibid., 145.

347 "crazy mandala": Roszak, Theodore, *The Making of a Counter Culture* (New York: Doubleday, 1968), 136.

347 "a hirsute, be-cowbelled contingent": Ibid., 140.

347 "What kind of America": Isserman, Maurice, and Michael Kazin, *America Divided: The Civil War of the 1960s* (New York: Oxford University Press, 2000), 164.

347 "She was outraged": McNamee, *Alice Waters and Chez Panisse*, 21.

348 "Alice was only twenty-one": Ibid., 25.

348 "I kept drinking": Ibid., 21.

348 "in other words, the French Revolution": Ibid., 23.

348 "She was never a shy person": Ibid., 23–24.

348 "plummy patrician accent": Barr, *Provence, 1970*, 52.

349 "The door was always": McNamee, *Alice Waters and Chez Panisse*, 23.

349 "health food nuts": Belasco, Warren J., *Appetite for Change: How the Counterculture Took On the Food Industry* (Ithaca, N.Y.: Cornell University Press, 2007), 16.

349 squirting a noisome fly with breast milk: Ibid., 36.

349 "organic's rejection": Pollan, Michael, *The Omnivore's Dilemma* (New York: Penguin, 2006), 143.

350 "White versus brown": Belasco, *Appetite for Change*, 48.

350 "I remember sitting around": McNamee, *Alice Waters and Chez Panisse*, 24–25.

350 "We'd get this great big old house": Ibid., 25.

351 "Alice would try anything": Ibid., 26.

351 "truly momentous disasters": Ibid.

351 "The whole trend of American family cooking": Kamp, David, "Cooking Up a Storm," *Vanity Fair*, October 2006.

351 "Those meals of ours": McNamee, *Alice Waters and Chez Panisse*, 27.

352 "she didn't even try": Ibid., 28.

352 "was the only one": Kamp, David, *The United States of Arugula* (New York: Broadway Books, 2006), 131.

352 "trying-to-be-French": McNamee, *Alice Waters and Chez Panisse*, 29.

353 "the restaurant fantasy": Ibid.

353 "There were a lot of artists": Gross, Terry, "Forty Years of Sustainable Food" (interview with Alice Waters), *Fresh Air*, NPR, August 22, 2011.

353 "Ingredients! Sure, you had": McNamee, *Alice Waters and Chez Panisse*, 32.

354 "Montessori went straight to my heart": Author interview with Alice Waters.

354 "It's an observation": McNamee, *Alice Waters and Chez Panisse*, 33.

355 "I'm your certified commie creep": Ibid., 34.

355 "You fed a space heater with shillings": Ibid.

355 "I nearly froze to death that winter": Waters, Alice, and Friends, *Forty Years of Chez Panisse: The Power of Gathering* (New York: Clarkson Potter, 2011), 20.

356 "Gypsy music": McNamee, *Alice Waters and Chez Panisse*, 36.

356 "A shy, big-eyed" and "The boy builds us": Waters, Alice, "Tea and Cheese in Turkey," in *The Kindness of Strangers*, ed. Don George (Melbourne: Lonely Planet, 2003), 42–43.

356 "finding grace in the unexpected": McNamee, *Alice Waters and Chez Panisse*, 52.

356 "practically nothing": Waters, *Forty Years of Chez Panisse*, 20.

356 "It was like a garden": Ibid., 37.

357 "a tightly wound cloche": McNamee, *Alice Waters and Chez Panisse*, 37.

357 "Martine was very important": Ibid., 38.

357 "Martine could feed": Author interview with Alice Waters.

357 "Film was the art form": McNamee, *Alice Waters and Chez Panisse*, 38.

358 "grew mistiest over": Kamp, *The United States of Arugula*, 138.

359 "an ugly, squat, two-story": Ibid.

359 "like a rundown hippie crash": McNamee, *Alice Waters and Chez Panisse*, 42.

359 "Nobody cared if you wanted": Ibid., 40.

359 "Well, of course, they were": Ibid., 42.

360 "I wanted it to be like": Ibid., 45.

360 "You ate what was there": Author interview with Greil Marcus.

361 "harmonious whole": McNamee, *Alice Waters and Chez Panisse*, 45.

361 "A couple could live": Author interview with Ruth Reichl.

362 "I hired them all": McNamee, *Alice Waters and Chez Panisse*, 4.

363 "It was exactly the right": Kamp, *The United States of Arugula*, 140.

363 "We were so happy": McNamee, *Alice Waters and Chez Panisse*, 5.

363 "No corners cut": Ibid., 49.

364 "Those of us who were working": Kamp, *The United States of Arugula*, 142.

364 "Everyone will have to be": McNamee, *Alice Waters and Chez Panisse*, 49.

365 "It was a train out of control": Ibid., 51.

365 "We were inventing": Author interview with Alice Waters.

365 "Such-and-such Meats": McNamee, *Alice Waters and Chez Panisse*, 52.

365 "nobody came before": Author interview with Alice Waters.

365 "She was a total workaholic": Kamp, *The United States of Arugula*, 143.

365 "We would have heated": McNamee, *Alice Waters and Chez Panisse*, 53.

366 "a notion of how to live": Marcus, Greil, "Chez Panisse: Seventies" (unpublished manuscript provided courtesy of Greil Marcus).

366 "It could not have happened": Ibid.

367 "She was very stubborn": McNamee, *Alice Waters and Chez Panisse*, 53.

367 "Scale down your attachments": Belasco, *Appetite for Change*, 26.

367 "'Natural' was a liberated": Ibid., 40.

367 "revolutionizing": Ibid., xi.

367 "No owner, no manager, no employees": Ibid., 20.

368 "There was a little cottage": Kamp, *The United States of Arugula*, 145.

368 "It was quite unremarkable": McNamee, *Alice Waters and Chez Panisse*, 54.

368 "I knew it was going to be hard": Ibid., 53.

369 "as big a percentage": Waters, *Forty Years of Chez Panisse*, 46.

369 "There has always been": McNamee, *Alice Waters and Chez Panisse*, 24.

369 "Once, I had this idea": Ibid., 190.

369 "I am uncompromising": Author interview with Alice Waters.

370 "We gathered watercress": Waters, Alice, "The Farm-Restaurant Connection," in *American Food Writing: An Anthology with Classic Recipes*, ed. Molly O'Neill (New York: Library of America, 2007), 561.

370 "the hunter-gatherer culture": McNamee, *Alice Waters and Chez Panisse*, 59.

370 *Terroir*: Ibid.

371 "They always had this local": Author interview with Alice Waters.

371 "Mediterranean reef": McNamee, *Alice Waters and Chez Panisse*, 86.

371 "one piece at a time": Ibid.

371 "Burgundy, the Pays d'Oc": Ibid., 87.

371 "But not here. Not in Berkeley" and "We were going to stalls": Author interview with Alice Waters.

372 "Station Wagon Way": Barr, *Provence, 1970*, 14.

372 *The Can-Opener Cookbook*: Ibid., 17.

373 "can claim the K-ration": Curtis, Olga, "Time-Saving Modern Kitchen Miracles," *Washington Post and Times Herald*, April 22, 1957, C2.

373 "how to talk and think": Barr, *Provence, 1970*, 9.

373 "revelation of taste": Ibid., 14.

373 "literary consideration": Ibid., 13.

373 "skimping": Ibid., 16.

373 "food acquired a chic": Ibid., 17.

374 "little family restaurants": Author interview with Alice Waters.

374 "heat-and-serve vacuum": McNamee, *Alice Waters and Chez Panisse*, 89.

375 "It was so dispiriting": Author interview with Alice Waters.

375 "If you wanted a peach": Ibid.

375 "When I traveled": McNamee, *Alice Waters and Chez Panisse*, 90.

375 "going to keep making bread": Ibid.

376 "They were doing it for": Author interview with Alice Waters.

376 "We were starting to reach": McNamee, *Alice Waters and Chez Panisse*, 90.

376 "I've always believed": Ibid., 60.

377 "Everyone was switched on": Waters, *Forty Years of Chez Panisse*, 151.

377 "if someone—most often": McNamee, *Alice Waters and Chez Panisse*, 62.

378 "There was scant division": Waters, *Forty Years of Chez Panisse*, 93.

378 "There are so many aspects": McNamee, *Alice Waters and Chez Panisse*, 67.

378 "Right now in an": Ibid.

378 "I never wanted to be chef": Ibid., 68.

378 "come one come all": Kamp, "Cooking Up a Storm."

378 "She wasn't Gallic": Kamp, *The United States of Arugula*, 145.

379 "a beef-stew-and-fruit-tart": Ibid., 146.

379 "roué": McNamee, *Alice Waters and Chez Panisse*, 76.

379 "Escoffier-style *grande cuisine*": Kamp, *The United States of Arugula*, 148.

379 "to roast barracuda"; "to smoke"; and "love boys": McNamee, *Alice Waters and Chez Panisse*, 76.

381 "I, of course, immediately": Ibid., 78.

381 "He was a perfectionist": Ibid., 79.

381 "could only be an asshole": Ibid.

382 "And it worked because": Kamp, *The United States of Arugula*, 154.

382 "wrote out these elaborate": Ibid.

382 "Jeremiah was Escoffier": McNamee, *Alice Waters and Chez Panisse*, 81.

382 "a better egg": Ibid., 82.

383 "Sometimes I'd bring in": Ibid., 93.

383 "and someone had hunted": Ibid.

383 "*l'entre-plat drogué*": Kamp, "Cooking Up a Storm."

383 "fascinating": Kamp, *The United States of Arugula*, 157.

383 "They were all so glamorous": Author interview with Ruth Reichl.

384 "There were magnums": Kamp, *The United States of Arugula*, 155.

384 "a big bag of blow": Ibid.

384 "I think the exuberance": McNamee, *Alice Waters and Chez Panisse*, 98.

384 "Drugs were easier": Kamp, *The United States of Arugula*, 156.

385 "Chez Panisse is joyously": McNamee, *Alice Waters and Chez Panisse*, 111.

385 "mostly dread" and "doubting, demanding": Ibid.

385 "I felt like I'd walked straight into": Waters, *Forty Years of Chez Panisse*, 73.

385 "That was when I really": McNamee, *Alice Waters and Chez Panisse*, 97.

385 "Lucien became my surrogate": Waters, *Forty Years of Chez Panisse*, 73.

386 "overripe, overrich": McNamee, *Alice Waters and Chez Panisse*, 114.

386 "If you read Jeremiah's": Kamp, *The United States of Arugula*, 159.

386 "a very closed place": Author interview with Greil Marcus.

386 "took a story that was rich": McNamee, *Alice Waters and Chez Panisse*, 114.

386 "Anybody who says that Jeremiah": Kamp, *The United States of Arugula*, 160.

387 "Jeremiah had a much": Author interview with Ruth Reichl.

388 "These new customers": McNamee, *Alice Waters and Chez Panisse*, 127.

388 "to use American ingredients": Kamp, *The United States of Arugula*, 161.

388 "We were doing some of": McNamee, *Alice Waters and Chez Panisse*, 124–25.

389 "We were really foraging": Ibid., 125.

389 "Forging a connection": Author interview with Alice Waters.

389 "One of them happened to be": Ibid.

390 "obsessive"; "health of the soil"; and "and which brought": Waters, *Forty Years of Chez Panisse*, 67.

390 "I had associated organic": Author interview with Alice Waters.

390 "It was a consistent pattern": Ibid.

391 "without a healthy agriculture": Waters, "The Farm-Restaurant Connection," 560.

391 "cream of fresh corn soup": McNamee, *Alice Waters and Chez Panisse*, 128.

392 "My idea of organic was to grow": Ibid., 136.

392 "We bought the seeds": Author interview with Alice Waters.

392 "I went to see Claude": McNamee, *Alice Waters and Chez Panisse*, 138.

392 "one endless meal": Ibid.

392 "Alice is a very loyal": Ibid., 199.

393 "In those days": Ibid., 142.

393 "I showed Alice": Ibid.

393 "Alice trusted me": Ibid., 143.

394 "I wanted to taste, travel": Ibid., 145.

394 "We could see the fire": Ibid., 148.

394 "no truffles, no foie gras": Ibid., 149.

395 "For eight years now": Ibid., 151.

395 "In some eight years" and "If you could eat": Ibid., 153.

396 "I remember distinctly": Ibid.

396 "a marriage of many": Ibid., 162.

397 "I remember very vividly": Ibid., 169.

397 "in that I kind of thought": Ibid.

397 "our responsibility reached": Waters, *Forty Years of Chez Panisse*, 101.

398 "I realized there was": McNamee, *Alice Waters and Chez Panisse*, 170–71.

398 "To see outside": Ibid., 172.

398 "her life, her family": Ibid.

398 "I was in my late thirties" and "And then, we met": Ibid., 173.
399 "I had a lot of sort of desperate": Ibid., 174.
400 "sensory overload": Ibid., 187.
400 "[Her] focus had been largely" and "about the earth": Ibid., 184.
400 "My sense of the ethics and politics": Ibid., 184–85.
401 "what food meant to the survival": Ibid., 190.
401 "clarity of flavor" and "I loved that he had": Ibid., 182.
402 "It's a fundamental": Waters, Alice, *Chez Panisse Menu Cookbook* (New York: Random House, 1982), 3.
402 "When I cook": Ibid., 6.
403 "This book is as much hers": McNamee, *Alice Waters and Chez Panisse*, 177.
403 "an ad hoc day-care center": Ibid., 189.
404 "eight kinds of heirloom": Ibid., 192.
404 "that nobody had ever seen": Author interview with Alice Waters.
405 "Nobody had ever done": McNamee, *Alice Waters and Chez Panisse*, 192.
405 "Sibella organized this very important": Author interview with Alice Waters.
405 "Alice's philosophy, and mine": McNamee, *Alice Waters and Chez Panisse*, 193.
405 "When you're around Alice": Ibid., 194.
406 "We buy everything Bob grows": Author interview with Alice Waters.
406 "an unruly child": McNamee, *Alice Waters and Chez Panisse*, 197.
406 "a revolution in American cooking" and "one of this country's most": Ibid., 204.
406 "More than any other single": Ibid.
407 "I just happened to see": Ibid., 205.
407 "a living exhibition" and "There's always money for good": Ibid., 208.
408 "if only for Fanny's sake": Ibid., 211.
408 "[Alice] had all the Pagnol fantasies": Ibid.
408 "an ideal reality where life": Waters, "The Farm-Restaurant Connection," 568.
409 "our responsibility to": McNamee, *Alice Waters and Chez Panisse*, 227.
409 "could and did do": Ibid., 216.
409 "As *famille Panisse*, we'd all take": Ibid., 219.
409 "Those of us who work": Waters, "The Farm-Restaurant Connection," 566.
410 "That's silly. That's like saying": Author interview with Alice Waters.
410 "a line of granola": Gopnik, Adam, "Annals of Gastronomy: The Millennial Restaurant," *The New Yorker*, October 26, 1998.
410 "Alice is the least elitist": Author interview with Ruth Reichl.

411 "What you have to understand": Author interview with Greil Marcus.

411 "organic" and "externalize": Author interview with Alice Waters.

412 "meat analogs" and "textured vegetable proteins": Belasco, *Appetite for Change*, 37–39.

412 "These guys find where": Author interview with Alice Waters.

412 "high-fructose corn syrup": McNamee, *Alice Waters and Chez Panisse*, 228.

413 "Food shouldn't be fast": Author interview with Alice Waters.

413 "While America's human population": Pollan, *The Omnivore's Dilemma*, 67.

414 "insanely high": McNamee, *Alice Waters and Chez Panisse*, 231.

415 "It was very distressing": Ibid., 232.

415 "You had to start somewhere": Ibid.

416 "the life of the restaurant": Author interview with Alice Waters.

416 "to come to his office": The account of the Edible Schoolyard on the next pages is based on my interview with Alice Waters, except as noted.

417 "seduction": Kalins, Dorothy, "Alice Waters: True Believer," *Town and Country*, January 2005.

417 "I think real food has a way": Author interview.

418 "It's so hard to remember": Author interview with Ruth Reichl.

418 "is the person": McNamee, *Alice Waters and Chez Panisse*, 261.

418 "a comprehensive solution": Ibid., 259.

418 "The tangerine peel": Ibid., 267.

420 "he would have been dead": Ibid., 262.

420 "They should have said to me": Ibid.

420 "I used to have a notion": Ibid., 263.

421 "It was always connected" and "This is the first generation": Author interview.

421 "Kids are disconnected": Ibid.

421 "I've been giving this talk": Ibid.

422 "epidemic": Pollan, *The Omnivore's Dilemma*, 90.

422 "unimaginable 655 percent": Taubes, Gary, "Big Sugar's Ally? Nutritionists," *New York Times*, January 13, 2017. For additional information see http://www.cdc.gov/diabetes/data.

422 "What's going on with children": Author interview.

422 "I've always believed": Ibid.

423 "They were afraid": Ibid.

424 "We began by teaching"; "so we designed a card"; "if they grow it"; and "It's why these models": Ibid.

424 "boys and girls together" and "They go there": Ibid.

425 "making visible the lines of connection": Pollan, Michael, *Cooked: A Natural History of Transformation* (New York: Penguin, 2013), 20.

425 "become so inured to the dogmas": Waters, "The Farm-Restaurant Connection," 561.

426 "Help us nourish": McNamee, *Alice Waters and Chez Panisse*, 268.

426 "The prospect of your second term": Ibid., 270.

426 "At best, we serve": Author interview with Alice Waters.

427 "It was right at the time": McNamee, *Alice Waters and Chez Panisse*, 274.

427 "He was so angry": Ibid.

428 "excruciating": Ibid., 276.

428 "Very painful": Ibid., 274.

428 "Alice didn't even want to cook for a while": Ibid., 275.

429 "It's only now": Author interview with Alice Waters.

429 "We haven't even begun": Ibid.

430 "And we have lots": Ibid.

430 "Well, do you want to change the food?": Ibid.

432 "Eating is an agricultural act": The conversation in the following pages is taken from my interview with Alice Waters.

433 "you'd follow her anywhere": Kalins, "Alice Waters: True Believer."

CHAPTER 5: HOPE IN THE SHADOWS

438 "28% of its electricity": Solnit, Rebecca, *Hope in the Dark: Untold Histories, Wild Possibilities* (Chicago: Haymarket Books, 2016), xxi.

438 "generates one-tenth": Tabuchi, Hiroko, "The Banks Putting Rain Forests in Peril," *New York Times*, December 4, 2016.

438 "fragile state": "Sustainable Land Use in the 21st Century" (study conducted by the United Nations, 2016), 5.

438 "protected areas": Ibid.

439 "To a populace": Wilson, E. O., afterword to Carson, Rachel, *Silent Spring* (New York: Houghton Mifflin, 2002), 358.

439 "Researchers were learning": Ibid., 357.

439 "the whip-hand over nature": Nash, Roderick, *Wilderness and the American Mind*, rev. ed. (New Haven: Yale University Press, 1973), 196.

440 "drowning in a sea": Isserman, Maurice, and Michael Kazin, *America Divided: The Civil War of the 1960s* (New York: Oxford University Press, 2000), 9.

440 "You may be hitched": Menand, Louis, "Books as Bombs," *The New Yorker*, January 24, 2011, 78.

441 "head and master": Coontz, Stephanie, *A Strange Stirring: The Feminine Mystique and American Women at the Dawn of the 1960s* (New York: Basic Books, 2011), 5.

441 "at least age 40": Ibid., 7.

441 "reckless assault": Marx, Leo, *The Machine in the Garden: Technology and the Pastoral Ideal in America* (1964; repr., New York: Oxford University Press, 2000), 369.

442 "thinking with all your senses": Klinkenborg, Verlyn, *Several Short Sentences About Writing* (New York: Vintage Books, 2013), 39.

442 "connection": Gould, Kira, and Lance Hosey, *Women in Green* (Washington, D.C.: Ecotone Press, 2007), 28.

442 "The change that counts in revolution": Solnit, *Hope in the Dark*, 26.

443 "practically and additively": Gould and Hosey, *Women in Green*, 30.

443 "essential linkages": Allen, Max, ed., *Ideas That Matter: The Worlds of Jane Jacobs* (Toronto: The Ginger Press, 1997), 207.

444 "Traditional conservation": McNamee, Thomas, *Alice Waters and Chez Panisse* (New York: Penguin, 2007), 315.

444 "exploitation and nurture": Berry, Wendell, *The Unsettling of America: Culture and Agriculture* (San Francisco: Sierra Club Books, 1977), 7.

445 "science-based technological progress": Marx, *The Machine in the Garden*, 369.

445 "nonviolence or nonexistence": King, Martin Luther Jr., "Remaining Awake Through a Great Revolution" (speech of March 31, 1968). The Martin Luther King Jr. Research and Education Institute, Stanford University.

446 "a total overhaul": Belasco, Warren J., *Appetite for Change: How the Counterculture Took On the Food Industry* (Ithaca, N.Y.: Cornell University Press, 2007), 26.

446 "quick death" and "slow strangulation": Ibid.

446 "A time comes when silence": Bromwich, David, "Martin Luther King's Speech Against the Vietnam War," Antiwar.com, March 16, 2008.

446 "In Vietnam that time": Ibid.

446 "the Stone Age": Sartre, Jean Paul, "On Genocide," in *America Since 1945*, ed. Robert D. Marcus and David Burner (New York: St. Martin's Press, 1981), 239.

446 "power and technology": Bromwich, "Martin Luther King's Speech."

446 "madness": Ibid., 7.

446 "I speak for the poor": Ibid., 8.

447 "We must realize": Belasco, *Appetite for Change*, 23.

447 "bid for green space"; "disdain for nature"; and "mass defoliation": Ibid., 21.

447 "the quality of our lives": Ibid.

448 "pointed away from violence": Ibid.

448 "Revolutionaries must begin to think": Ibid.

449 "the tyranny of the immediate": Shabecoff, Philip, *A Fierce Green Fire: The American Environmental Movement* (New York: Hill and Wang, 1993), 254.

450 "founding text": Marx, *The Machine in the Garden*, 381.

INDEX